Represent[ing the] UK valve a[nd] actuator industry

- A meeting place for industry
- *VALVEuser MAGAZINE*
- Training by industry, for industry
- HR and Health & Safety Service
- Exhibitions inside customers' premises
- *VALVEuser.COM*
- Industry web portal and data resource
- Tenders Service - introducing customers and suppliers
- Technical Advice & Publications
- Standards Development & European Directives Guides
- UK and Overseas Promotion
- Marketing & Statistics
- Conferences & Seminars

Business Shield

www.bvaa.org.uk

VALVES MANUFACTURED TO YOUR INDIVIDUAL NEEDS

ADANAC
VALVE SPECIALITIES LIMITED
www.adanac.co.uk

ADANAC MANUFACTURE AND MODIFY PIPELINE VALVES FOR SPECIALIST APPLICATIONS WITH THE BACKING OF ACCREDITED QUALITY MANAGEMENT SYSTEMS AND MAJOR MANUFACTURER APPROVALS.

PED 97/23/EC
CE 56828

COMMON VALVE MODIFICATION SERVICES:
- CRYOGENIC CONVERSION
- HEATING JACKET
- DEGREASING, PREPARATION FOR OXYGEN DUTY
- PIPE STUB ATTACHMENT
- END CONNECTION CONVERSION
- FUGITIVE EMISSION BONNET
- TRIM CHANGE
- STEM EXTENSION
- OPERATING LINKAGE
- ACTUATION

ATaCS
ADANAC TESTING AND CRYOGENIC SERVICES

THE ATaCS FACILITY AT ADANAC OFFERS SPECIALIST PRESSURE & PERFORMANCE TESTING TO USER AND NATIONAL STANDARDS INCLUDING CRYOGENIC AND FUGITIVE EMISSIONS TESTING

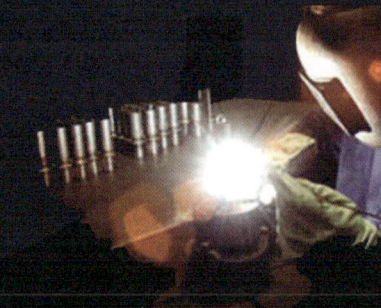

ADANAC VALVE SPECIALITIES LTD, WOOLPIT BUSINESS PARK, WOOLPIT, SUFFOLK, UK IP30 9UP
TEL: +44(0)1359240404 FAX: +44(0)1359240406 EMAIL: SALES@ADANAC.CO.UK

THE VALVE AND ACTUATOR USERS' MANUAL

6th Edition

Edited by Martin Greenhalgh CEng FIMechE

The British Valve & Actuator Association
9 Manor Park, Banbury, Oxfordshire OX16 3TB (UK)
Tel: (0)1295 221270, Fax: (0)1295 268965
Email: enquiry@bvaa.org.uk
www.bvaa.org.uk

DISCLAIMER

The publishers endeavour to ensure the accuracy of the contents of The Valve & Actuator Users' Manual. However, the Publishers do not warrant the accuracy and completeness of the material in The Valve & Actuator Users' Manual and cannot accept responsibility for any error and subsequent claims made by any third parties. The contents of The Valve & Actuator Users' Manual should not be construed as professional advice and the publishers disclaim liability for any loss, howsoever caused, arising directly or indirectly from reliance on the information in The Valve & Actuator Users' Manual.

COPYRIGHT

© All rights reserved. All material (including without limitation photographs and graphics) in The Valve & Actuator Users' Manual, unless clearly indicated to the contrary, may not be reproduced in any format and in any circumstances without the prior written consent of the publishers.
Published by The British Valve & Actuator Association Ltd, 2010.

ISBN 978-0-9567450-0-2

Control Valve
SOLUTIONS
The Smart Solution

Aberdeen - Liverpool - Lincoln

- **Supply:** We supply any type and style of control valve and related products.

- **Service/Repair:** Guaranteed 48 hour site response (UK).

- **Installation/Commissioning:** Carried out by qualified technicians.

- **Survey:** Engineers to carry out site surveys and reports.

- **On/Offshore Maintenance:** Offer OPITO registered technicians to work in National and International waters.

- **Solutions:** Tailor Made to suit your needs.

- **Free Help Line:** 24 hours.

- **Simple uncomplicated rates.**

Aberdeen - 01224 583116
Liverpool - 0845 2269975
Lincoln - 0845 2269975

e-mail - sales@controlvalvesolutions.co.uk
web - www.controlvalvesolutions.co.uk

A proud member of 'The United Kingdom Oil and Gas Industry Association Ltd'

List of Contents

Foreword	7
Acknowledgements	9
List of Advertisers	11
The History of Valves & Actuators	13
Terminology	19
Valve Basics	25
Corrosion & NACE	39
Standards & Directives	45
Fluid Sealing Materials	55
Valve Selection Techniques	61
Linear Valves	67
Gate Valves	69
Globe Valves	75
Diaphragm Valves	79
Pinch Valves	83
Rotary Valves	85
Plug Valves	87
Ball Valves	93
Butterfly Valves	101
High Performance Butterfly Valves	105
Check Valves	109
Safety Valves	115
Pressure Regulators & Control Valves	123
Recent Developments	135
Valve Operating Torques	143
Actuators – Introduction	151
Pneumatic Actuators	155
Electric Actuators	163
Hydraulic Actuators	171
Control Valve Actuators	179
Installation & Operation	183
Maintenance & Repair	189

Appendices

List of Standards	199
SI Units for Valves	210
Flow Data	211
Steam Tables	213
Index	215

Best product for process!

Engineered solutions for flow control

Weir has the capability and experience to design, deliver and maintain project-wide solutions in many of the world's most technically challenging industrial environments: engineering our products to satisfy the most stringent process requirements.

Our valves are industry renowned brands, each with an established reputation for quality engineering and reliability.

Our valve aftermarket solutions are based on our engineering heritage, applying our OEM knowledge and expertise to maintenance strategies, through to life extension and upgrade projects.

We have extensive references and a proven track record in the supply of valves across a number of key industries, including:

- Nuclear power generation
- Fossil-fired power generation
- Desalination
- Offshore oil & gas production
- Gas storage
- Refining and petrochemical
- Pulp & paper
- Intensive industrial processes

For further information please visit:
www.weirpowerindustrial.com

Excellent Power & Industrial Solutions

ALLEN STEAM TURBINES
Single and Multi-stage Steam Turbines

ATWOOD & MORRILL
Engineered Isolation & Check Valves

BATLEY VALVE
High Performance Butterfly Valves

BLAKEBOROUGH
Control & Severe Service Valves

HOPKINSONS
Parallel Slide Gate & Globe Valves

HYDROGRITTER
De-Gritting Machines

MAC VALVES
Ball & Rotary Gate Valves

ROTO-JET PUMP
High Pressure Pitot Tube Pumps

SARASIN-RSBD
Pressure Safety Devices

SEBIM
Nuclear Valves

TRICENTRIC
Triple Offset Butterfly Valves

WEMCO PUMP
Extra Heavy Duty Vortex Pumps

WEMCO SELF PRIMER
Solids Handling Self Primer Pumps

WEIR POWER & INDUSTRIAL SERVICES

Weir Power & Industrial
Divisional HQ

Pegasus House
Bramah Avenue
Scottish Enterprise Technology Park
East Kilbride, Lanarkshire
G75 0RD, UK

Tel: +44 (0) 141 308 2800
Fax: +44 (0) 141 308 2856

Foreword

The British Valve & Actuator Association was formed in 1939 and serves the interests of manufacturers, distributors and repairers of valves, actuators and related products. Its mission is to provide the means for collective representation on matters affecting the British valve and actuator industry, to promote and represent members and their products and services with end users and customers at UK and international levels, and to facilitate the growth and profitability of the industry.

The BVAA is a not-for-profit organisation, with a highly professional staff. It provides its members with a number of invaluable services. These include technical and commercial support, technical committees for the development of UK and international standards and directives, training courses on all aspects of valves and actuators and related topics, exhibitions and conferences, a HR and health & safety support service, global market forecasts, promotional DVDs, R&D projects, the industry web portal, the internationally renowned 'Valve User' magazine and website, and of course the Valve & Actuator Users' Manual, which the Association has been publishing in various guises since 1964.

BVAA members have a consistent record of quality, innovation and advanced manufacturing technology, putting them at the forefront of the global flow control industry. They also have the wealth of experience, knowledge and skills necessary to ensure that the specification and purchase of valves and actuators from a BVAA member is your route to safe and reliable flow control solutions.

I would like to thank Martin Greenhalgh, BVAA's Director Rob Bartlett, the staff at the Association and all the BVAA members who have assisted in the development of this latest issue of the Valve & Actuator Users' Manual, and I commend it to you.

Bill Whiteley
BVAA Chairman

Pro-Kits Ltd
Professional Solutions for the Valve Industry

Everything for the valve and actuator industry except valves and actuators

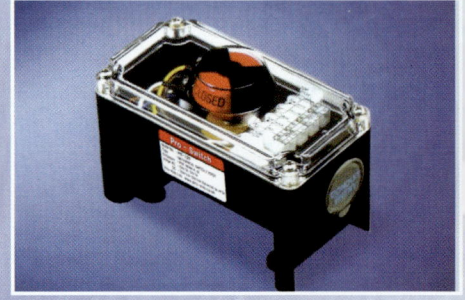

Pro-kits Ltd are ISO 9001-2008 Approved

The UK's largest independent supplier of valve related products to the industry

The Sidings, Off Debdale Lane,
Mansfield Woodhouse,
Notts NG19 7FE

Tel: +44 (0)1773 860629
Fax: +44 (0)1773 860672
E-mail: enquiries@pro-kits.co.uk

The better alternative

Acknowledgements

I was very pleased when the BVAA asked me to write a new version of the User Manual. It has given me an opportunity to put on paper some of the knowledge I have gained in over 30 years in this industry, hopefully to improve the general understanding of those who use valves and actuators. Knowledge sharing is one of the key roles of the BVAA and I have always been very grateful to the many friends and colleagues I met through the association, in particular Bob Cleary and Alan Scriven, who over the years made the time to discuss with me many important design issues.

The preparation of the Manual involved the help of many people who have provided technical review of the text and again I most appreciative of all this assistance freely given. I am also very grateful to the team of BVAA training course lecturers for their kind assistance. Two individuals deserve special mention; Rob Bartlett, BVAA Director, who has added all the pictures and illustrations to my text and Peter Churm, BVAA Technical Consultant, who has reviewed much of the content.

Our grateful thanks also go to the many BVAA members and non-members who graciously allowed us to use their photographs and schematics in this edition. Many contributors went to the trouble of staging specific photographs, drawing new figures etc., often at extremely short notice with pressing deadlines. Regrettably there were space limitations and we were not able to include every contribution proposed or received, but we thank everyone who, through their own strenuous efforts, helped make this manual the attractive and readable publication it has become.

Martin Greenhalgh CEng, FIMechE
Editor
December 2010

Your Cryogenic Valve Partner

Leading manufacturers of Cryogenic Valves Worldwide for 50 years

Bestobell Valves are one of the world's leading suppliers of cryogenic valves to the industrial gas and downstream LNG industries.

Bestobell has 50 years' experience in the design and manufacture of cryogenic valves for the processing, storage and transportation of cryogenic fluids. Bestobell Valves are in operation all over the world where safety and reliability are critical. Our innovative range of products handle numerous industrial gases such as nitrogen, oxygen, carbon dioxide, argon, ethylene and liquid natural gas (LNG).

The product range includes;
- Globe Valves
- Gate Valves
- Ball Valves
- Emergency Shut Off Gate Valves
- Manifold Fill Assemblies
- Lift and Swing Check Valves
- Safety Relief Valves
- Pressure Regulators
- Flow Divertors
- Strainers

SAFETY   QUALITY   RELIABILITY   LONG LIFE

www.bestobellvalves.com

Bestobell Valves
Your Cryogenic Valve Partner

T: +44 (0)114 224 0000
E: sales@bestobellvalves.com
Sheffield, S. Yorkshire S4 7UR UK

President Engineering Group Ltd

List of Advertisers

Inside front cover	Blackhall Engineering
Title page	Adanac Valve Specialities
Contents page	Control Valve Solutions
Foreword	Weir Power & Industrial
Acknowledgements	Pro-Kits
List of Advertisers	Bestobell Valves

Chapters

The History of Valves	Heap & Partners
Valve Terminology	Solent & Pratt/Farris Engineering, Div. of Curtiss Wright Flow Control
Valve Basics	Peter Smith Valve Company
Corrosion & NACE	valveuser.com
Standards & Directives	Flowserve Flow Control
Valve Selection Techniques	Econosto UK
Fluid Sealing Materials	BVAA Training
Linear Valves	D&D International Valves
Gate Valves	HH Valves
Globe Valves	Koso Kent Introl
Diaphragm Valves	Crane ChemPharma Flow Solutions
Pinch Valves	Linatex
Rotary Valves	Hobbs Valve
Plug Valves	BVAA Exhibitions
Ball Valves	Apollo Valves
General Purpose Butterfly Valves	eTec Engineering Services
High Performance Butterfly Valves	Hindle Cockburns
Check Valves	Goodwin International
Safety Valves	Dresser Flow Technologies
	SGS United Kingdom
Pressure Regulators & Control Valves	Dresser Flow Technologies
	Schubert & Salzer UK
Recent Developments	BEL Valves
Double Block and Bleed	Colson Industries
Valve Operating Torques	Turner Workflow Manufacturing/Doig Springs
Actuators, Introduction	Rotork Controls
Actuators, Pneumatic	Emerson Process Management
Actuators, Electric	Rotork Controls
Actuators, Hydraulic	Advanced Component Technology
Actuators for Control Valves	Rotork Process Controls Ltd
Installation & Operation	BVAA Exhibitions
Maintenance & Repair	Comid
	Steam Plant Engineering
	Excel Engineering Services
Index	St Gobain PAM UK
Inside back cover	Shipham Valves

Heap & Partners Ltd

Fluid Control Engineers Since 1866

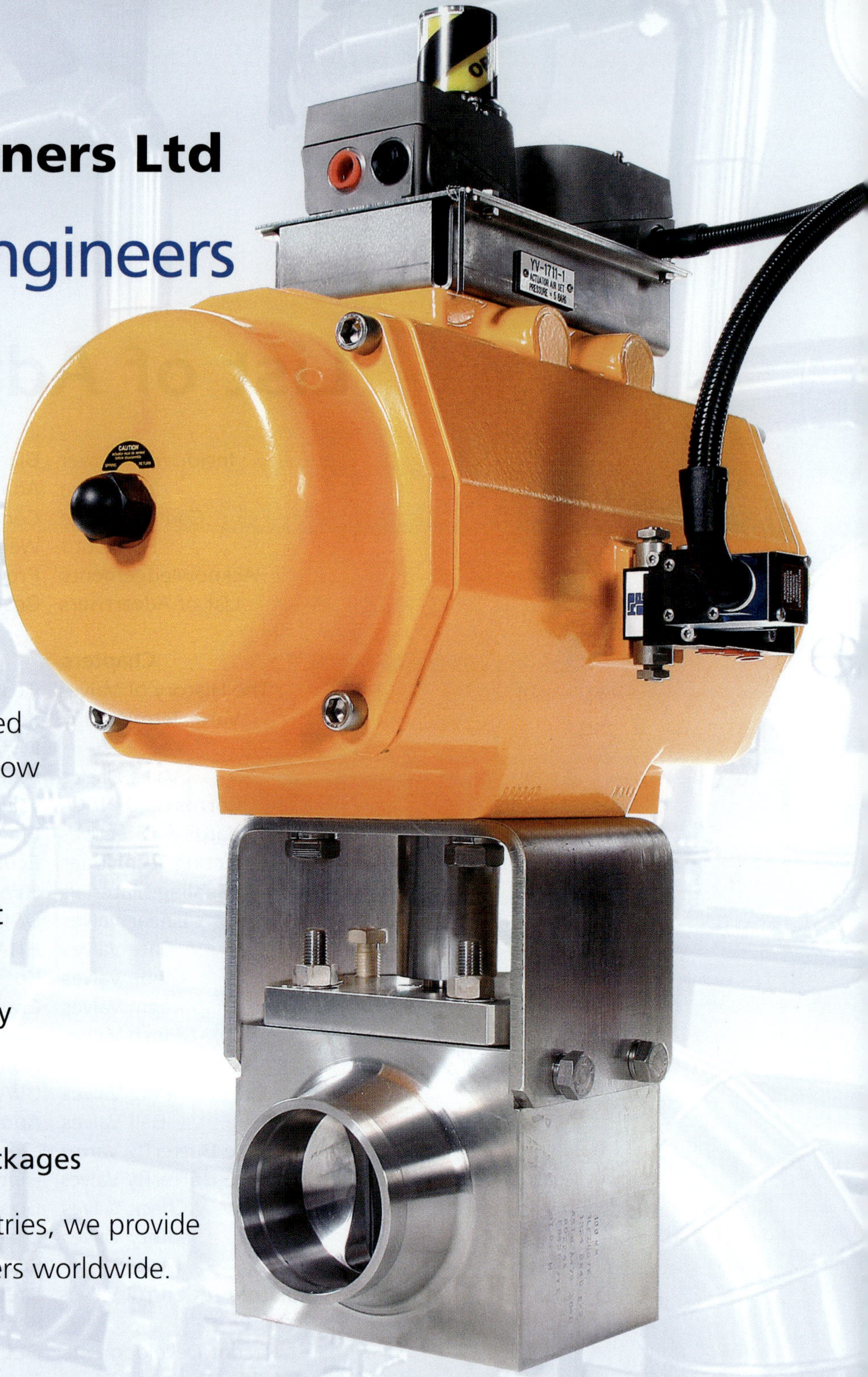

With over 140 years experience in fluid control engineering, Heap & Partners not only represent manufacturers of some of the very best flow control products available, we also provide a range of specialised services and bespoke solutions for flow related issues all over the world.

- Project Management
- Valve Customisation
- Design & Consultancy
- Manufacturing
- Distribution
- Valve & Actuator Packages

With exports to more than 80 countries, we provide engineering support to our customers worldwide.

Pharmaceutical

Chemical

Power Generation

Oil & Gas

Food & Beverages

Water

The Only Source of Knowledge is Experience

www.heaps.co.uk

The History of Valves and Actuators

Figure HV1: Bronze Cock Valve C. AD25, discovered at the ruins of the Palace of Tiberius, Capri (Sandia).

Need is very often the mother of invention and this is largely true for valve innovation.

Prior to Greek and Roman times, little is known about the methods used to control the flow of fluids. Some form of sluice gate was obviously used to hold and retain water in irrigation channels, and we know there was some knowledge of the principles of flow because of the water clocks made by the early Egyptians.

The First Valves

The Greek and Roman periods saw the development of many mechanical and hydraulic machines and the first use of valves of sophisticated design. In the case of the plug cock valve, the design remained virtually unchanged until the 19th century.

Flap valves and coin valves were the forerunners of the present swing and lift check valves and were used in the water force pumps. Bronze and brass plug cocks were in common use as stop valves on water mains and supply pipes to public and domestic buildings during the Roman period. The large bronze cock valve shown in Figure HV1 was found on Capri among the ruins of the Palace of Tiberius, built around AD 25.

The 18th and 19th Centuries
The early years of the 18th century marked the start of the industrial revolution and the arrival of the steam engine as a practical and commercial proposition. In 1698, Thomas Savery had patented his engine for the 'raising of water' and in 1705 Thomas Newcombe introduced his advanced version of Savery's engine, the atmospheric beam engine.

James Watt provided the decisive step forward in the development of the steam engine when he patented the separate condenser in 1769. His single-acting steam engine included lift valves, flap valves and a plug cock, and the catalogue from 1788 illustrating his steam engine is shown in Figure HV2.

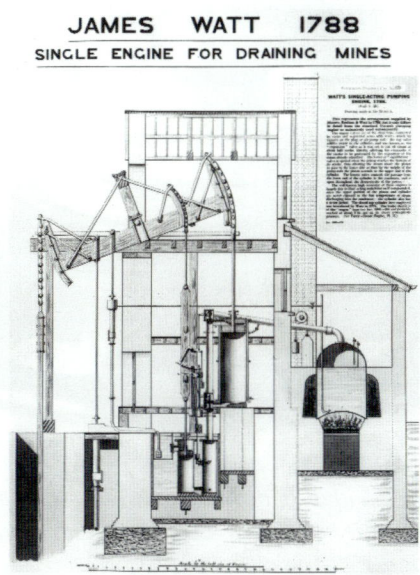

Figure HV2: James Watt's single-acting pumping engine, © Science & Society Picture Library

During the Nineteenth Century, a number of eminent engineers directed their attention to valves, notably Timothy Hackworth, who introduced adjustable springs instead of weights to the steam safety valve, as shown in Figure HV3. This valve is preserved in the Science Museum in London. Another major innovation was the introduction of the groove plug cock by Dewrance & Co. in 1875. The grooves were packed with asbestos which made the valve easier to operate and far more effective in achieving tight shut off in steam service. In 1886, Joseph Hopkinson introduced the parallel slide valve, in which the sealing of the valve was effected by the line pressure on the downstream disc, a design which is still being manufactured today and widely used in power stations for high pressure steam service.

Figure HV3: Spring arrangement of Hackworth's early steam safety valve, © Science & Society Picture Library

The Modern Era

Throughout the twentieth century and continuing today developing industrial processes have required innovative designs for valves and the industry continues to respond to these needs. New materials both metallic and non metallic have helped engineers to provide successful solutions to the range of applications we have today. New production processes and production machinery allow manufacture of designs on a volume basis that was unimaginable for the nineteenth century pioneers of this industry.

Figure HV4: Turn of the century Globe Valves from 'Peglers' – a brand that still exists today

Steam and the availability of power drove the industrial revolution. Valves for steam were therefore the first to receive the attention of the designers. Then came developments in water supply, the use of oil and gas as energy sources and finally chemicals for the many products we consume today and nuclear power. All these industries needed valves and presented manufacturers with almost daily challenges to their innovative skills.

Mostly these developments were based on existing valve types. The lubricated taper plug valve was developed during World War I by Sven Nordstrom, a Swedish engineer, who was trying to overcome the excessive leakage and sticking of ordinary plug valves (see Figure HV5). The ball, or spherical plug valve, is a relative newcomer to the valve family. Initially developed for fuel systems on aircraft during World War II, the valve was further developed in the post-war years to produce the first industrial range of ball valves. Reference had been made earlier to James Watt. He made use of a butterfly valve in his steam engine, and the first Mercedes car built around 1901 introduced a butterfly valve in the fuel intake linked to the accelerator pedal.

A completely new type, the diaphragm valve, was developed by a South African engineer named P. K. Saunders, who, working in the gold mines, was faced with excessive leakage of compressed air from the glands of the valves he was using. In 1928 he developed a valve using a diaphragm both to isolate the valve operating mechanism and also to act as the closing member, which proved a great success (see Figure HV6).

Figure HV5: The Taper Plug Valve (Flowserve Audco), which bears a remarkable resemblance to the bronze cock valve in Figure HV1

Figure HV7: Proving the versatility of the design. A 2½" Diaphragm Valve that successfully endured 36 years of continuous service (Heap & Partners / Crane Saunders)

The accidental discovery in 1938 by Dr Plunkett in Du Pont's Jackson Laboratory in New Jersey that he had polymerised tetrafluoroethylene (TFE) to a white waxy solid now known universally as PTFE, was of huge importance to the valve industry. PTFE has proved to be a remarkable material. It is resistant to almost every chemical and solvent and its surface is so slippery that virtually nothing will stick to it. Moisture does not affect it, nor does it degrade after prolonged exposure to sunlight. In the valve industry PTFE became an instant success for seals and seats not only in ball valves but in virtually all valve types. Later derivatives and modifications of PTFE continue to have a profound effect on valve engineering and application capability, for instance in fully lined products (Figure HV8).

Figure HV6: C. 1970, 4" and 2" 'Saunders' non-rising handwheel Diaphragm Valves, complete with a valve interlock system (Heap & Partners) and (below) a modern-day descendant of P K Saunders' Diaphragm Valve (Crane Saunders)

Figure HV8: A 6" ASME Class 150, PN16 PTFE Lined Ball Valve (Crane Xomox)

Over the years there has been competition between the different valve types for different applications. This has also been a driver for innovation. Many valve manufacturers directed their attention to the ball valve and a variety of new, improved designs were introduced. This has led to a much wider diversification and expansion of the capabilities of the ball valve and its use in applications in practically all sections of the valve market. The first butterfly valves used metal to metal seats, but after 1945, improvements in modern synthetic rubbers for the sealing members extended the application of the butterfly valve into many industrial fields. In recent years, the butterfly valve has been developed further to handle much higher pressures and temperatures than previously envisaged with the appearance of first the double offset (high performance) design and latterly the triple offset design.

order by individual valve makers and it wasn't until the beginning of the 1950s that industry's widespread desire to use electricity translated into the creation of standardised products for all valve types, manufactured by independent actuator companies.

Figure HV10: Large, multi-turn actuators with bolt-on electric motors, designed for rising stem gate valves. These were en-route to a refinery in Holland in 1957, where they operated for over 40 years (Rotork Controls)

Figure HV9: George H Pearson (right) photographed here in 1935, was the author of the first book on the Design of Valves and Fittings published in 1953

As process plants became more sophisticated and pressures increased the need grew for actuator development, encompassing increasingly sophisticated electric designs as well as hybrid fluid power innovations such as gas-over-oil and high pressure gas designs for pipeline applications and electro-hydraulic actuators for critical failsafe duties. Actuator development also responded to new materials and the availability of new industrial processes, machinery, etc. but the biggest driver in this business has been computer technology.

The valve is an important item of pressure equipment. Good design is important. As the industry grew good practice was shared and developed. One of the first books on the subject was by George Pearson (see Figure HV9). Today there are many books and standards, which are intended to help both the designer and the user to ensure that designs are safe and perform satisfactorily. It must always be remembered that a partnership is required between the user and the manufacturer/designer if the best results are to be achieved.

Actuators

A similar history can be told of the actuator industry although this is mostly confined to the twentieth century. The Industrial Revolution pioneered the use of water as a source of hydraulic power and many of these ingeniously actuated valves remained in service until the twentieth century. By the 1920s the use of pneumatic actuation was growing and by the 1940s was well established as a standardised product design. By this time the use of hydraulic oil to power actuators was also well established. By comparison, most electric actuators were still built to

Figure HV11: An Electric Actuator C.1957 mounted on a secondary worm gearbox and its modern-day, multi-turn counterpart with quadrant worm gearbox (Rotork Controls)

The use of this technology allows the control of the actuator in complex process plant but equally it allows the actuator to be a source of data on both its own performance and on the performance of the valve to which it is attached. Many actuators can be controlled via a single pair of wires or even no wires at all as wireless technology is being introduced. For the valve manufacturer the availability of actuators with computer technology has had the greatest impact on control valve capability.

Figure HV12: An Electric Actuator mounted to a secondary spur gearbox being commissioned on a 34" Gate Valve on an oil pipeline in France, C. 1950s (Rotork Controls)

Market Growth & Consolidation

The valve and actuator industries grew in those countries with strong domestic markets. The companies were entrepreneurial in nature, such as Joshua Hindle (Figure HV13) very often promoting a specific design that they had patented. In the 1980s and 1990s the size of the market and the needs of their customers, who were starting to operate globally, saw amalgamations and groupings of valve companies across national boundaries.

In 2009 estimated sales world wide of new valves including those fitted with actuators were US$45 billion of which over 40% was to oil and gas production and refining and the chemical industry. 36% of total sales came from the 12 largest companies, but there are estimated to be in excess of 10,000 companies in the valve industry world wide. China has the largest industry with over 6,000 manufacturers. The industry remains diverse for many reasons not least of which, is the continuing need for innovation and the opportunity that gives an entrepreneur with a good idea.

Figure HV13: A view of the yard at Joshua Hindle & Sons, C.1950s (Tyco Hindle)

Standardisation

Standardisation has always been an important part of the industry and manufacturers have a long history of such co-operation. Initially this was on a national basis. The United Kingdom trade association, then grandly titled the British Iron & Steel Valve Manufacturers Association (or BISVMA, nowadays know as BVAA) had its first meeting on 23rd November 1939, at a time when increasing production, standardization and control of exports became a matter of increasing national importance. It is interesting to note the names and companies represented, one of whom was the aforementioned Mr P. K. Saunders for the Saunders Valve Co., and the author's own grandfather, Mr J Greenhalgh. The surnames of many of these founding fathers continue today as valve brands known around the globe.

Similar trade associations helped to formulate national standards like DIN in Germany, NF in France, BS in the UK and ANSI in the USA. Today the same trade associations actively co-operate to formulate European (EN) and International (ISO) standards. There is a fuller description of these activities in the chapter on Standards & Directives.

BVAA

The British Valve & Actuator Association is the organization that eventually evolved from BISVMA. Despite contraction and consolidation in the industry in general, in 2011 BVAA has more members than ever before in its 70 year history. This perhaps illustrates the current strength of the modern day UK valve and actuator industry. Standardisation is still a core activity for BVAA, but the organization now devotes much of its time to marketing and business development for its members. BVAA is also the publisher of the widely circulated 'Valve User' magazine. You can find out more about the BVAA at www.bvaa.org.uk and Valve User at www.valveuser.com

Not all days are this perfect.

That's why our products are engineered for the harshest environments.

- Pressure Relief Valves
- Triple Offset and High Performance Butterfly Valves
- In-House Automation & Control Specialists
- Asset Management and Valve Repair Service

Two great companies. One convenient location. Over a century of service combined. Curtiss Wright Flow Control UK, Ltd. provides product and service solutions that transform the way our customers work. Solent & Pratt redefines the boundaries of isolation and butterfly control valves in design and performance. Farris Engineering sets the standard, offering pressure relief valves and PRV management solutions that support a facility's entire lifecycle. From our premier facilities located in the UK and globally, we can manufacture, set and test valves for critical service needs.

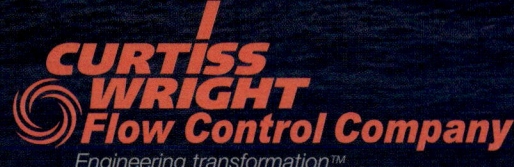

Solent & Pratt   Farris Engineering

For more information visit us on the web at http://farris.cwfc.com or http://solentpratt.cwfc.com

Terminology

Figure TM1: 'Getting into valves' will require a thorough understanding of valve and actuator terminology in order to avoid misinterpretation (Flowserve)

A clearly defined and understood language is an essential requirement within any industry so that all participants including manufacturers, engineering consultants, purchasers, distributors, users, installers and repairers can safely and effectively communicate.

The valve industry in particular has a huge diversity of products and applications. It is therefore important that its language is defined, understood and used by everyone working in the industry.

Valve and Valve Types

A Valve is a piping component which influences the fluid flow by opening, closing or partially obstructing the passage of the fluid flow or by diverting or mixing the fluid flow. There are five basic types of valves, which are distinguished by the direction of motion of their obturator (the closure device more commonly called disc, plug or ball) and the direction of flow in the seating area.

The valve types are illustrated in Figure TM2.

Gate valve - A valve in which the obturator moves in a linear motion and, in the seating area at right angles to the direction of flow.

Globe valve - A valve in which the obturator moves in a linear motion and, in the seating area in the direction of flow.

Diaphragm and Pinch valve - A valve in which the fluid flow passage through the valve is changed by the deformation of a flexible obturator. The obturator moves in a linear motion.

Plug and Ball valve - A valve in which the obturator rotates about an axis at right angles to the direction of flow and, in the open position the flow passes through the obturator.

Butterfly valve - A valve in which the obturator rotates about an axis at right angles to the direction of flow and, in the open position, the flow passes around the obturator.

Valve Functions

A valve can also be categorized relative to its function as follows:-

Isolating valve - A valve intended for use only in the closed or fully open position.

Regulating valve - A valve intended for use in any position between closed and fully open.

Control valve - A power operated device which changes the fluid flow rate in a process control system. It consists of a valve connected to an actuator with or without a positioner that is capable of changing the position of the obturator in the valve in response to a signal from the controlling system.

Safety valve - A valve which automatically, without the assistance of any energy other than that of the fluid concerned, discharges a certified quantity of the fluid so as to prevent a

Figure TM2: The Five Basic Valve Types					
Operating motion of the obturator	Linear motion		Deformation of a flexible component	Rotation about an axis at right angles to the direction of flow	
Direction of flow in the seating area	At right angles to the operating motion of the obturator	In the direction of the operating motion of the obturator	Depends on design	Through the obturator	Around the obturator
Schematic figures					
Basic types	Gate valve	Globe valve	Diaphragm and Pinch valves	Plug and ball valves	Butterfly valve
⇨ Direction of fluid flow					
→ Operating motion of the obturator					

Figure TM2: The Five Basic Valve Types

predetermined safe pressure being exceeded, and is designed to re-close and prevent further flow of fluid after normal pressure conditions of service have been restored. Additional terminology relating specifically to safety valves is given in the later chapter on Safety Valves.

Check (non-return) valve - A valve which automatically opens by fluid flow in a defined direction and automatically closes to prevent fluid flow in the reverse direction.

Diverting valve - A valve intended to influence the proportion of two or more output flows from a common input flow by changing the position of the obturator.

Mixing valve - A valve intended to influence the proportion of two or more input flows to produce a common output flow by changing the position of the obturator.

Valve Components

The various components of the valve are defined as follows:-

Shell - This is the collective term for the components forming the pressure containing envelope of the valve. It normally comprises the body and when included in the design a bonnet or cover and the body bonnet or body cover joint and associated bolting. A shell tapping is a threaded hole in the wall of the shell.

Body - The main component of the valve which provides the fluid flow passageways and the body ends. Important terms associated with the body are the body end defined as that part of the body provided with the means of connection to the piping component and the body end port defined as the fluid flow opening in the body end.

Bonnet - Component of the shell which closes an opening in the body and contains an opening for the passage of the operating mechanism. An important term associated with the bonnet is the Bonnet bushing which is a separate component mounted in the bonnet which serves as a stem guide and can also provide a back seat.

Cover - Component of the shell which provides a closure for an opening in the body (it does not have an opening for the passage of the operating mechanism).

Union nut - A threaded ring which connects the union bonnet or cover to the body.

Body bonnet/cover gasket - Gasket which seals the body bonnet/cover joint.

Pressure seal joint - Body bonnet/cover joint in which the internal fluid pressure increases the compressive loading on the bonnet/cover gasket or pressure seal ring. For high pressure valves this type of seal significantly reduces the required bolting and body bonnet/cover dimensions.

Pressure seal ring - Ring which acts as the sealing component in a pressure seal joint.

Seal weld - Weld which provides a seal between two parts, for example body and bonnet/cover.

Trim - Functional components of a valve excluding the shell components which are in contact with the fluid inside the valve. In general the relevant product standard defines the trim components but as a minimum it would be body seating surface, obturator and stem or shaft.

Seating surface - Contacting surfaces of the obturator seat and the body seat which effect valve closure. The seating surfaces can be integral or a separate seat ring assembled into the body or obturator. Hard facing defined as a deposit of molten metallic material intended to provide wear resistance or soft seat defined as a part made of soft material are commonly specified seating surface options.

Obturator - Movable component of the valve whose position in the fluid flow path permits, restricts or obstructs the fluid flow. The term disc continues to be more widely used. The ball in ball valves and plug in control valves are words, which have the same meaning as obturator, when used in association with these valve types.

By-pass - Piping loop provided to permit fluid flow from one side to the other side of the main valve obturator in its closed position.

Operating mechanism - Mechanism, which translates the motion of the operating device to the motion of the obturator. The Operating element is the component of the operating device by which the mechanical power is introduced. For manually operated valves the operating element is a handwheel, lever or chainwheel.

Actuator – Operating element which uses electrical, hydraulic or pneumatic power.

Stem - Component extending through the shell which transmits the motion from the operating device to the obturator which has a linear motion. Depending on the detail of the valve type and operating mechanism design stems can variously be described as either rising or non rising, rotating or non-rotating. The operating thread of the stem can be either in contact with the fluid, inside screw, or not in contact with the fluid, outside screw.

Figure TM3: Being able to identify and understand valve components and their functions is a very important skill (Pegler & Louden)

Stem nut - Component of the operating mechanism mounted on the obturator which together with the thread of the stem converts rotary motion into linear motion. A valve design incorporating a stem nut has an inside screw, non-rising stem.

Yoke - Component of a valve which supports the yoke sleeve, yoke bushing or the actuator. It can be a separate component or an integral part of the bonnet or actuator.

Yoke bushing - Fixed component of the operating mechanism mounted on the yoke which together with the thread of the stem converts rotary motion into linear motion. A valve design incorporating a yoke bushing has an outside screw, rising stem.

Yoke sleeve - Rotating component of the operating mechanism mounted on the yoke which together with the thread of the stem converts rotary motion into linear motion. A valve design incorporating a yoke sleeve has an outside screw, non-rotating, rising stem.

Shaft - Component extending through the shell which transmits the motion from the operating device to the obturator which has a rotary motion.

Packing chamber - Chamber of the shell provided to contain the packing.

Packing - Component made of deformable material which provides the seal of the passage of the operating mechanism through the shell.

Packing gland - Component used to compress the packing.

Gland flange - Flange bearing against a packing gland used to compress the packing.

Lantern ring - Rigid spacer used in the packing chamber to separate two packing sets.

Bellows seal - Component utilizing a bellows which provides the seal of the passage of the operating mechanism through the shell.

Soft seal - Component utilizing a resilient seal ring which provides the seal of the passage of the operating mechanism through the shell.

Seal ring bushing - Bushing designed to accept the seal ring(s) of a soft sealed operating mechanism sealing.

Diaphragm seal - Component utilizing a diaphragm which provides the seal of the passage of the operating mechanism through the shell.

Back seat - Contacting seating surfaces on the bonnet or bonnet bushing and the stem or corresponding component when the stem is fully retracted. Applicable to gate and globe valves in the open position.

Figure TM4: Left to right - 'Wafer Lugged Through Drilled,' 'Double Flanged Long Pattern' and 'Wafer non-lugged' are just a few of the many Butterfly Valve body styles available (Hobbs Valve)

Body Styles and Body Ends

The main component of the valve is the body. There are many options for different body styles and end connections to the pipe and the main ones are defined as follows:-

Straight pattern body - Body having two body end ports and where the axis of the bonnet or cover is parallel to the faces of the body end ports.

Angle pattern body - Body having two body end ports and where the faces are at right angles.

Oblique pattern body - Body having two body end ports and where the axis of the bonnet or cover is not parallel to the faces of the body end ports.

Lug type body - Body designed with threaded or unthreaded holes for bolting to the adjacent flange(s) of the pipeline.

Wafer type body - Body designed to be installed by clamping between flanges.

Multi end body - Body with more than two body end ports.

Capillary end - Body end prepared for connection to a tube by soldering or brazing.

Compression end - Body end prepared for connection to a tube by the compression of a ring or sleeve on to the outside surface of a tube by a tubing nut.

Flanged end - Body end provided with a flange for mating with a corresponding flange. Connecting flanges are usually in accordance the international standards, either ASME, EN or JIS. The dimensions of the flange in terms of thickness, mating dimensions and the different face types are precisely defined.

Socket end - Body end prepared for connection to a spigot end.

Spigot end - Body end prepared for insertion in a socket.

Threaded end - Body end provided with internal or external thread for mating with a corresponding threaded component.

Welding end - Body end prepared for welding to a corresponding end of a component. Such body ends can be of the butt welding or socket welding type. A **butt welding end** is prepared for welding to a component by abutting the ends and welding within the groove formed between the prepared ends. A **socket welding end** is prepared for insertion of a component end into the socket and joining and sealing by fillet welding.

Valve Terms

Maximum allowable pressure, PS - The maximum pressure for which the valve is designed, as specified by the manufacturer.

Maximum allowable temperature, TS - The maximum temperature for which the valve is designed, as specified by the manufacturer.

PN - A numerical designation of pressure for flanged components, which is a convenient round number for reference purposes.

Class - Alphanumeric designation used for reference purposes related to a combination of mechanical and dimensional characteristics of a component of a pipework system. It comprises the word Class followed by a dimensionless whole number. The number following the word Class does not represent a measurable value and should not be used for calculation purposes.

Allowable differential pressure - Maximum allowable static differential pressure at a given temperature of a valve when it is in the closed position.

DN (nominal size) - A numerical designation of size which is common to all components in a piping system other than components indicated by outside diameters or by thread size. It is a convenient round number for reference purposes and is only loosely related to manufacturing dimensions. The nominal size is designated by DN followed by a number.

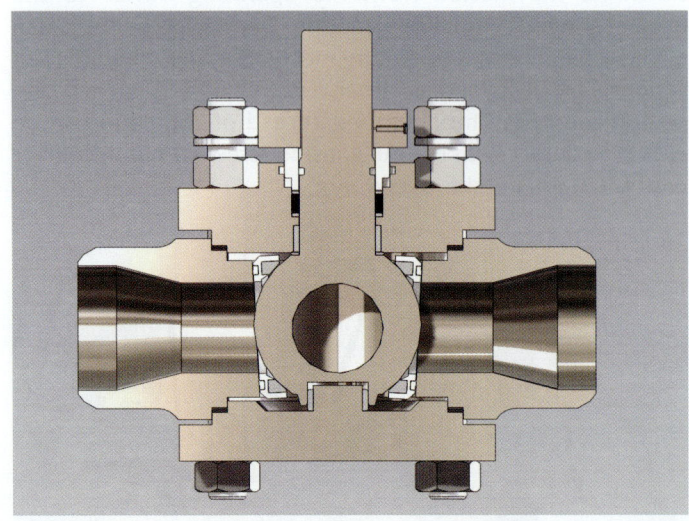

Figure TM5: DN50, Reduced Bore, Class 600, Stainless Steel, Top Entry Ball Valve with Butt Weld Ends (Truflo Marine)

NPS - Alphanumeric designation of size for components of a pipework system, which is used for reference purposes. It comprises the letters NPS followed by a dimensional number which is indirectly related to the physical size of the bore or outside diameter of the end connections. The number following the letters NPS does not represent a measurable value and should not be used for calculation purposes.

Face-to-face dimension, FTF - The distance, expressed in millimetres, between the two planes perpendicular to the valve axis located at the extremities of the body and ports or as may be specified in the relevant valve product standard.

Centre-to-face dimension, CTF - The distance, expressed in millimetres, between the plane located at the extremity of either body end port and perpendicular to its axis and the axis of the other body end port.

End-to-end dimension, ETE - Distance expressed in millimetres between the ends of the body for straight pattern valves other than those with flanged ends.

Centre-to-end dimension, CTE - Distance expressed in millimetres between one of the ends of the body and the centre of the body for angle pattern valves other than those with flanged ends.

Travel - Displacement of the obturator from the closed position.

Maximum travel - For valves with mechanical end stops, the total displacement of the obturator between these mechanical end stops. The mechanical end stops may be in the body, the bonnet or cover, the operating device, etc.

Full bore valve - A valve with the seat diameter not less than 90% of the nominal inside diameter of the body end port. Alternatively a **Reduced bore valve** has a seat diameter less than 90% and not less than 60% of the nominal inside diameter of the body end port and a **Clearway valve** has an unobstructed flow way which allows the passage of a theoretical sphere with a diameter which is not less than the nominal inside diameter of the body end port. The nominal inside diameter of the body end port for the particular valve type is generally specified in the corresponding product or performance standard.

Anti-blow out design - A valve design which ensures that the shaft or stem cannot be blown out of the shell when the valve is under pressure:

by disassembly of any external part; or

by failure of the connection between obturator and shaft or stem even when external parts are removed. External parts are parts which are not included in the bare shaft valve e.g. bracket, lever, actuator.

Anti-static design - A valve design which ensures electrical continuity between all the components in contact with the fluid and the shell.

Lining - Non replaceable part made of plastomer and/or elastomer, designed to protect a component from the fluid.

Coating - A protective layer applied to a valve component or the valve itself to provide a protection against corrosion and/or to prevent contamination of the fluid by the valve.

Operating torque - The torque applied to the operating device which is necessary to operate the valve against specified working conditions, between the open and closed positions.

Strength torque - The torque applied to the operating mechanism or when fitted, the operating device, for which the valve has been designed to resist.

Type test - A test carried out on one or more valves representative of the design and the manufacturing process to confirm conformance of the production with specified requirements.

Production test - A test carried out on valves during the manufacturing process to confirm conformance of the production with the specified requirements.

Acceptance test - A test carried out in accordance with the technical specifications of the order.

Related Organizations and websites

ASME	American Society of Mechanical Engineers; http://www.asme.org
ANSI	American National Standards Institute; http://www.ansi.org
API	American Petroleum Institute; http://www.api.org
ASTM	ASTM International, formerly American Society for Testing and Materials; http://www.astm.org
BSI	British Standards Institution; http://www.bsi-global.com
BVAA	British Valve and Actuator Association; http://www.bvaa.org.uk
CEN	Comité Européen de Normalisation; European Committee for Standardization; Europäisches Komitee für Normung; http://www.cen.eu
DIN	Deutsches Institut für Normung; http://www.din.de
EU	European Union, formerly EEC; European Economic Community; http://ec.europa.eu
EN	Europäische Norm; European Standard; Norme Européenne; See CEN website
IEC	International Electrotechnical Commission; http://www.iec.ch
ISA	International Society of Automation; http://www.isa.org
ISO	International Organization for Standardization; http://www.iso.org
JIS	Japanese Industrial Standards; http://www.jsa.or.jp
MSS	Manufacturer's Standardization Society of the Valves and Fittings Industry; http://mss-hq.org
NACE	NACE International, formerly the National Association of Corrosion Engineers; www.nace.org
VMA	Valve Manufacturers Association of America; http://www.vma.org

The Peter Smith Valve Company Ltd

An independent Private Company that is proud to trace it's history back to Sidney Smith, the inventor of the Steam Pressure Gauge in 1847, specialising in the manufacture of small to medium size valves for the control of Fluids and Gasses in most pipe conveying systems.

Manufactured in Cast Iron, Cast Steel and Gunmetal for use in major industries, Power, Process, H&V, Water and Marine. The various types include Cast Steel Parallel Slide, Globe and Angle valves, Product Approved by BSI.

- **Parallel Slide Valves**
- **Globe and Angle Stop Valves Combined Stop and Non-Return Valves**
- **Boiler Crown Valves and Lift Check Valves**
- **Swing Check Valves**
- **Equilibrium Ball Float Valves**
- **Spring Relief Valves**
- **Strainers**
- **Oblique Double Regulating Valves and Metering Stations**

The Company has it's own Design Department with an on-going development programme, planned expansion to increase the volume of production and is BSI assessed and registered for Quality Assurance to BS EN ISO 9001 Registered No. FM 1392 and are approved by BSI to apply the CE mark in compliance with European Community Pressure Equipment Directive 97/23/EC, to manufactured products at our works.

The Company operates an Occupational Health and Safety Management System and has been registered by BSI to BS OHSAS 18001, Certificate No. 512384.

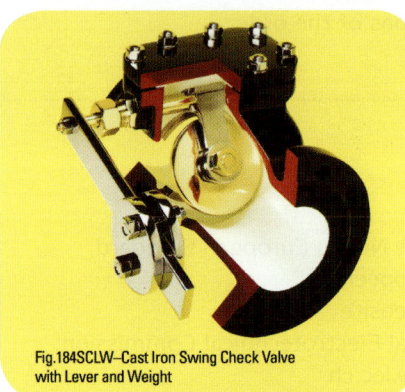

Fig.184SCLW–Cast Iron Swing Check Valve with Lever and Weight

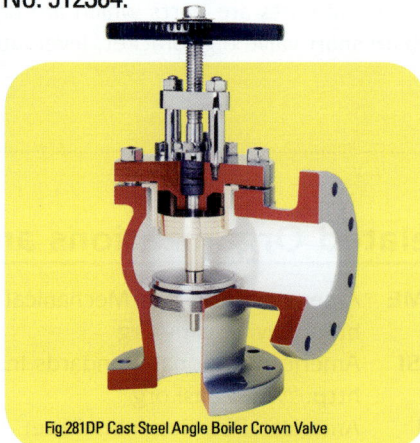

Fig.281DP Cast Steel Angle Boiler Crown Valve

Fig.501 Cast Iron Equilibrium Ball Float Valve

Fig.650 Cast Steel Parallel Slide Valve

Fig.BV203 Cast Iron Double Regulating Valve

The Peter Smith Valve Company Ltd

MANUFACTURERS OF INDUSTRIAL AND MARINE VALVES

Occupation Road, Cinderhill Road, Bulwell, Nottingham NG6 8RX

Tel: 0115 927 2831 • Fax: 0115 977 0233

www.petersmithvalve.co.uk • email sales@petersmithvalve.co.uk

H&S : BS 18001
CERT : OHS 512384

QA : BS 9001
CERT : FM 1392

Valve Basics

Figure VB1: 20" (500mm) class 150lb rated Swing Check Valve in Nickel Aluminium Bronze material for handling corrosive sea water (Shipham Valves)

The valve has two essential roles to fulfill. Firstly it is a component of a piping system and must maintain the pressure integrity and fluid containment of that system. Secondly it must control the fluid flow either in an isolating (on/off) or fully modulating mode.

The user has a huge variety of options available all of which provide those two fundamental functions. However there are a number of basic features that are common to all valves and this section will help with understanding the importance of these to a specific application.

Figure VB2: Pouring of a stainless steel two piece valve body (Apollo)

Body, Bonnet and Cover Materials

The pressure envelope comprises the body, the bonnet and any other cover, the removal of which would result in gross release of the fluid (see Figure VB3). The material selected must be strong enough to withstand all the anticipated loadings, be capable of resisting the corrosive nature of the fluid and of sufficient quality to prevent leakage through the shell wall.

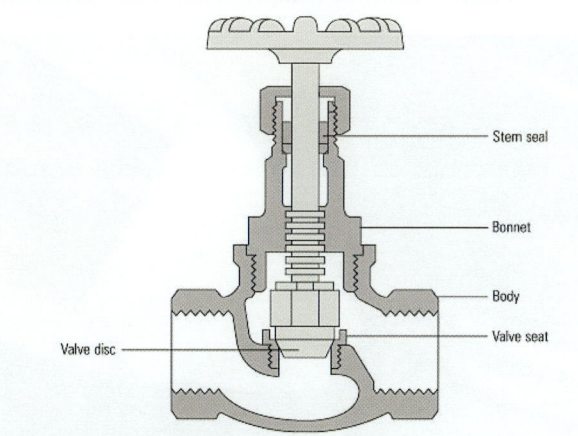

Figure VB3: Composition of a basic Globe Valve (Spirax Sarco)

Selection of the appropriate material is typically a function of minimum cost to suit the fluid and the design temperature. Usually the material is chosen to match the user's selection of pipe line material. However other factors e.g. weight may be critical and can lead to alternative choices such as titanium. Valves for water at ambient temperature and 10 bar could use grey cast iron whereas the selection for valves for Hydrofluoric Acid (HF) at 250°C is likely to be a nickel alloy Grade CY40.

BVAA VALVE & ACTUATOR USERS' MANUAL - 6th Edition

25

Figure VB4: Titanium body casting of a Swing Check Valve (Titanium Castings UK / Shipham Valves)

Material properties of strength, ductility and corrosion resistance are a function of both chemical composition and heat treatment. The need for correct heat treatment must be recognized. The latest technology for Positive Material Identification (PMI) can rapidly and accurately confirm the percentage composition of most alloying elements but this is only half the story. It is therefore very important that the supplier of pressure envelope materials is one that has demonstrated competence and the capability to produce reliable quality. It is for this reason that the European Pressure Equipment Directive 97/23/EC Essential Safety Requirements specifies that the material manufacturer must have an appropriate Quality Assurance system, certified by a competent body established within the European Community.

Figure VB5: PMI testing which identifies the composition of the metal for quality purposes (Econosto)

Cast Iron - contains Carbon (C) between 2 and 4%. It is a material easy to cast and easy to machine and therefore still an important material for example in water and gas supply applications. In its original as cast form, known as Grey Iron, it has high strength but almost zero ductility due to the presence of laminar graphite flakes. A variation is SG Iron, short for Spheroidal Graphite Iron, or sometimes known as nodular iron. By adding magnesium the graphite in SG Iron is encouraged to form in circular nodules. These nodules give the material its ductility as they help to prevent the propagation of cracks. SG Iron can be produced with a wide range of properties with increasing strength offset by decreasing ductility.

Low Alloy Steels - are the work horse materials of the valve industry. Carbon steel containing approximately 0.2% carbon and 1% manganese (Mn) can be used from -29°C to 425°C for many oil and gas applications. By heat treating to produce a fine grain structure the impact properties are improved and the material can be used down to -46°C, although the upper temperature limit is then reduced to 350°C. For temperatures above and below these limits additional alloying elements are added in small quantities; Nickel (Ni) to improve the ductility for low temperatures and Chromium (Cr) and Molybdenum (Mo) to provide resistance to creep and oxidation at higher temperatures. Creep is time dependent strain at constant stress. The material stretches as a function of load and time and at temperatures over 370°C design stresses are based on the stress to cause rupture after 100,000 hours of loading. The alloys 1¼%Cr ½%Mo and 2¼%Cr 1%Mo were developed to provide oxidation resistance and creep strengths at 520°C and 565°C respectively. Correct heat treatment is an essential part of achieving the mechanical properties of all low alloy steels. The material is heated above its austenising temperature and then cooled to ambient temperature where it returns to a ferritic structure. This cooling can be rapid using water or oil (called quenching) or it can be slower using still air (called normalising). Depending on the material a second heat treatment known as tempering may be required to provide ductility and reduce hardness.

The latest power station steam temperatures are 600°C and this has lead to the development of the material known as modified 9% Cr. This material contains 9% Cr 1% Mo and 0.05% Nitrogen (N). The nitrogen increases its yield and tensile strength, the chrome gives it high temperature oxidation resistance and the chrome and molybdenum its creep resistance. The significantly higher design stresses allow thinner wall sections to be used, which reduces thermal stresses during start up and shut down.

Stainless Steels - were developed originally for their ability to resist corrosion by the cutlery industry in Sheffield. The two main alloying elements are Chromium and Nickel. The percentage of each element changes the nature of the material either ferritic, martensitic, austenitic or duplex (a mixture of ferritic and austenitic). Pressure envelope stainless steels are usually austenitic nominally 18%Cr and 8% Ni. They typically have excellent corrosion resistance to air and water making them ideal for clean service applications in the food industry. They also find wide application in the chemical industry. The addition of molybdenum in amounts of 2 to 3% increases pitting corrosion resistance in the presence of chlorides e.g. salt. This is the Grade 316 and without molybdenum it is Grade 304.

Mechanically austenitc stainless steels have excellent notch toughness with no brittle transition temperature and are therefore used extensively in cryogenic service down to 196°C. The duplex stainless steels have high strength but correct heat treatment is very important if the two phases are to be uniformly distributed to achieve the expected combination of strength and corrosion resistance without loss of ductility.

The super austenitic stainless steels add more nickel (18 to 30%) and more molybdenum (typically 6%). The increase in these elements is intended to reduce the susceptibility of the material to stress corrosion cracking and pitting corrosion in specific chemical services (see Chapter on Corrosion and NACE).

Carbon is still an important element in stainless steels. In the ferritic and martensitic grades it contributes to hardness and tensile strength. In the austenitic grades the percentage is nominally much lower than in ferritic steels but it still provides an important contribution to tensile strength. In the austenitic grades carbon and chromium can combine together to form chromium carbide, which can be seriously detrimental to corrosion resistance. Therefore within the various austenitic grades there are low carbon e.g. CF3 with 0.03% C max and high carbon e.g. CF8 with 0.08% C max.

Nickel alloys - were developed to provide high strength at high temperature in turbine blades. They also have important applications as valve pressure envelope materials in many of the severe chemical service conditions, where corrosion resistance coupled with good mechanical strength at high temperature is required. The original nickel chrome materials (known very often by the original trade name of Inconel®) have been expanded to include the nickel chrome molybdenum series (known by the original trade name Hastelloy®).

Figure VB6: A 138" Cast Iron Butterfly Valve for power station service (Curtiss Wright Flow Control, Solent & Pratt)

Figure VB7: Three-piece, socket weld end Ball Valves in Stainless Steel and Carbon Steel (Apollo Valves)

Figure VB8: An 8" API 6A (5000psi) Choke Valve manufactured from solid Inconel 625 material with solid Tungsten Carbide trim (Koso Kent Introl)

These materials were originally developed by specialist alloy supply companies such as Special Metals Corporation (formerly Inco Alloys International), Haynes International and Carpenter Technology Corporation. While the alloys are still available from these companies under their original trade names, materials with very similar chemical analysis and mechanical properties are also available under UNS numbers using ASTM material standards.

Additional alloying elements for Inconel® are iron, molybdenum, niobium and cobalt and for Hastelloy® are iron and tungsten. An important issue with the Hastelloy® group is the very low level of carbon in the original alloys. This is possible in the wrought material by refining of the melt to remove carbon. In the cast material equivalents for example CW12MW the maximum permitted carbon is 0.12% and this increase in carbon can result in a different corrosion performance to the wrought equivalent.

The earliest nickel alloy group was developed in 1906. The alloy contains copper at approximately 30% and is known by the trade name of Monel®. An important application is hydrofluoric acid service. Additional alloying elements are iron and for the precipitation hardened grade UNS N05500 aluminium and titanium.

Copper alloys - are also an important group particularly for small bore valves ≤ DN50. It is possible to cast relatively thin pressure tight sections so that even though the per kg cost of the alloy is much higher than steel it is still competitive to make valves out of copper alloy. The most common alloying element is zinc between 30% and 50% and the resulting alloy is known as brass and is used extensively in the building services industries for water, gas and air where temperatures do not exceed 200°C. By reducing the zinc percentage to 5% and adding 5% tin and 5% lead the famous copper alloy known as gunmetal (red brass in USA) is produced. As the UK name suggests this alloy was originally developed for naval cannon but today is used in low pressure steam valves up to 260°C.

Copper alloys become stronger and more ductile as temperature goes down and they also retain excellent impact resistance to -250°C making them suitable for cryogenic service.

A third important copper alloy group is the nickel aluminium bronze group containing approximately 9% aluminium, 4.5% nickel and 4.5% iron. This material group has excellent sea water corrosion resistance in flow rates up to 4.5 m/s and good mechanical properties comparable with the austenitic and super austenitic stainless steels. However it is not always easy to produce pressure tight castings and foundry experience should be assessed when considering this material. The alloy is complex and small variations in composition can make significant changes to corrosion resistance. Therefore uniformity of analysis on large castings is important.

Trim Materials

Trim materials comprise typically the seating surfaces and the stem or shaft. The requirement for corrosion resistance on seating materials is higher than for the pressure envelope materials because significant seat leakage can result from small amounts of corrosion. Shafts and stems are usually required to be high strength as well as corrosion resistant. Another essential property for most seating face materials is resistance to sliding wear.

Many of the materials listed under body, bonnet and cover materials are suitable trim materials. Another factor to be taken into account is that the quantity of trim material is small therefore the cost benefit of using the higher strength and higher corrosion resistant materials in terms of valve life and performance is high. The manufacturer usually has a standard set of trim materials associated with each pressure envelope material. The user may need to consider having these upgraded in critical applications.

Figure VB9: Tungsten carbide coated balls for ball valves (Hardide)

Many trim materials obtain their mechanical strength and properties due to heat treatment after machining. It is helpful to the machining process to have a soft material. Once the component has been machined it can be either quenched and tempered as in the martensitic stainless steels or precipitation hardened for example for 17% Cr 4% Ni material (17-4ph). The differences in chemical composition between martensitic and ferritic stainless steels are quite subtle. A martensitic steel has lower chromium and higher carbon. If nickel is added up to 2% then chromium can be increased to 16% while still retaining the martensitic nature of the material. This particular material is Type 431 stainless and its high strength and wear resistance makes it a very useful trim material.

Another important group of trim materials particularly for seating faces is the coating group. The material may be applied by welding, spraying, plating or impregnation. The coating can be a few millimeters thick or tens of microns thick. The coating must prevent wear and corrosion of the base material and it must not be brittle or potentially initiate cracks that can then propagate into the base material.

Cobalt alloys - are usually deposited by the welding process and typically have a thickness of 3mm. Chromium is the main alloying element with additions of tungsten and carbon to give hardness between HRC35 and HRC50. The replacement of tungsten with molybdenum gives a softer deposit typical HRC33 but a higher corrosion resistance to reducing or complex environments such as sulphuric acid or sour gas. This group of alloys are commonly known by the Deloro Stellite name Stellite®. They have excellent resistance to galling wear even under high contact loads at temperatures between 500°C and 700°C depending on grade and an acceptable coefficient of friction making them very good for sliding surfaces. The thickness of the coating allows for a number of repair cycles which can recover the surface to the as new condition.

Chemical Vapour Deposition – CVD is a process by which the component surface is infused with a very hard inorganic chemical such as tungsten carbide. For valve applications the coating thickness is around 50 microns thick with hardness of Hv1200. It can be applied to all the different types of stainless steel materials. It can be applied to Titanium, which allows the use of this light weight high strength material in sliding applications without galling wear even under high contact loads. The coating process gives a uniform thickness with good surface finish allowing coating to be performed as the final operation. Small components can also be made out of solid tungsten carbide.

Figure VB10: Bush assembly, with 'as coated' tungsten carbide CVD to left, and lightly polished post-coating right (Hardide)

Figure VB11: Solid Tungsten Carbide plug and cage sleeve (Total Carbide)

Thermoplastic Polymers - have become a very important class of seating materials. The most commonly known grade is the fluoropolymer PTFE. The material is inert to most fluids and has low friction. It is therefore widely used as a seating face material and its resilient nature allows it achieve bubble tight isolation. In its pure form PTFE has a maximum service temperature of 230°C due to lack of rigidity. The addition of carbon or ceramic improves rigidity allowing these filled PTFE's to be used up to 260°C. At the other end of the temperature scale below -50°C pure PTFE starts to loose resilience and is no longer as capable of providing bubble tight isolation. Another fluoropolymer ETFE provides greater resilience at low temperatures and for temperatures up to 280°C polymers such as PEEK can be used.

Figure VB12: Thermoplastic polymers are often used for ball valve seats (Econosto)

Ceramics - are used for seating components in very severe abrasive and corrosive applications. The usual materials are aluminium oxide (alumina) and zirconium dioxide (zirconia).

End Connections

The choice of end connections for connecting a valve to its associated pipe work is dependent upon the pressure and temperature of the working fluid and the frequency of dismantling the pipeline or removing the valve from the line.

The types of end connection in general use are as follows: -

Screwed end - Male threads of various forms may be used for special purposes, but as a rule screwed end valves have female pipe threads, either tapered for assembly to taper threaded pipe, or parallel for assembly to taper or parallel threaded pipe. In taper-to-taper and in taper-to-parallel connections, the pressure-tight joint is made on the threads. In parallel-to-parallel connections, the pressure-tight joint is made by compressing a grummet or gasket against the end face of a valve. Screwed ends are confined to valve sizes ≤DN150 and are widely used for bronze valves and, to a lesser extent, in iron valves. The Class 800 pressure temperature rating is an intermediate rating applied to forged low carbon and stainless steel valves with screwed NPT female ends in sizes typically ≤DN40.

which do not require frequent dismantling. An important consideration with valves that are welded into the pipe line is that the chemical composition of the body material is often modified to either reduce the need for welding pre heat or make the post weld heat treatment easier. For example in carbon and low alloy steel valves the maximum carbon content and allowable carbon equivalent are restricted. For austenitic stainless steel the low carbon or stabilized grades are preferred.

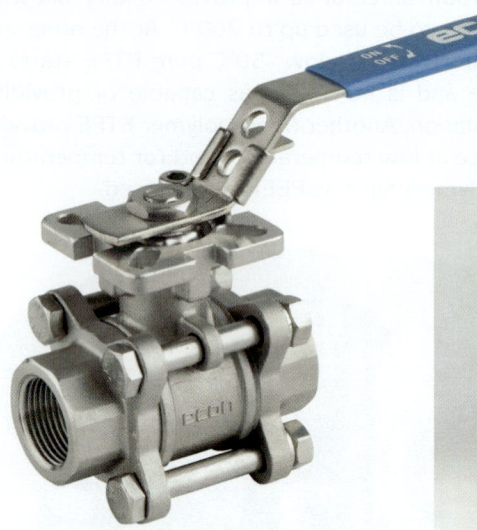

Figure VB13: Stainless steel, three piece Ball Valve with Screwed Ends (Econosto)

Flanged end - Flanged-end valves are easy to install or remove from the pipeline, being bolted to the mating pipe flanges. To ensure a tight seal, a gasket is usually fitted between the machined facing of the flanges. The type of gasket, which can be non-metallic, metallic, or a combination of both, depends upon service conditions and upon the type of flange facing employed. Bronze and iron valves are normally supplied with plain (flat) facings, and steel valves with raised faces. Gasket flange surface finish is crucial to satisfactory joint performance. Modern practice is to reduce gasket thickness and surface finish roughness. The normal surface finish is known as smooth and is in the form of a spiral groove with Ra 3.2 to 6.3 roughness. Alternative types of flange faces such as male, female, tongue, groove, or ring-joint types are also specified in both the ASME and EN flange systems. Flanged end valves are made in sizes ≤DN15.

Figure VB14: A Flanged End ball valve (Tyco Hindle)

Socket welding end - In this type, the ends of a valve are machined with a toleranced internal bore and depth to receive a plain-end pipe. A circumferential fillet weld is made on the outside of the pipe attaching the valve to the pipe. Socket-weld ends are used only on steel valves, normally in sizes ≤DN50 for higher pressure/temperature applications in pipelines not requiring frequent dismantling. A common pressure rating for this type of end connection is Class 800.

Butt welding end - In this case, the ends of the valve are bevelled to match the wall thickness and machined bevel at the end of a mating pipe. A circumferential weld is made into the groove created by the abutted mating bevels. Butt-weld ends are used only on steel valves, normally in sizes ≥DN50, for the higher pressure/temperature applications in pipelines

Figure VB15: Stainless Steel Gate Valve with Socket Weld Ends (Bestobell Valves)

Compression end - This type of valve end has a socket to receive the pipe and is fitted with a screwed union nut. The joint is made by the compression of a ring or sleeve on to the outside of a plain-end pipe, or by compressing a preformed portion of the pipe end. As a rule compression ends are used with copper tubing and steel tubing up to 65 mm diameter and are used for low pressures or where pipes may require frequent dismantling.

Capillary end - These valves are soldered to the mating pipe. The ends of the valve have a socket, machined to close tolerances, to receive the plain-end pipe. The joint is made by the flow of solder by capillarity along the annular space between the socket and the outside of the pipe. Capillary ends are commonly used with copper tubing and confined to sizes ≤DN65. The high temperature use of capillary end valves is limited due to the comparatively low melting point of the solder.

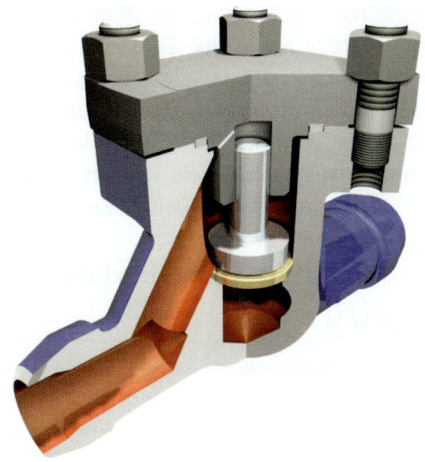

Figure VB16: Lift Check Valve with Butt Weld End connection (Weir)

Socket end - The ends of the valve are machined with a toleranced internal bore to receive the plain spigot end of the pipe, the seal being made by the insertion of a yarn ring joint, corked with lead. Other forms of socket ends use a rubber sealing ring. These ends are either in the style of flange and socket adaptors for bolting to flanged end valves, or incorporated in the valves. Socket ends are normally associated with cast iron valves for water services in sizes ≥DN50.

Spigot end - The type of socket used in the coupling or on the pipe end determines the form of the spigot ends. For cast iron pipes with lead joints the spigot end is provided with a raised band; for screwed and bolted gland and other forms of mechanical joint the spigot end is prepared to suit the joint. These spigot ends are normally associated with cast iron valves for water services in sizes ≥DN50.

Pressure Temperature (p/t) Ratings

The range of allowable service conditions that can be applied to a valve is an important and essential piece of information for the valve user. This tells the user whether the valve is suitable for his design conditions or not. The curve plotting the allowable conditions is known as the p/t curve. The allowable pressure reduces as the service temperature increases. User design conditions, which are below or on the curve, are within the capability of the valve. Conditions above the curve are beyond the capability of the valve.

The p/t curve of the valve is usually that of the end connection for flanged valves i.e. the body and bonnet are always stronger than or of equal strength to the flange. In the case of socket weld or butt welding end valves the p/t curve is that of the body and bonnet. P/t rating curves for other types of end connections are either published in specific valve product standards or manufacturers catalogues.

Flanged End p/t Ratings - ASME and EN standards publish allowable p/t ratings for the different flange designs. In both systems materials are divided into groups and each group has a defined p/t curve. Each flange design and associated p/t curve is given an alpha numeric designations e.g. Class 300, PN40. The allowable pressure is generally proportional to the material mechanical properties at temperature. For low alloy and carbon steel the yield strength or 0.2% proof stress is the mechanical property used up to 375°C. Above this temperature the creep stress to cause rupture after 100,000 hours is the applicable mechanical property. For austenitic and super austenitic materials creep stress is used at temperatures above 510°C. Both the Class and the PN flange systems also take into account elastic deformations and impose limits on the allowable stresses of the higher strength materials. In the Class system there are ceiling allowable pressures, which cannot be exceeded therefore there is no advantage to using higher strength materials at certain temperatures. In the PN system the p/t curve shows a horizontal line truncation at the maximum allowable pressure at 50°C. For PN40 the maximum allowable pressure irrespective of material strength is 40 bar. For low strength materials this truncation has an impact on allowable pressures up to 100°C whereas for the high strength materials allowable pressures at temperatures up to 450°C are truncated. The manufacturer's catalogue should always be checked to make sure that he has not imposed any de-rating or truncation on the standard flange p/t rating due to use of particular trim materials for example PTFE seats or packing or body bonnet joint gasket material.

Figure VB17 compares the Class 300 p/t rating curves in EN1759 for three material Groups 1C1 (ASTM A216WCB, A350 LF2), 1C9 (ASTM A217 WC6) and 2C1 (ASTM A351 CF8).

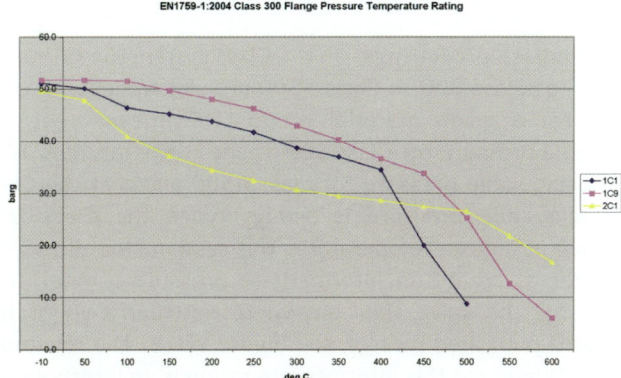

Figure VB17: p/t Rating Curves in EN1759

Figure VB18 compares the PN40 p/t rating curves in EN1092 for the same three material groups. The figures clearly show that at temperatures below 400°C the austenitic stainless steel flanges have a lower allowable pressure than the carbon and low alloy steels whereas above 500°C they have the highest allowable pressure.

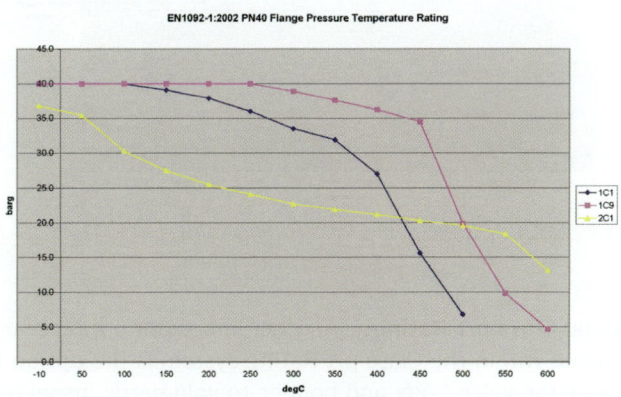

Figure VB18: p/t Rating Curves in EN1092

Butt Welding End p/t Ratings - For valves with welding ends the valve designer is no longer constrained by the contiguous series of flange pressure temperature ratings. In EN standards he can either design for the specific user design conditions or he can design to a pressure temperature rating identical to a flange rating or he can design to his own pressure temperature rating known as an intermediate rating. If he chooses to design to a pressure temperature rating identical to one of the flange ratings the standard describes the method by which he establishes the minimum body wall thickness. This wall thickness is independent of material and the allowable pressure temperature curve for the valve shell is tabulated in the standard and is the same as the equivalent Class flange rating for the same material group.

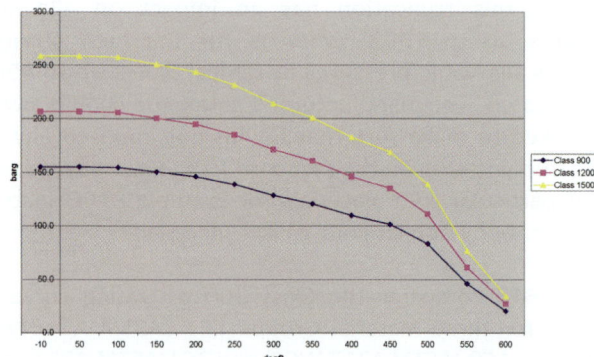

Figure VB19: p/t Rating Curves in EN12516-1, Class 1200

Intermediate p/t Ratings - In the Class system the p/t rating curves for Class 300 to Class 4500 are all parallel with each other for a given material group. This means that the allowable pressure for Class 300 at any temperature is 1/15 of the allowable pressure for Class 4500 and that the Class 600 allowable pressure is always twice the Class 300 pressure at a given temperature. This system also allows the designer to create an intermediate pressure temperature rating for the valve shell anywhere between two of the standard published ratings, see Figure VB19. The minimum wall thickness for this intermediate rating is an interpolation between the minimum thickness required for the standard rating above and below.

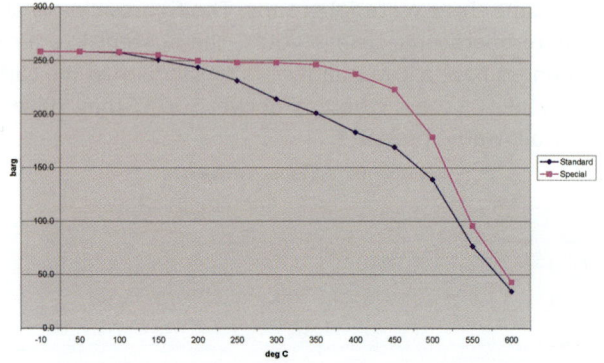

Figure VB20: p/t 'Special Class' Rating Curves in EN12516-1, Class 1500

Special Class p/t Ratings - Another important option that is available within the Class p/t Rating system is the option to subject the valve body and bonnet to volumetric inspection either by radiography or by ultrasonic testing to confirm that there are no casting defects greater than those defined in the relevant mandatory appendix. Bodies and bonnets that meet this acceptance criteria can be used for higher pressures at a given temperature than bodies and bonnets that have not been volumetrically examined, see Figure VB20. These higher p/t ratings are known as Special Class ratings and are only applicable to valves with welding ends.

Body Bonnet Joint and Stem/Shaft Seal integrity

Fluid leakage to the environment is most likely to occur from either the joint between the body and bonnet or past the packing intended to seal around the stem or shaft as it is moved during operation of the valve. It is important that the user considers these particular features of the valve design and assesses whether they will match their expectations in terms of reliability and maintainability.

The manufacturer has many options available in terms of the design of both the body bonnet joint and sealing around the stem or shaft. No manufacturer produces a new valve that leaks to the environment from these areas. The user therefore has to try to assess the future operating performance of the valve based on a data set that does not differentiate between the good and the bad. An understanding of the different options that are available will help the user select the most suitable and cost effective solution for the particular application whether it be drinking water, or toxic or flammable fluids.

Many of these sealing options involve the use of a Fluid Sealing Material and the issues concerned with the selection and use of the appropriate material are discussed in a later chapter.

Body bonnet joint or body cover joint is a static seal involving no moving components. The simplest form is to mate the two components together using a screw thread, which seals either on taper threads or on flat mating surface between the two components, if the threads are parallel. This type of joint is very common on small bore valves ≤DN25 in all pressure classes. It is also used on copper alloy valves from -196°C to 260°C. It works best when the body and bonnet are of identical materials so that differential thermal expansion is impossible. It is generally not an option on larger top entry valve designs because of the forces required to mate the two components together. For linear motion valves very often this type of joint is used with an inside screw stem design. The stem thread mates with a thread machined in the bonnet. If the stem thread seizes in the bonnet for any reason then the action of turning the handwheel may unscrew the body bonnet joint. To prevent this happening the bonnet should be locked to prevent it unscrewing. The union type bonnet is an alternative solution to overcoming this problem. A separate union nut pulls the body and the bonnet mating faces together.

The most common form of joint is to make a seal between the two components using a gasket made of either a resilient material or a resilient material reinforced with metal or all metal. This type of joint requires the two components to be held together with sufficient force to compress the gasket and to withstand the pressure acting on the area inside the gasket. This force is provided by bolting. In order for the joint to work satisfactorily the bolting must produce the desired operating stress on the gasket, relative to the operating conditions. The most common way of estimating the applied gasket stress is for the bolts to be tightened to a defined torque. This torque produces a stress in the bolt the value of which can be calculated based on assumptions. This sounds simple but in practice there are a number of potential pit falls to be considered. Firstly torque should be applied using a torque wrench so that the magnitude is known. Secondly the condition of the bolting threads and the body or bonnet or washer surface under the nut must be taken into account. Friction at all these surfaces plays a large part in determining how efficient the bolt thread is in converting torque applied to the nut to tensile stress in the bolt. Lubricating everything does improve the efficiency but that might just result in the bolt being over stressed if the torque figure being used was intended for a non-lubricated situation. Remember that torque only gives an indication of the turning power applied to the nut, not the actual load to the fastener. For small bolting sizes there is generally a big margin of safety in the design but for larger sizes used on high pressure valves it is very important that correct procedures are applied should it ever be necessary to do maintenance work on the body bonnet joint.

Figure VB21: Gaskets come in a multitude of different types (James Walker)

The grade of bolting should always be stamped on the end of the bolt and on the nut. When the valve is first assembled these should be visible when viewed from the operating device. When doing maintenance replacing the bolting with the identical grade and size is best practice.

Each type of gasket and each thickness requires potentially different bolt loading to ensure its correct performance. Much research has been done on the huge variety of gasket materials that are available to determine the suitability for a particular application. The valve manufacturer works closely with the seal supplier to ensure that the design of joint will be satisfactory and generally performance is very reliable. Issues such as thermal cycling and fluid corrosion attack on the inside of the joint or on the body or bonnet mating surfaces can still lead to leakage if there is poor detailed design.

On large diameter bonnets and high pressure applications the bolting required is large. The forces within the bolts must be reacted by the flange of the body and of the bonnet. Not only must these flanges not be overstressed they must also not be overstrained. This requires the flange to be stiff and not to rotate under the bolt load.

An alternative design is the pressure seal bonnet (Figure VB22) which uses fluid pressure to push the bonnet upward against a seal significantly reducing the bolt loading required and hence the flange thickness. In the pressure seal design the bonnet is placed inside the body and then the seal and a segmented ring is placed on top of the bonnet. Studs fitted to the bonnet are then used to jack the bonnet upwards until it has pre loaded the seal by the desired amount. The segmented ring when assembled is larger than the body ID and therefore prevents the seal and bonnet from being blown out of the body. This type of joint offers a number of advantages for high pressure high temperature applications namely lower weight more compact design, better reliability as the smaller bolting is less influenced by thermal cycling.

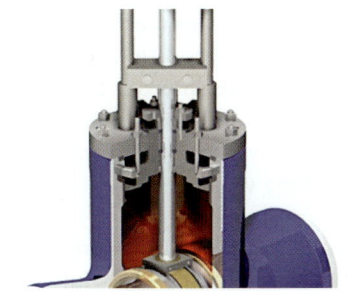

Figure VB22: Pressure Seal Bonnet (Weir)

Stem shaft seal is the place where environmental leakage is most likely as the seal needs to accommodate movement. The traditional method of sealing is to surround the stem or shaft with a flexible packing which is compressed against the shaft or stem by applying load to the gland. Rotary shaft motion for the butterfly, ball and plug valve is the easiest to seal. Pure linear motion of the gate valve and of the non-rotating stem globe valve is more difficult to seal and the most difficult of all is the helical motion of the rotating and rising stem globe valve. Good stem sealing results from good detailed design and tight tolerances in manufacture none of which can easily be confirmed by looking at manufacturers catalogues. For this reason type testing is becoming very important in assessing and comparing gland sealing performance. See the chapter on Standards & Directives.

Stem and shaft diameters need to be circular and have a surface finish compatible with the resilient packing material providing the seal. The packing chamber needs to be concentric with the shaft or stem so that the packing is loaded equally around 360°. For linear motion valves the stem diameter must not vary along the length of the stem and it must also be straight.

The stem or shaft seal did not leak in the new valve. Reliability of this seal in service is a function of how well the manufacturer's design will maintain the as new conditions. In general there are two causes of loss of as new

Figure VB23: A 6" Class 300 Globe Valve undergoing emission testing of the stem seal by sniffing method (Adanac)

conditions. The first is wear of the resilient packing due to repeated shaft rotation or stem movement. To compensate for this wear designs incorporate springs, i.e. live loading or pressure energized seals in order to maintain a constant as new loading between the packing and the stem or shaft. High cycle isolating valves or control valves are most likely to suffer from packing wear.

The second cause is contamination/damage of the seal and this can be created either by the fluid or the ambient environment or a combination of the two. Linear motion valves suffer more from contamination/damage than do rotary as a part of the stem moves from the fluid into the packing and back out again on every open and close cycle. Equally part of the stem exposed to the ambient environment enters the top section of the packing during closing. Either of these movements can do damage to the packing and lead to leakage.

Thermal cycling can also be damaging due to expansion-contraction of the stem and packing chamber, which can cause additional internal compaction of the packing, and unless there is sufficient resilience, any additional compression set leads to a loss of overall gland tightness. There is also the issue of stem taper created by different temperatures inside and outside the valve particularly on high temperature steam. Live loading is often provided to accommodate thermal cycling, perhaps more so than with frictional wear.

A common means of providing additional shaft or stem seal integrity with a traditional packing is to provide two seals (see Figure VB24) usually separated with a spacer or lantern ring. The difficulty with this type of design in linear motion valves is maintaining sufficient compressive loading on the lower packing. This can lead to early leakage from the bottom seal although the leakage would be detected through a hole drilled into the lantern ring area.

An important factor that must be remembered in compressing the packing is that the friction force between stem or shaft and packing needs to be overcome by the operating torque. It has been known that valves become inoperable because too much compression is applied to the packing.

Bellows seal is an alternative means that is available to provide a stem seal in linear motion valves (see Figure VB25). The bellows is a thin wall usually multi layer metallic element that is welded to the stem and either welded to the bonnet or to a metal gasket forming part of the body bonnet joint. As the stem moves the bellows stretches or contracts maintaining the fluid seal.

The bellows is a highly stressed component subjected to cyclic compressive and tensile loading. These conditions generate fatigue type failure and bellows have a finite life as a result. The life is a function of the length to movement ratio, the external pressure, the material and design of the bellows and the method of manufacture. Valve manufacturers quote a cycle life and it is important to understand the assumptions behind the cycle life. Sometimes this cycle life is based on actual test results or alternatively on calculations.

Figure VB24: Double Shaft Seal arrangements (Metso Jamesbury)

It is usual in bellows seal isolation and control valves to provide a secondary seal in the form of a traditional flexible packing which acts as back up should the primary bellows seal fail.

Bellows are used on safety valves, globe isolation valves, globe control valves and gate valves. Each type of valve has different requirements for the number of open closed movements and length of stroke and therefore the bellows for each is a different design. Bellows can be made from a series of cones that are circumferentially welded together around the mating diameters or they can be made from two, three or four nested thin wall tubes that are hydro formed into the convoluted shape.

The length of the bellows can add substantially to the overall height of the valve if the design cycle life and/or stroke length is large for example in control valves or gate valves.

The bellows material ranges from type 321 (stabilized type 304) stainless steel to UNS10276 (Hastelloy® C276). They are expensive components and add significantly to overall valve cost, but when the fluid is radioactive, toxic, or expensive to produce the cost of plant shutdown time, clean up costs, high insurance premiums, and loss of public confidence in the plant, is likely to be significantly higher than the investment in this type of stem seal.

Flow Characteristics

It is important for the design of the overall piping system that the pressure loss across the valve in the full open position is comparable between different designs and types of valves. Manufacturers make decisions in their designs about valve body shape, seat bores, method of body manufacture, the shape and position in the flow path of internal components all of which affect the pressure loss as fluid flows through the valve.

Figure VB25: Bellows Seal Globe Valve with auxiliary PTFE packed gland (Shaw Valve)

The coefficient of flow is the universally recognized measure of comparison between different valve designs. Manufacturers will provide this coefficient for each of their designs. This coefficient may be a calculated value or one obtained by actual flow testing.

Two flow coefficients are commonly quoted, C_v based on USA imperial units and K_v based on metric units.

$$C_v = Q\sqrt{\frac{SG}{\Delta p}} \qquad C_v = 1.17 K_v \qquad K_v = Q\sqrt{\frac{SG}{\Delta p}}$$

Where C_v = flow coefficient of the valve
  Q = flow rate in US gal/min
  Δp = pressure drop across the valve in lb/in²
  SG = specific gravity of fluid (water =1)

where K_v = flow coefficient of the valve
  Q = flow rate in m³/hr
  Δp = pressure drop across the valve in bar
  SG = specific gravity of fluid (water =1)

The C_v and K_v are the flow rates of pure water in US gal/min or m³/hr that will generate respectively 1psi or 1 bar pressure loss across the valve.

Pressure Testing

Pressure testing is the performance test that each new valve undergoes to prove that it is capable of fulfilling the two roles of maintaining the pressure integrity and fluid containment of the system into which it is to be installed and providing the specified level of leak tightness for isolating valves or fluid flow control for control valves.

The testing procedures and acceptance criteria are laid down in international standards both within ASME, API or EN and these standards are used by all manufacturers.

There are small differences between the different standards but each describes a test method to be used, a test pressure, a test fluid, a test duration time and an acceptance criteria.

When considering the selection of a valve for a particular application the user needs to understand the nature of the testing that the manufacturer routinely carries out on the products he manufactures. The testing the manufacturer does is part of a comprehensive manufacturing quality control process that should ensure that 100% of the valves meet the final criteria. Failures at the final test reflect on the robustness of the manufacturers' processes to achieve the goal of 100% first time acceptance. It is not in the interests of the manufacturer for sub standard valves to leave the factory and so he is rigorous in the level of testing that is applied. Those involved in the testing are able easily to spot valves that do not perform as they should do because of the experience they have from testing many valves.

It is common practice for users to choose a manufacturer and then to impose their own set of pressure tests on the manufacturer. It is not clear as to whether this is done so that the valves they purchase can be considered of better quality than those normally supplied by the manufacturer or whether they doubt the standards that the manufacturer has adopted.

Figure VB26: Pressure testing on a valve (Emerson)

There are three fundamental pressure tests that have to be carried out on every new valve: -

The strength test applies a pressure usually 1.5 x PS at 20°C for a short period of time to make sure that the valve shell is strong enough. The acceptance criteria is no gross leakage i.e. bursting.

BVAA VALVE & ACTUATOR USERS' MANUAL - 6th Edition

The shell leakage test applies an internal pressure to the assembled valve. The magnitude of this pressure is dependent on the test fluid and the test time is dependent on the test fluid, the size of valve and the required shell leakage integrity. Test fluids can be water, air, nitrogen, helium or a helium/nitrogen mix. Test times vary from tens of seconds to hundreds of seconds. Pressures vary from 1.5 x PS at 20°C to 1bar. The acceptance criteria is no visible or detectable leakage using a technique appropriate to the testing fluid. If water is the test fluid the external surface is examined visually. If air or nitrogen is the test fluid the body can be immersed in water to show bubbles escaping from any leak site or the surface sprayed with leak detection fluid. If helium is the test gas the surface is sniffed with a mass spectrometer. The lower the viscosity of the test fluid, the higher the test fluid pressure, the longer the test time and the more sophisticated the leak detection equipment the higher will be the achieved standard of shell leakage. The test adopted by the EN industrial valve product standards is testing with water at 1.5 x PS at 20°C or air at 6 ± 1 bar. Specified test times are 15 seconds for ≤DN50, 60 seconds DN65 to DN200, 180 seconds ≥DN250.

The seat leakage test applies a pressure to the closed valve. The magnitude of this pressure is dependent on the test fluid and the test time is dependent on the test fluid and the required seat leakage integrity. Test fluids can be water, air, nitrogen, helium or a helium/nitrogen mix. Test times are typically tens of seconds and are shorter than shell leakage test times except on small valves ≤DN50. Pressures vary from 1.5 x PS at 20°C to 1bar. The acceptance criteria is selected from a range specified in the standard varying from Rate A no visible or detectable leakage to Rate G 2.0xDNmm^3 liquid (6000xDNmm^3 gas) in the specified time. The different leakage rates are there to accommodate the capability of different types of valves and to a much more limited extent, different manufacturers' capabilities. Even the largest allowable leakage rate is equivalent to only 0.002% of the nominal flow in the pipe. Rate A is the rate most commonly specified in isolating valve standards and the rate that most manufacturers achieve. It must be remembered that the criteria is no visible leakage over a specified time and that in many valve types direct visual access to the seating surfaces on small valves is not possible. Manufacturer's therefore use different techniques to try and detect the seat leakage and each technique has potential limitations. For example, when testing with liquid, falling movement of the pressure gauge reading would be an indication of leakage due to the incompressibility of the fluid being used. However the volume of the system supplying the valve and the graduation and size of pressure gauge will influence the sensitivity of this method. When testing with air leakage can be collected at the outlet flange and fed via a tube into a container filled with water. Zero bubble leakage is probably the specified criteria. The sensitivity of this test is influenced by the bore of tube, the volume of air between the seating faces and the water meniscus in the tube and the testing time.

Other tests that may be required depending on the valve are back seat test and operation test against the specified differential pressure. This latter is very relevant to actuated valves particularly those with a spring to close option. Valves that have been back-seat tested using water as the test medium may have caused the packing to become wet. This can lead to pitting corrosion problems on austenitic stainless steel stems if the test water used or the packing material has a high chloride content. Cleaning and drying the gland before fitting the packing and placing into storage will mitigate against this problem.

Marking

It is absolutely essential that all valves are clearly marked with the pressure rating for which they are suitable. International standards specify the nature and location of the marking. It is very important that before any valve is installed that it is checked to see that it is correctly marked and that this marking confirms that it is suitable for the intended operating conditions.

Pressure temperature ratings are a function of dimensions and material. Therefore the most important information that must be on the valve is the material of the body and bonnet. This can be either cast or stamped onto the body. It can be in the form of an alpha numeric group e.g. WCB, LF2, AB2 or a numeric group e.g. 1.0619, 1.4408.

The second most important information is the PN or Class rating. For a flanged valve this may be the same as the flanged end connection. It may also be that the body has a higher rating than the end connection so both should be checked. Again this should be cast or stamped onto the body although sometimes in small valves this is not practical and it has to be on a plate attached to the body.

Other mandatory information that should be marked either on the body or if size does not permit on a separate label, is size, either DN or NPS and the manufacturer's name or trade mark. In the EU valves supplied in accordance with the Pressure Equipment Directive 97/23/EC must also be marked with the date of manufacture, the manufacturer's address or means of identification and model or type identification of the valve.

If the valve is uni-directional the correct flow direction shall also be marked by an arrow stamped or cast or permanently attached to the body.

Valves that have a truncated pressure temperature rating, due to materials of construction e.g. soft seats or because the actuator is only suitable for a limited differential pressure, shall have this information on a name plate securely fixed to the valve. The name plate should also give details of trim materials and any applicable international product standard to which the manufacturer is claiming that the valve complies. The melt identification of the material should be stamped on the relevant shell component.

Figure VB27: Clear marking is essential (Schubert & Salzer)

Figure VB28: Gunmetal Ball Valve for marine applications, with a flange one end and a Storz fire hose connection the other. Marking is cast directly into the body, end connections and handle (Econosto / Brooksbank)

BVAA® 70 years of service to industry

VALVEuser.com

Valve User readers, you now have your own dedicated and official valveuser.com website!

Valveuser.com is the new, online presentation of the entire back catalogue of Valve User magazine. Every editorial, press release, YourVU comment, MasterClass article and news story is now preserved forever in a readily accessible format.

Valveuser.com - *the easiest search engine imaginable*

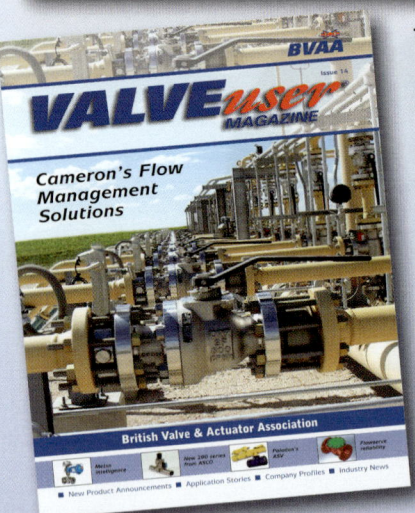

The website is keyword sensitive, enabling any topics of interest to be easily found within the thousands of articles in our archive. To make it even easier to use and search, articles are also categorised by topic and company, making it the most user-friendly valve industry resource on the web.

You can even download the latest issue of Valve User magazine, or order yourself a hard copy!

The 'upgrade' option gives a direct link to your article from the valveuser.com homepage

ADVERTISING OPPORTUNITIES

There are many advertising opportunities available on Valveuser.com, such as:

- Alternating Headline Banner Ad opportunities (viewable on every web page).
- Homepage scrolling side-bar ads
- Sponsoring of the specific article categories is available at an insanely low rate
- BVAA Members can opt to have their article 'upgraded' to 'feature' (direct hotlink) on the homepage

As BVAA is a not-for-profit organisation, our advertising fees are typically 1/10th of commercial rates, with an additional 50% discount for BVAA members.

Don't miss out on this fantastic opportunity to get your company seen by thousands at very low cost. For advertising opportunities, media information and rate card please contact BVAA on enquiry@bvaa.org.uk

Corrosion and NACE

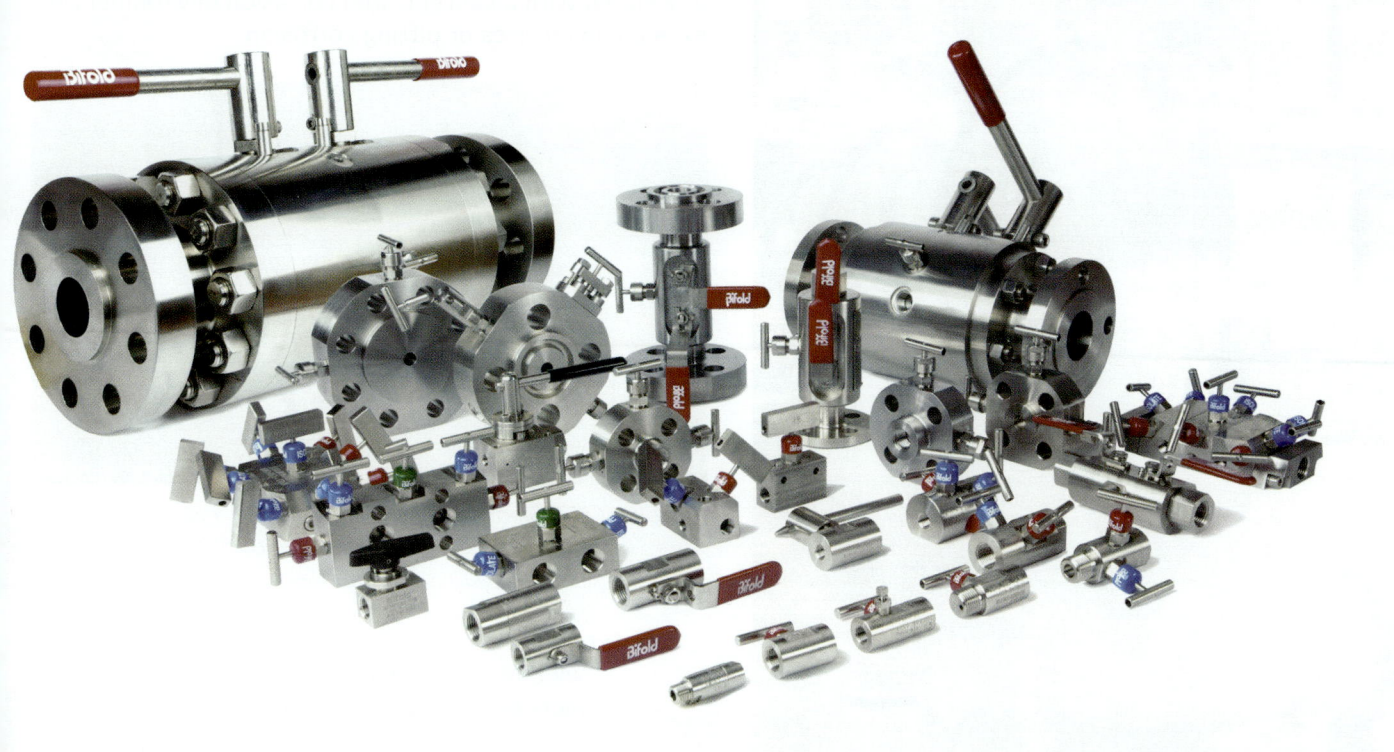

Figure CN1: A wide range of valves are designed so that their wetted metallic parts meet NACE Standard MR-01-75 (Bifold)

Corrosive Attack

Corrosive attack can be either uniform or localised. Uniform attack is relatively easy to design for either by the provision of additional material in the form of a corrosion allowance or by selection of a suitable material that suffers almost negligible uniform corrosion in the service conditions. It is important having selected a supposedly corrosion resistant material that there is an understanding of the mechanism by which uniform corrosion is inhibited.

For example stainless steel forms a passivating layer of chromium oxide on the surface so low levels of oxygen in the service conditions of stainless steel will create corrosion issues. Equally passivating surfaces may be broken down by flowing fluids resulting in significant increases in corrosion when flow rates exceed threshold levels.

Intergranular corrosion

This is normally confined to the non-stabilized varieties of austenitic stainless steel containing in excess of 0.03% carbon. When alloys of this type are heated in the range 500 to 800°C, for example if welding is required, the carbon in excess of 0.03% precipitates preferentially as chromium carbide around the grain boundaries. This robs the regions adjacent to the grain boundaries of Chromium and renders them susceptible to inter-granular attack.

To reduce the susceptibility of the material to this form of attack it is always solution annealed and water quenched. By water quenching from the solution temperature the carbon is fixed in solution and thereby prevented from forming a carbide with chromium. This heat treatment must be undertaken after welding and or post weld heat treatment unless resistant grades are used.

The susceptibility is further reduced by the carbon content being 0.03% or less. This requires additional decarburisation of the melt, which adds to the cost of castings. Forged or wrought material is generally low carbon but has the mechanical properties of the higher carbon material. Very often forged or wrought material is supplied dual certified and the user gets the advantage of the low carbon without paying the price of reduced mechanical strength.

The final method is to add either niobium or titanium to the base analysis. These elements have a fundamentally greater affinity for carbon than has chromium and so any carbides formed or precipitated will be based upon niobium or titanium. These are the stabilised austenitic stainless steels grades.

If re-heated above 500°C, sensitisation will occur again. For high temperature applications stabilised grades are normally used. The low carbon grades have insufficient creep strength. It should be noted that the sensitisation is only a problem at low temperatures where wet corrosion occurs. Many refinery operators allow sensitisation and deal with corrosion problems by nitrogen purging and caustic washing.

be carefully assessed against the risks of SCC. Fracture mode is normally trans granular but may be inter granular if the material is sensitised along grain boundaries or a mixture of both. The cracking is usually very localised and associated with a concentrated corrosion environment for example in crevices or pitting corrosion.

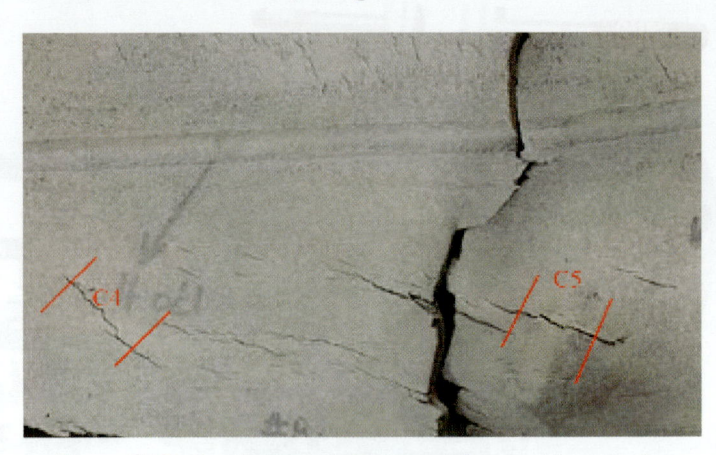

Figure CN3: Small cracks joining together and intersecting circumferential cracks on the exterior of a pipeline (© Dr. Rust Inc., courtesy of Bob Heidersbach/Wiley/NACE) R. Sutherby, T. Hamre, and J. Purcell, "Circumferential Stress Corrosion Cracking Rupture Case Study," NACE Northern Area Western Conference, February 6-9, 2006, Calgary, Alberta, Canada.

Figure CN2: Choke Valves for sour service application (NACE) for the offshore industry (Weir)

Stress Corrosion Cracking (SCC)

SCC is a well documented phenomenon. The most common cases of SCC are chloride stress corrosion cracking of austenitic and duplex stainless steel in the presence of chloride ions, season cracking of brass in the presence of ammonia and caustic cracking of steel in the presence of a strong alkali. In all cases the material must be susceptible, there must be an anodic corrosion environment present and the material must be under stress either a residual stress resulting from welding or cold working or an applied loading. The severity of attack increases with temperature. Carbon steels have the advantage of a relatively high threshold stress for most environments, consequently it is relatively easy to reduce the residual stresses to a low enough level. In contrast, austenitic stainless steels have a very low threshold stress for chloride SCC. This sometimes makes it a good choice to fit carbon steel body bonnet bolting in stainless steel valves. The material usually fails quickly and at a value of externally applied stress, much lower than would be expected from its normal mechanical properties. The most vulnerable areas are likely to be associated with welds and valves with structural welds as part of their design should

Sulphide Stress Cracking (SSC)

SSC is a form of hydrogen embrittlement. Hydrogen dissolves in all metals to a moderate extent. It is a very small atom, and fits in between the metal atoms in the crystals of the metal. In particular it is attracted to regions of high tensile stress where the metal lattice structure is dilated. Thus, it is drawn to the regions ahead of cracks or notches that are under stress leading to embrittlement of the metal.

The bcc (body-centred cubic) crystal structure of ferritic iron has relatively small holes between the metal atoms, but the channels between these holes are relatively wide. Consequently, hydrogen has a relatively low solubility in ferritic iron, but a relatively high diffusion coefficient. In contrast the holes in the fcc (face-centred cubic) austenite lattice are larger, but the channels between them are smaller, so materials such as austenitic stainless steel have a higher hydrogen solubility and a lower diffusion coefficient.

Consequently, it usually takes very much longer (years rather than days) and much higher amounts of hydrogen are required for austenitic materials to become embrittled by hydrogen diffusing in from the surface than it does for ferritic materials, and austenitic alloys are often immune from the effects of hydrogen.

The presence of H_2S in the oil or gas leads to the generation of hydrogen atoms from acid corrosion. If the hydrogen atoms do not combine to form gaseous hydrogen they are available to diffuse into the metal surface. The sulphides resulting from the corrosion process poison the formation of molecular Hydrogen and so encourage this diffusion.

The embrittlement process is associated with low ductility so high strength metallic materials and hard weld zones are prone to SSC. Inclusion stringers, particularly manganese sulphide, absorb hydrogen and this may also initiate or propagate cracks.

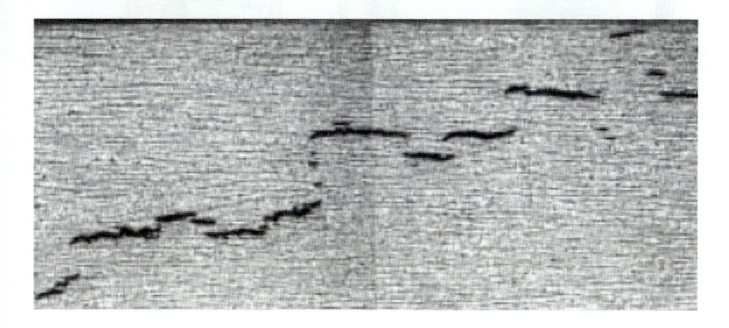

Figure CN4: Hydrogen-induced cracking (HIC) in pipeline steel (© Dr. Rust Inc., courtesy of Bob Heidersbach/Wiley/ASM International)
R. D. Kane, Corrosion in Petroleum Refining and Petrochemical Operations, in Metals Handbook, Volume 13C--Environments and Industries,
ASM International, Metals Park, Ohio, 2003, pp. 967-1014.

Pitting corrosion

A localized type of attack occurring in chloride solutions due to a localized loss of surface passivity. Environments which favour pitting such as stagnant solutions containing some dissolved oxygen usually enhance stress corrosion cracking. Conditions giving a high alkalinity, turbulence and low temperatures are factors which help mitigate against this type of failure. Alloys containing high nickel content or the additions of molybdenum help in combating this form of corrosion.

Corrosion resistance tables

There are a number of tables published in the general literature, which summarise the most reliable laboratory data available on the corrosion resistance of the metals commonly used in the construction of valves. It must be noted, however, that the performance of a material in actual plant service can be influenced, sometimes quite significantly, by many factors, such as the design and operating characteristics of the valve itself, the type of process involved, fluctuations in concentration and temperature of the fluid, presence of other chemicals, and so on.

Materials selection will usually be based upon expected performance and cost. A gain in one value of materials selection will often mean a loss of another property. This means that it is usually necessary for the user to decide upon the relative merits of a particular selection of materials.

It must be remembered that the valve manufacturer may not be a metallurgical expert in the particular field involved. Experience of the corrosive behaviour of a material in the particular fluid is the only reliable guide to materials selection and this information is generally best known by the user, particularly when the fluid or fluid mixture is unique to their process. Tabulated data should, therefore, be used only as a guide to selection of the most likely materials and must not be construed as a guarantee of their suitability. In all cases where there is any doubt, further advice should be sought from the material supplier using all available information including corrosion tests carried out on site if at all possible.

Chemical Resistance of plastic and rubber linings

The selection of the appropriate lining material for a lined valve is equally as complex as the selection of a metal for corrosion resistance. In many cases more than one material will be suitable, offering a chance to consider economy, availability and rationalization. Use should be made of manufacturers experience and expertise as well as consulting published materials data. Materials selection can be extremely complex due to the combined effects of temperature, pressure, concentration, and additives in the fluid. Wherever possible field tests should be conducted if fluid conditions are complex.

Figure CN5: A 6 inch, Class 150, PN16 Lined Ball Valve (Crane Xomox)

NACE MR0175 Materials for Use in H$_2$S Containing Environments in Oil and Gas Production

Introduction

NACE is an abbreviation for the National Association of Corrosion Engineers. The MR0175 standard addresses all mechanisms of cracking that can be caused by the presence of hydrogen sulphide (H$_2$S) in oil and gas, so called sour service. The requirements of MR0175 are intended to help prevent catastrophic failure of and costly corrosion damage to equipment used in oil and gas production. They are intended to supplement not replace requirements specified in design codes, standards and national regulations.

MR0175 lists materials which have already been qualified for defined H$_2$S service environments together with the relevant requirements. Part 2 of the standard is for carbon steel, low alloy steel and cast iron. Part 3 is for corrosion resistant alloys and other alloys.

If a particular material is not listed in MR0175, the standard describes the protocol that must be followed in order to qualify the material for use either in a specific sour service application or a range of sour services.

An essential and very important feature of MR0175 is that it recognises that providing equipment that will operate satisfactorily in an H$_2$S environment requires co-operation and collaboration between user and manufacturer. It is the user's responsibility to determine the operating conditions, to specify when the standard applies and to ensure that a material will be suitable in the intended environment. It is the equipment manufacturer's responsibility for meeting the specified metallurgical requirements.

Lining and internal coating is not permitted by MR0175 as protection against sour service (H$_2$S).

Sour environment

Fluids containing water as a liquid and H$_2$S are sour environments. Hazards associated with sour environments in addition to cracking of metallic materials include: -

Toxicity - Hydrogen sulphide is a highly toxic gas and has been responsible for many industrial poisoning accidents, a number of which have been fatal. The maximum allowable concentration in the air for an 8-hour work day is 10 parts per million (ppm), well above the level detectable by smell. However, the olefactory nerves can become deadened to the odour after exposures of 2 to15 minutes, dependent on concentration, so that odour is not a completely reliable alarm system. Being heavier than air, it tends to accumulate at the bottom of poorly ventilated spaces. Concentrations over 1000 ppm cause immediate collapse with loss of breathing, even after inhalation of a single breath.

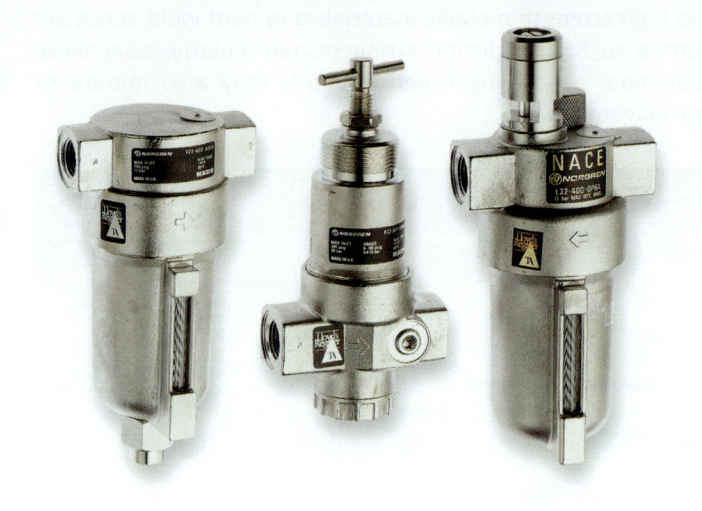

Figure CN6: The metallic parts of these pneumatic Filters and Pressure Regulators meet NACE Standard MR-01-75 (Norgren)

Fire and explosion hazards - Hydrogen sulphide is a flammable gas, yielding poisonous sulphur dioxide as a combustion product. In addition, its explosive limits range from 4 to 46 percent in air. Appropriate precautions should be taken to prevent these hazards from developing.

Carbon and Low Alloy Steels

The behaviour of carbon and low alloy steels in sour environments is affected by the complex interaction of parameters such as: -

1. Metal chemical composition, strength, heat treatment;
2. Acidity (pH) of the water phase
3. H$_2$S partial pressure or equivalent concentration in the water phase;
4. Chloride ion concentration in the water phase;
5. Presence of sulphur or other oxidants
6. Tensile stress;
7. Fluid temperature;
8. Exposure time.

The severity of the sour environment is a function of the acidity and the H$_2$S partial pressure. The more acidic the fluid and the higher the H$_2$S partial pressure the greater is the risk of SSC. Three levels of severity are defined with Region 3 being the most severe. Region 0 has a H$_2$S partial pressure of less than 0.3kPa and normally no precautions against SSC are required when selecting materials for this application.

For equipment operating in each of the three SSC regions MR0175 specifies the requirements of chemical composition, heat treatment and hardness of the materials that can be used. The maximum permitted average hardness level is 22 HRC (Rockwell C) except for A105 forgings where the maximum is 187 HBW (Brinell). The nickel content must be less than 1%. In general all the carbon and low alloy steels can be used provided they are heat treated in accordance with one of the specified conditions.

Where welding is used in the fabrication or repair of a valve for sour service it is essential that the welds be given a stress relieving heat treatment at a minimum temperature of 620°C to produce a maximum individual hardness of 250Hv or less in the heat affected zone (HAZ).

Corrosion Resistant Alloys and Other Alloys

The behaviour of corrosion resistant and other alloys in sour environments is affected by the complex interaction of parameters such as: -

1. Metal chemical composition, strength, heat treatment;
2. Acidity (pH) of the water phase
3. H_2S partial pressure in the gas phase or equivalent concentration in the water phase (calculated as the partial pressure at last separation from gas);
4. Chloride or other halide ion concentration in the water phase;
5. Presence of oxygen, sulphur or other oxidants
6. Tensile stress;
7. Fluid temperature;
8. Exposure time:
9. Galvanic effects
10. Pitting resistance of the material as defined by the PREN (pitting resistance equivalent number) i.e.

F_{PREN} = Cr +3.3 (Mo +0.5W) +16 N where Cr, Mo, W, and N are the % by mass of the element in the material.

The various materials within the classification of corrosion resistant and other alloys are broken down into different groups. Within the group materials are identified by type or by individual alloy. Each environment is then considered in detail in terms of temperature, H_2S partial pressure, chloride content. pH and sulphur resistance and recommendations made as to the suitability of each type or individual alloy for specific equipment types in the different environments.

In the complete absence of oxygen (below about 5ppbv O_2) and H_2S pitting corrosion and SCC of stainless steel does not occur. However H_2S is also an oxidising medium and pitting corrosion and SCC of stainless steel may occur with no O_2 but small amounts of H_2S.

There are a large number of tables covering all the various materials and ensuring that the right material has been selected requires experience and expertise.

Figure CN7: Body to bonnet bolting on a Gate Valve (Velan)

Again welding is of great importance and correct pre heat and/or post weld heat treatment is necessary to ensure that any weld will not be a potential crack initiation site.

Bolting

It should not be forgotten that most valves have bolted body to bonnet connections. While nominally this bolting is not in contact with the fluid it is potentially exposed to an H_2S environment in the atmosphere surrounding the valve. The bolting is typically the most highly stressed component in the valve and therefore susceptible to SSC if suitable conditions exist. It is therefore good practice to fit recommended MR0175 bolting in valves where the materials are subject to the requirements of MR0175.

Recommended grades for carbon and low alloy steel valves are ASTM A193 Grade B7M and A320 Grade L7M for bolts and A194 Grade 2HM and 7M for nuts. These materials have been specifically heat treated to reduce the hardness to maximum 235HBW. They are also heat treated after the completion of the thread forming process.

An issue that valve manufacturers must address when using these bolting grades is that they have reduced yield strengths. The design of the body bonnet or cover joint must be reassessed to confirm it's suitability for the specified operating conditions.

For more information about NACE International, see www.nace.org

Second best – one option we don't provide

Whatever your flow control requirement, Flowserve has the systems, the products and the experience to help your processes run smarter, safer and more efficiently.

Flowserve supplies contractors, OEMs, and end users with the industry's most complete range of valves, actuators, positioners, controls and switches. Complemented by our comprehensive R&D, engineering and global support services, we provide every Flowserve customer with a one-stop solution for some of the most demanding flow control needs of users around the globe.

Professional partnerships with some of the leading distributors mean that Flowserve customers have on-the-spot availability, experienced technical support and unmatched service while rugged and reliable products ensure extended service life, with on-site and off-site service and rebuild options providing some of the lowest life cycle costs in the industry.

And, backed by Flowserve, you also have access to one of the leading pumps and seals manufacturing organisations in the world.

To find out how Flowserve can benefit your business, contact:
Flowserve Flow Control
Burrell Road, Haywards Heath,
West Sussex RH16 1TL
T: 01444 314400
F: 01444 314401
E-mail: ukfcinfo@flowserve.com

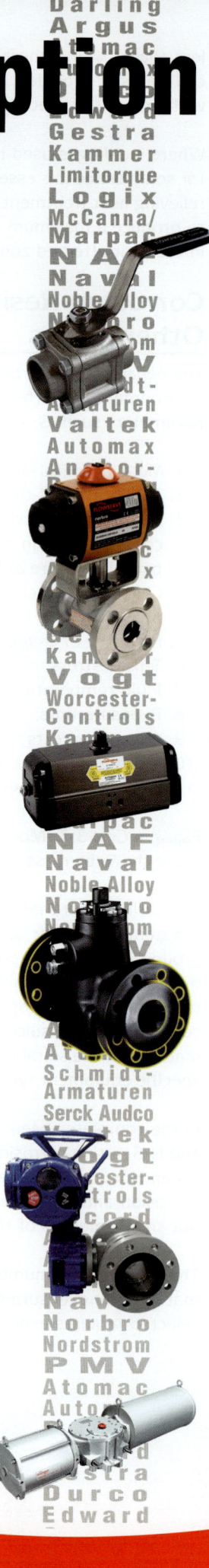

FLOWSERVE

Anchor-Darling
Argus
Atomac
Edward
Gestra
Kammer
Limitorque
Logix
McCanna/Marpac
NAF
Naval
Noble Alloy
Norbro
Nordstrom
PMV
Schmidt-Armaturen
Valtek
Automax
Anchor-Darling
Kammer
Vogt
Worcester-Controls
Kammer
Marpac
NAF
Naval
Noble Alloy
Norbro
Nordstrom
PMV
Schmidt-Armaturen
Serck Audco
Valtek
Vogt
Worcester-Controls
Accord
Atomac
Naval
Norbro
Nordstrom
PMV
Atomac
Automax
Edward
Gestra
Durco
Edward

Experience In Motion

flowserve.com

Standards and Directives

Figure SD1: Standards and Directives might initially appear burdensome, but can prove extremely beneficial.

Standards

Standards make life safer, healthier and easier for people, organizations and enterprises all over the world. They enable and simplify communication and trade, while allowing resources to be used more efficiently.

All areas in the valve sector can benefit from adopting standards, from global manufacturers to small local firms carrying out valve modification and repair work. For a business involved in any area of the valve sector, being compliant with the relevant standards can save time, effort and expense, while giving added peace of mind that some of their many legal responsibilities have been met.

Obeying regulations can be much more challenging when diversifying into new markets or trading in non-domestic territories. Although practices and regulations can and do vary, standards can provide the information necessary to make the trading of valves and their associated services in new markets much less difficult, time consuming and costly.

Industrial valve standards can be used to reduce the time, effort and money that has to be invested in the research and development of new products, while increasing their likelihood of success in the marketplace. In addition, industrial valve standards can provide best-practice guidance helping businesses to assess their processes, allowing them to take steps to increase efficiency and become more profitable.

Standards enable innovation by defining and measuring valve performance, leaving the user free to use a standard without divulging intellectual property. Arguably, the most innovative businesses have the most to gain from the strategic use of standards. Standards provide a reliable benchmark against which performance can be judged, enabling businesses to demonstrate product performance. Being able to claim compliance with internationally recognized and respected standards is an effective way of supporting the quality of goods, services or processes. If compliance cannot be supported in this way, customers only have the manufacturers/suppliers word regarding the quality of their goods, products or services.

To claim compliance with a standard the manufacturer must ensure that all the requirements are implemented in the product or all relevant areas of the business for standards such as ISO 9001. In the case of ISO 9001 a manufacturer's claim for compliance is supported by auditing by a third party organisation. For companies supplying valves and actuators that require to be CE marked, the third party organisation must be approved by a national body confirming that it is capable of performing this work to the required standard. In the UK this approval body is UKAS, the United Kingdom Accreditation Service and each approved organisation has a unique UKAS number. For added weight, some businesses also choose to have their product compliance verified by an outside auditor or test organization e.g. the BSI Kitemark scheme.

Appendix 1 'British & International Standards for Valves, Actuators, Seals and Related Products' gives a comprehensive list of the standards that are widely used in the valve industry. The BVAA is a significant contributor to the work on EN and ISO standards through the British Standards Institution (BSI) and its membership of committee PSE/18.

Figure SD2: The familiar BSI Kitemark© logo (BSI Global)

Performance Standards

Most standards in the valve industry relate to dimensions, interchangeability, and comparative data documenting how each valve has performed under the test conditions. A small but important number of standards relate to performance type testing of representative samples. The samples are subject to a defined testing protocol and documented comparative data as to how each valve performs under the test conditions is produced, very often involving witness by an independent third party. The standard also defines the range of sizes and pressure ratings that are considered to be qualified by the type test.

One of the oldest of these performance standards is that for fire testing. Another standard that has been introduced recently is that for fugitive emissions.

Figure SD3: A valve undergoing Fire Testing (Colson)

Testing of Valves - Fire type testing requirements EN ISO 10497

Fire testing is generally required for valves with soft seating surfaces and thermo plastic polymer packings, which might be destroyed in the event of a fire. The test is intended to demonstrate an acceptable level of leakage after the valve has been exposed to the fire. Since valves are often the first line of control for flammable liquids it could be hazardous to install a valve, which has a design that has not been fire type tested.

It is impossible to have a single definition of a fire. A fire which may occur in a refinery would certainly be different to one in a chemical plant. Some fires will burn longer or hotter, others will spread faster. Therefore, standards have evolved which derive from the theoretical assumptions of what a fire is and how a valve should operate in such a situation.

In EN ISO 10497 the closed valve, completely filled with water under pressure, is enveloped in flames with an environmental temperature in the region of the valve of 750°C to 1000°C for a period of 30 minutes. The objective is to completely envelop the valve in flames to assure that the seat and sealing areas are exposed to the high burn temperature. During this period the internal and external leakage is recorded. After cool-down from the fire test, the valve is hydrostatically tested to assess the pressure containing capability of the valve shell, seats and seals.

Figure SD4: Cooling down stage of a fire test on a 3" Class 150 Cryogenic Ball Valve (Apollo Valves / Adanac)

The standard specifies the maximum through seat and external leakage rates in ml/min for valve sizes DN8 through DN200.

Instead of testing each nominal size and nominal pressure rating of a given valve design, additional valves of the same basic design as the test valve may be deemed to have been fire-tested as defined in Clause 7 of the standard. Some examples of what this means are given below: -

- Where the test valve is DN50 Class 150 all valves <DN50 and all valves ≤DN100 of Class 150 and Class 300 are also qualified

- Where the test valve is DN80 Class 1500 all valves ≤DN150 of Class 1500 and Class 2500 are also qualified.

Other fire testing standards exist namely API 607. This was specifically for soft seated, quarter turn valves. ISO adopted API 607 5th Edition as the basis to create the latest ISO 10497 but the title and scope of the latest ISO 10497 applies to all valves and is not limited to soft-seated quarter-turn valves. It is important to note that the API 600 series standards for gate, globe, check, ball and butterfly valves used in refineries all require API 607. Since ISO adopted API-607 to create ISO 10497, they are the same for all intents and purposes.

In the past manufacturers also tested to BS6755 Part 2. Older designs may well have fire type test certificates bearing this number.

Industrial valves - Measurement, test and qualification procedures for Fugitive Emissions (EN ISO 15848)

Increasing attention is being paid by regulator authorities to emissions from process equipment in general and valves in particular. Continuous improvements from manufacturers are required to meet these demands. To support this process users and manufacturers came together to create a type test which would provide comparable performance data on this important issue. Fugitive emissions is the collective term for all leakage from a valve to the environment. The standard specifies testing procedures for evaluation of external leakage of valve stem (or shaft) seals and body joints of both isolating valves and control valves intended for application in volatile air pollutants (hydrocarbons) and hazardous fluids.

The standard is in two parts. Part 1 details the classification system and qualification procedures for type testing of valves. Part 2 details the production acceptance test of valves. A performance class is defined by the combination of the following criteria: -

- tightness class
- endurance class
- temperature class.

Tightness classes are defined only for stem (or shaft) sealing systems. Leakage from body bonnet or cover joints shall be ≤50 ppmv in every case. Three classes are defined Class A, typically achieved with bellow seals or equivalent stem (shaft) sealing system for quarter turn valves, Class B typically achieved with PTFE based packings or elastomeric seals and Class C typically achieved with flexible graphite based packings. Endurance classes for isolating valves are identified 'CO' and relate to the number of closed open closed cycles, which the valve has under gone during the test programme. The higher the numeric value after the letters CO the greater the number of test cycles the valve has seen. Endurance classes for control valves involve a significantly higher number of mechanical cycles however these are performed at 50% of stroke/angle with an amplitude of ± 10% of the full stroke/angle. Designation for control valve endurance classes are 'CC.'

Figure SD5: Ball Valve being tested for emissions following a cryogenic test (Tyco / Hindle Winn)

The temperature class is the temperature to which the valve is cycled during the endurance test. If the temperature class is Room Temperature 'RT' then no thermal cycling is performed. Both low temperature classes and high temperature classes are specified which cover a range of potential application temperatures. For these low and high temperature classes the valve must be thermally cycled a number of times during the mechanical cycling tests.

The test pressure inside the valve is that appropriate to the material and pressure rating of the body.

Upon the successful completion of the test program as defined in Part 1 the qualification may be extended to untested sizes and classes of valves of the same type if the criteria specified in Clause 8 are met.

Part 2 is to establish standard production tests for valves that have been successfully type tested in accordance with Part 1 in order to demonstrate that the production valves are likely to perform to the level of the type tested valves. Samples of production valves are tested using helium at 6 bar and a sniffing method to a tightness class agreed with the purchaser.

Directives

Compliance with a particular standard is not a requirement of the law. It may be a contract requirement or it may be the manufacturer's choice to work in accordance with a particular standard.

Within the European Union (EU) it was recognised that while standards had the potential to be beneficial they could also be detrimental in that by adopting different standards particularly on safety issues member states of the EU would effectively create barriers to trade. On the 1st July 1987 the Single European Act came into force, which included the concept of the Single European Market. The basic purpose of a Directive is to support the concept of the single market by harmonising national legislation as the means of removing technical barriers to trade by producing one harmonised set of Essential Safety Requirements (ESRs) across the EU.

The EU Directives are subject to the force of law through legislation in the individual member states. For example UK companies must comply with the UK Statutory Instrument enacting a particular directive into UK law. Directives apply to equipment that is put onto the market within the countries of the EU.

Figure SD6: The ubiquitous 'CE' Mark, the application of which means that the product to which it is fixed conforms to all the European Directives that apply to it.

Directives have a number of common features and it is important to understand these: -

- Each Directive has an objective and a definition of the equipment and applications to which it applies and to which it does not apply. It is just as important to know what is excluded, for example the Pressure Equipment Directive excludes well control equipment and equipment associated with water distribution.

- The Directive specifies the ESRs that must be satisfied. It is the responsibility of the manufacturer to ensure conformity with the ESRs.

- The Directive specifies procedures that the manufacturer has to undergo as part of the process of conformity to the ESRs. These procedures may involve verification by a third party called a Notified Body. The Notified Body is authorised by a member state to provide verification activities for a particular directive not necessarily every directive. Each Notified Body is allocated a unique number and this number will appear alongside a CE mark if the Notified Body has been involved in the verification process. An important aspect of the Notified Body's work is to assess the competency of the manufacturer to meet his responsibilities as specified in the Directive.

- The manufacturer must prepare a technical file detailing his design, the hazards that he has considered and the means by which he has satisfied the ESRs. The manufacturer makes a Declaration of Conformity signed by a competent person in the organisation.

- Harmonised Standards are European Norms (EN) Standards that have been listed in the Official Journal of the EU and which provide a means of satisfying specific ESRs as defined in the Annex to the EN standard.

- The CE mark is applied to the equipment indicating that the requirements of all relevant directives have been met.

- It is illegal to CE mark equipment under a particular Directive if that equipment is excluded from that Directive.

- The manufacturer must put his name on the equipment and the year of manufacture.

- The manufacturer must provide operating and maintenance instructions in the language of the user of the equipment.

- The BVAA has consulted widely with experts and prepared detailed Guidelines on interpretation of the various Directives relevant to the valve and actuator industry. These directives are listed below with some of the key points taken from the Guidelines.

Figure SD7: The BVAA has produced numerous guidelines on the European Directives relevant to the valve and actuator industry

Pressure Equipment Directive (PED) - 97/23/EC

This is the most important directive for the valve industry. The Directive applies to the design, manufacture and conformity assessment of pressure equipment and assemblies with a maximum allowable pressure PS greater than 0.5 bar above atmospheric pressure.

'Maximum allowable pressure PS' means the maximum pressure in bar for which the equipment is designed, as specified by the manufacturer.

NOTE: For valves it is the allowable working pressure PS at room temperature. For a safety device it is the maximum set pressure for which the safety device is designed.

'Pressure equipment' means vessels, piping, safety accessories and pressure accessories.

'Safety accessories' means devices designed to protect pressure equipment against the allowable limits being exceeded. Such devices include: -

- Safety valves
- Bursting disc safety devices
- Controlled safety pressure relief systems
- Limiting devices, which either activate the means for correction or provide for shutdown and lockout, such as:-

- Pressure switches
- Temperature switches
- Fluid level switches
- Safety related measurement and regulation devices.

'Pressure accessories' means devices with an operational function and having pressure-bearing housings.

NOTE: *Valves and fluid pressurised actuators are pressure accessories.*

Valves are pressure accessories and have to meet the requirements for piping as set out in Article 3, paragraph 1.3 of the PED. Safety devices are safety accessories and have to meet the requirements of Article 3, paragraph 1.4 and 1.1 or 1.2 or 1.3 of the PED, i.e. 'Vessels' or 'Fired or otherwise heated pressure equipment' or 'Piping', whichever they protect.

Valves, safety devices, fluid pressurised actuators and assemblies, are classified by category in accordance with Annex II of the PED according to ascending level of hazard. There are four tables two for dangerous (Group 1) fluids, either gas or liquid, and two for non-dangerous (Group 2) fluids, either gas or liquid. The user must provide the manufacturer with information on the intended fluid, pressure and temperature so that the manufacturer can determine by consulting the correct chart the applicable category for the pressure equipment.

Figure SD8: One of the Charts from the PED, for Group 1 'Dangerous' gases

This is essential for the manufacturer because both the conformity assessment module the manufacturer may choose and the applicable Essential Safety Requirements are based on the equipment category he determines.

The pressure equipment may fall below Category I for reasons of size or of fluid or service pressure. In this case the equipment must be designed and manufactured in accordance with the sound engineering practice (SEP) of a Member State in order to ensure safe use. Such pressure equipment must be accompanied by adequate instructions for use and must bear markings to permit identification of the manufacturer. Such equipment must not bear the CE marking.

Equipment that is Category I may be CE marked by the manufacturer without the involvement of a Notified Body but the manufacturer must still complete the technical file and the Declaration of Conformity. Equipment that is Category II or higher, must involve a Notified Body in the assessment procedure.

Equipment and Protective Systems Intended for Use in Potentially Explosive Atmospheres Directive (ATEX) - 94/9/EC

The Directive covers equipment and protective systems, which may be used in areas endangered by potentially explosive atmosphere created by the presence of flammable gases, vapours, mists or dusts.

"Equipment" is any item which contains or constitutes a potential ignition source and which requires special measures to be incorporated in its design and/or its installation in order to prevent the ignition source from initiating an explosion in the surrounding atmosphere. Also included in the term "equipment" are safety or control devices installed outside the hazardous area but having an explosion protection function. A wide range of products comes within the definition of equipment, including valves, actuators and ancillary equipment, which will include mechanical, electrical & electro/mechanical equipment.

The ATEX directive also applies to non-electrical equipment such as valves, pneumatic and hydraulic actuators and gearboxes.

All Industrial Valves must be subjected to an ATEX Risk Assessment to identify any potential ignition source. This includes those originating from the static charge build-up, arising from the throughput of the flow media or system vibrations, which should be considered an "own ignition source" for a valve.

The European ATEX Guidelines 2nd edition (July 2005) introduced the term 'simple product' stating that a manually operated valve is a 'simple product' and is outside of the scope of the ATEX 94/9/EC directive. This statement is based upon the assumption that a manually operated valve, as a 'simple product' does not have an ignition source.

The BVAA view is that having undergone an ATEX Risk Assessment by the Manufacturer and having been proven to have no own ignition risk in normal operation as well as expected and rare malfunctions, such an industrial valve is outside of the scope of the ATEX Directive. However, where an ATEX risk assessment shows that an industrial valve does have its own ignition sources, including electrostatic charges build up, such a valve does fall within the scope of the ATEX Directive.

Explosion protection is the technique of preventing or controlling the effects of explosions, which might otherwise occur where flammable materials are handled, stored or processed. It is widely recognised internationally by the following symbol: -

Figure SD9: European Directives usually require CE Marking, but ATEX requires an 'Ex' mark of its own.

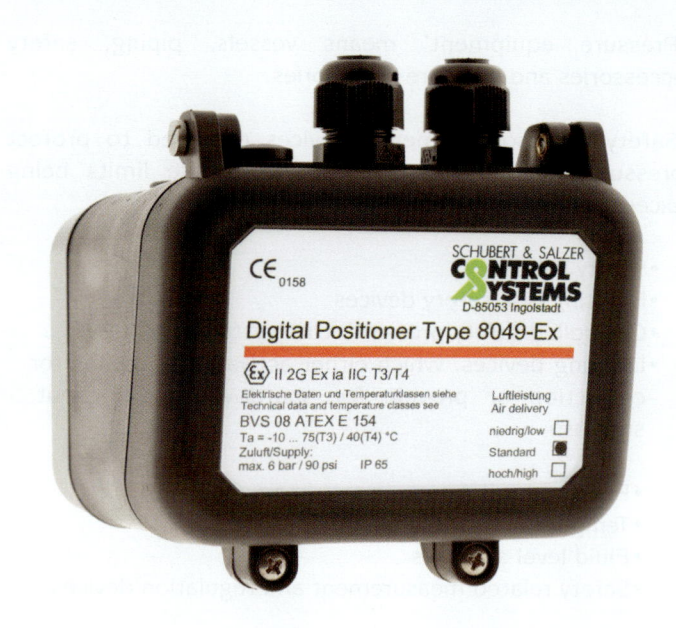

Figure SD10: A Digital Positioner for pneumatically controlled actuators, bearing both the CE and Ex markings (Schubert & Salzer)

Where formation of explosive atmosphere cannot be ruled out, measures must be taken to prevent their ignition. The measures will be related to the risk involved. The area classification system commonly used for non-mining (Group II) applications identifies: -

- Zone 0 - Explosive atmosphere present continuously or for long periods of time. Atex Equipment Group II Category 1 which would include equipment designed to the Intrinsic Safety 'ia' standard.

- Zone 1 - Explosive atmosphere present occasionally under normal operating circumstance. Atex Equipment Group II Category 2, which would include equipment designed to Intrinsic safety 'ib'; Flameproof enclosure 'd'; Increased safety 'e'; Purged and pressurised 'p'; Encapsulated, 'm'; Oil filled, 'o'; Powder filled, 'q'.

- Zone 2 - Explosive atmosphere not normally present and then only for short periods. Atex Equipment Group II Category 3, which would include equipment designed to the non-incendive standard 'n'.

Annex II of the ATEX Directive, 94/9/EC specifies the Essential Health and Safety Requirements (EHSRs) relating to the design and construction of equipment and protective systems that are necessary in order to prevent explosions or to control the effects of incipient explosions. In addition to design and construction, Annex II also provides concepts on marking, instructions for use and replacement parts and the level of detail requirements are specified therein. The EHSR are divided into three groups: -

- Common requirements for equipment and protective systems
- Supplementary requirements for equipment
- Supplementary requirements for protective systems

Machinery Directive - 2006/42/EC

Article 1.2 of the Directive states that for the purposes of the Directive "machinery" means an assembly of linked parts or components, at least one of which moves, with the appropriate actuators, control and power circuits etc., joined together for a specific application, in particular for the processing, treatment, moving or packaging of a material.

Article 1.3 of the Directive contains a list of exclusions, the first one of which is "machinery" whose only power source is directly applied manual effort, unless it is a machine used for lifting or lowering purposes.

The BVAA interpretation is that manual valves are not "machinery" as defined in the Directive as the power source of manual valves (handwheel, lever or gear operated) is directly applied manual effort. Power operated valves, actuators and valve/actuator assemblies do not in themselves process, treat, move or package a material and so are not "machinery" as defined in the Directive but are only components or parts of "machinery", which requires that the manufacturer provides a Declaration of Incorporation.

Article 1.2 of the Directive also defines a "safety component" as "a component placed on the market to fulfil a safety function when in use and the failure or malfunctioning of which endangers the safety or health of exposed persons". Further, a safety component is also interpreted as a component fulfilling a safety function and having no reason for being used for any other purpose. Its absence should not impair the productive function of machinery.

The BVAA consider some valves, actuators and valve/actuator assemblies, when used in specific applications (e.g. emergency shut down (ESD) or SLAMSHUT), meet the above definitions. They are therefore "safety component" which requires the manufacturer to prepare and issue a "Declaration of Conformity"

It is the manufacturer who declares his product to be a "safety component". He may be advised by his customer that his product is to be used as a "safety component" or he may decide this himself based on the function for which the valve has been designed e.g. a fire control device, an ESD.

For those products declared to be "safety components" the essential steps to preparing the Declaration of Conformity are: -

- Fulfil the ESR's of ANNEX I of the Directive
- Prepare the Technical File of ANNEX V of the Directive
- Prepare the EC Declaration of Conformity of ANNEX II.C of the Directive.

The manufacturer must identify all hazards that apply to their "safety component" and having identified the hazards define the preventative measures that have been applied to each. The order of priority that must be use in determining the applicable preventative measure is 1. Elimination of hazard, 2. Risk reduction, 3. Technical protective measures and 4. Information.

Paragraph 2 of Article 8.1 refers to CE marking of "machinery alone". The manufacturer or his authorised representative in the Community must affix to the machine the CE Mark referred to in Article 10. Neither a valve, an actuator nor a valve/actuator assembly is "machinery" as defined by the Directive. It is therefore not allowed to be CE mark under this Directive. Safety components are also not allowed to bear the CE mark but they must have a Declaration of Conformity.

Figure SD11: Heavy duty, intrinsically safe, low power pilot valve with 'Ex' certification (Asco Numatics)

Electromagnetic Compatibility Directive (EMC) - 2004/108/EC

This Directive is of major importance for electrically powered apparatus. Often EMC is considered something which is only important to audio and video equipment. However, the functioning of many products may be affected by electromagnetic disturbance or may cause the functioning of other products to be affected. The Directive covers both of these aspects, immunity and emission.

Apparatus is defined as all electrical and electronic appliances together with equipment and installations containing electrical and/or electronic components. The BVAA interpretation of Apparatus is that the Directive typically applies to the following valve/actuator industry products where "intolerable" electromagnetic disturbances might occur: -

- Electric motor driven actuators
- Solenoid valves
- Servo valves
- Electro pneumatic positioners
- Electrical (analogue or digital) positioners
- Electro pneumatic transducers
- Electrical transmitters
- Linear/rotary variable differential transformers

The above list of typical items is not intended to be exhaustive but is given only as an example for purposes of clarification. The Directive defines the following exclusions: -

- Exports to third countries
- Spare parts (components)
- Electromagnetically "benign" apparatus
- Apparatus in a sealed electromagnetic environment
- Amateur radio apparatus
- Military equipment

The BVAA interpretation of Exclusions is that typical excluded apparatus includes the following: -

- Limit switches
- Isolating switches
- Proximity switches (non-inductive type)
- Thermostats
- Potentiometers
- Junction boxes
- Conduit/cable glands

The above list of typical items is not intended to be exhaustive but is given only as an example for purposes of clarification. The examples given of excluded apparatus are on the basis that they are "benign" according to the Directive.

There are two principal routes to follow to enable valve and actuator manufacturers to demonstrate conformance of their products with this Directive: -

a) Standards route – the product is manufactured and tested in accordance with a relevant Harmonised European standard and the manufacturer assesses conformity of his product with this standard and produces an EC declaration of conformity. There is no requirement for independent third party involvement and the manufacturer can consequently "self certify" his product.

b) Technical Construction File route – the manufacturer prepares a full Technical Construction File giving details of how his product has to be manufactured to be in conformity with the EMCD. This Technical Construction File should be in accordance with the Directive and has to be submitted to a competent body for assessment (i.e. non-self certification).

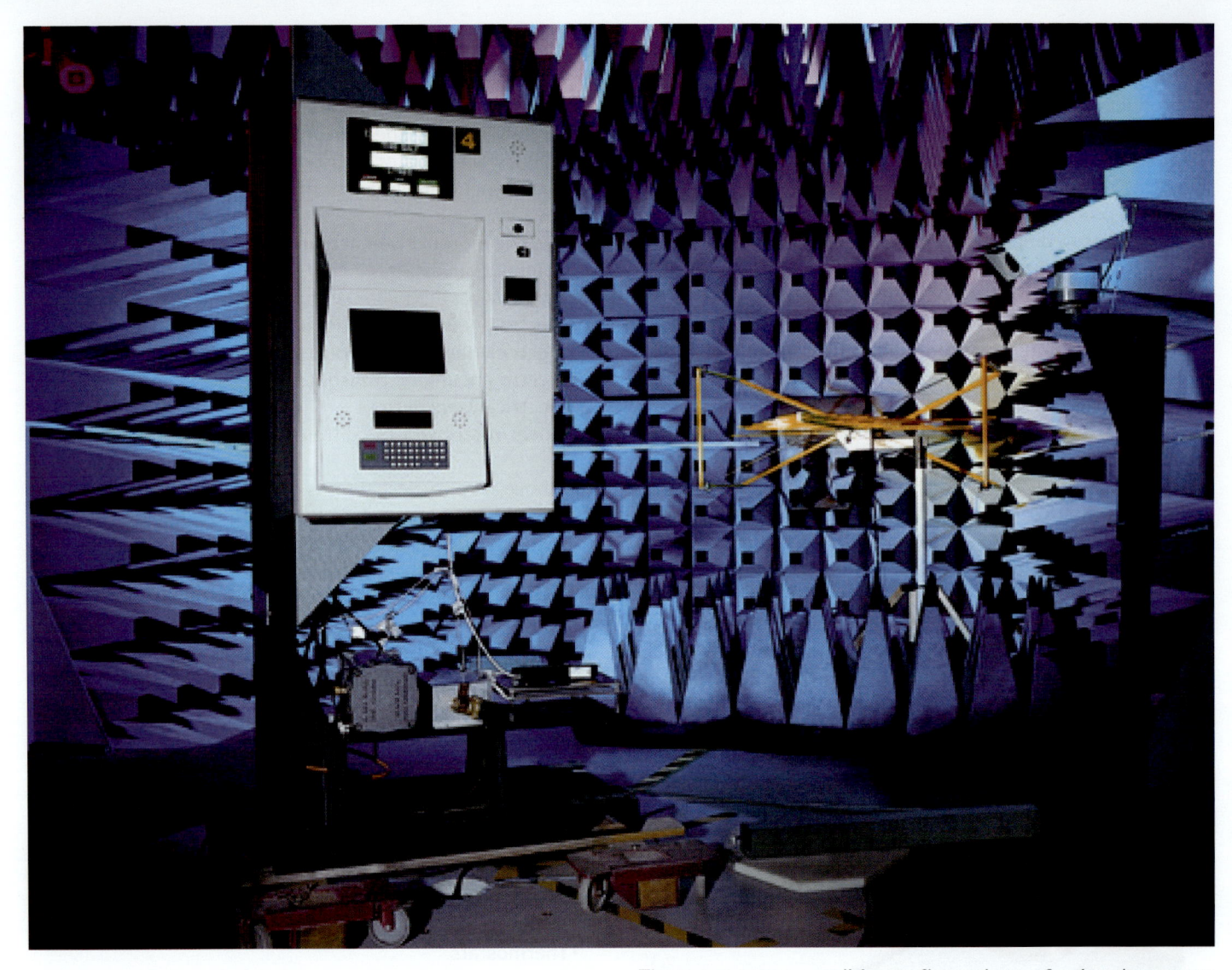

Figure SD12: EMC Test Laboratory (York EMC Services Ltd)

Low Voltage Directive (LVD) - 2006/95/EC

This Directive is of major importance for electrically powered equipment operating within the limits of 50 volts a.c. to 1000 volts a.c. or 75 volts d.c. to 1500 volts d.c. It is principally concerned with the safety of persons, domestic animals and property when the equipment is used in applications for which the equipment was manufactured and installed as instructed. It is important to consider any warnings of misuse in appropriate instruction manuals/leaflets.

The term "electrical equipment" for valve/actuator products is synonymous with the term "apparatus" as given above under the Electromagnetic Compatibility Directive. Certain valve/actuator industry equipment is excluded from the Directive, e.g. 24 volt supply equipment.

To comply with the LVD, it is necessary to design and manufacture in accordance with the Electrical Equipment Safety Regulations 1994. Electrical equipment which complies with the safety provisions of a Harmonised European standard will be presumed to comply with the safety requirements of the 1994 regulations. Where no suitable Harmonised European standard exists then reference can be made to an ISO standard or, where none exists, to a national standard.

There are many possible configurations of valves/actuators with or without electrical apparatus/equipment. When the valve, actuator and all ancillary electrical apparatus forms one complete unit, the supplier of the complete unit should have available the individual declarations of conformity for all non-benign ancillaries covered by this Directive and the Electro Magnetic Compatibility Directive.

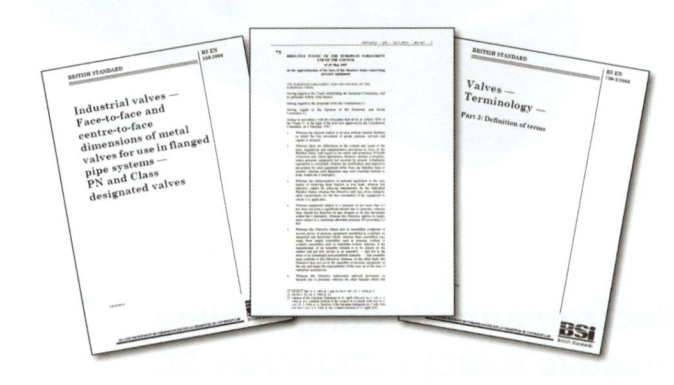

Figure SD13: Due to the long and inclusive development processes, the final content of most Standards and Directives is widely known prior to publication.

Who are 'They'?

Whenever you have a conversation about standards, you will inevitably hear 'They' mentioned...

- '*They* wrote this standard...'
- '*They* tested and decided...'
- '*They* met recently and discussed...'

But just who are '*They*'?

It's us! BVAA Members have thousands of years of collective experience of standardisation, and we regularly participate in over 50 technical and standards committees around the world. For every standard being developed, you can be sure there is a BVAA group monitoring and contributing to the work.

Faceless people? Not us!

If you would like to participate in standards making, just contact the BVAA.

BVAA

British Valve & Actuator Association
9 Manor Park, Banbury, Oxfordshire OX16 3TB
Tel: 01295 221270 Fax: 01295 268965
Email: enquiry@bvaa.org.uk
www.bvaa.org.uk

BVAA
Valve Training Courses

30% OFF

30% Discount For BVAA Members

"Good and pitched at right level"
- Shaw Valves

"Very good with lots of experience and knowledge" - BP

"Professionally done"
- British Energy

"Good - clear, concise and knowledgeable"
- Titanium International Ltd

"Friendly and funny, explained concepts in laymen's terms, making the course easy to understand"
- AMEC

These courses are a **MUST** for those involved in the engineering industry who need to know more about valves and actuators. BVAA valve courses are delivered by our lecturers who have tremendous knowledge and experience of the industry. The sessions always result in comments of the highest praise.

Introduction to Valves
Introduction to Valve Actuators
Control Valves
Safety Valves
Basic Fluid Sealing
Intermediate Fluid Sealing
Safety Integrity Levels (SILs)
PED & ATEX Directives

BVAA Training sessions are held twice annually at our Banbury (UK) Headquarters. Subject to sufficient numbers, courses can also be arranged at a venue of your choosing.

British Valve and Actuator Association Limited
9 Manor Park • Banbury • Oxfordshire OX16 3TB
Telephone: +44 (0)1295 221270
Fax: +44 (0)1295 268965
Email: enquiry@bvaa.org.uk www.bvaa.org.uk

For full details on each course, see http://www.bvaa.org.uk/training

Fluid Sealing Materials

Figure FS1: A selection of gasket materials and styles (James Walker)

Fluid sealing materials are vital to the satisfactory operation of many items of process equipment and valves in particular. The importance of good quality fluid sealing materials should not be overlooked and the temptation to consider these items as commodities rather than critical components should be resisted. Failure of the body and bonnet seal or leakage past the packing material intended to seal around the stem or shaft will result in environmental leakage the consequences of which can vary from minor inconvenience to explosions, loss of life and catastrophic destruction of plant.

Elastomer seals are also used in many valves and on actuators to ensure reliable sealing. These too can be subject to aggressive chemicals and wide ranges of service temperatures and pressures.

Bonnet joints and line flanges

Gaskets are used to seal the valve bonnet and the connecting flanges in the pipe and come in a wide variety of materials and types (see Figure FS1). A gasket has to be compliant enough to conform to the flange surface to accommodate any irregularities and surface defects, whilst also being strong enough to withstand the compressive stresses from the bolt loading. Gaskets can be categorised into three main types:

a) Non metallic gaskets – for example rubber, compressed fibre, expanded graphite and PTFE
b) Semi metallic types with spiral-wound gaskets, kammprofiles and metal clad joints
c) Metallic – solid metal API ring joints, lens rings and various other designs such as those employed on some compact flange connectors.

As a general rule, the higher the operating temperature and/or pressure becomes, then the more metallic types of gasket tend to be employed, as there is the need to resist the likely overall joint forces as well as having to use materials resistant to deterioration by the effects of high temperature. If compressed non-asbestos fibre materials are being used, then care should be taken to use the thinnest material that will accommodate the potential flange irregularities and misalignment. Thinner materials are less prone to creep effects and have better stress retention performance. The tightness also tends to be better as there is less thickness for leakage due to gas permeation. The use of jointing pastes is generally not recommended as these can facilitate creep of the gasket across the flange face. The smoother the flange surface finish the greater is the potential for this to occur.

Expanded graphite material has played an increasingly significant part in sealing flange joints over the years. It is used both in sheet form for cut joints, and as a filler material in spiral-wound gaskets, as well as a facing for kammprofile type gaskets. It is soft and conformable, with a wide range of chemical compatibility although it is not suitable for use with strong oxidising media. It is available in standard grades (98%) for general duties as well as high purity grades (>99.5%) for nuclear duties.

Spiral-wound gaskets are commonly used as a valve bonnet seal as well as in the pipeline joints. They are manufactured using a continuous strip of metal wound with a soft material filler alongside, which is usually graphite or PTFE, though mica and vermiculite are used for very high temperatures. When used with a raised face flange in the pipeline these gaskets have inner and outer supporting rings to locate them within the bolt holes and act as a compression stop. When a sealing element alone is used as a valve bonnet seal, care should be taken to ensure the correct clearances on the inside and outside diameters, as these can affect the load-compression and tightness characteristics, and it is often wise to seek the manufacturer's recommendations on this matter.

Whilst the majority of valve bonnets may typically use a cut joint or a spiral-wound gasket, the higher pressure oilfield valves often use metal ring joints. Oval type ring type joints (RTJs) were developed in the 1930s as oil well pressures increased, in order to give a more robust gasket design compared to sheet material types. These were originally housed in a round bottomed groove, though later this was modified to the 23° bevel groove that we still use today, and the octagonal type joint was a logical derivative of the original oval joint.

Even higher pressure flanges, though of a more compact design, were produced in the 1950s as the BX type ring joint – and these are used in high pressure flanges, rated at 10,000 psi, 15,000 psi, and 20,000 psi. For hydro-testing of valve bodies with RTJ flanges, it is possible to use rubber-covered ring joints in order to avoid damaging the ring groove before final assembly.

In order to get the best results from any flange gasket, care and attention to the installation and bolting is essential, in order to achieve the correct levels of initial and operating stress on the sealing element for the operating conditions. Also bolted joints on valve bonnets and pipeline flanges can see significant external forces which need to be accommodated, for example on high pressure API oilfield valves.

As already discussed in the Valve Basics chapter some valves are constructed with a pressure seal bonnet. These use a sealing ring which is energised by the pressure within the valve acting on the underside of the bonnet and forcing the wedge-shaped seal tightly into the housing. These are sometimes known as Bridgeman or Bredtschneider joints, and can be a soft plated metal gasket, or alternatively made from moulded expanded graphite, usually with some form of metallic anti-extrusion element in the corners.

Gland sealing

The first compression packing materials were little more than greasy ropes, lubricated with natural products such as tallow and animal fats. However, today's materials can be produced from a wide variety of natural and synthetic yarns, with products such as exfoliated graphite being widely employed in many high temperature steam applications as well as on searching gases in the hydrocarbon processing sector (Figure FS3). The subject of fugitive emissions and environmental pollution has received a lot of attention from this sector. To provide users with comparative performance data a variety of test methods have evolved such as the ISO 15848 for valves (see chapter on Standards and Directives) and the API 622 standard test for packing. API 622 evaluates packing performance in a standardized fixture, outside the influence of a particular valve's design. A number of packing formats have been produced by manufacturers as specifically low emission grades, mostly comprising graphite based materials.

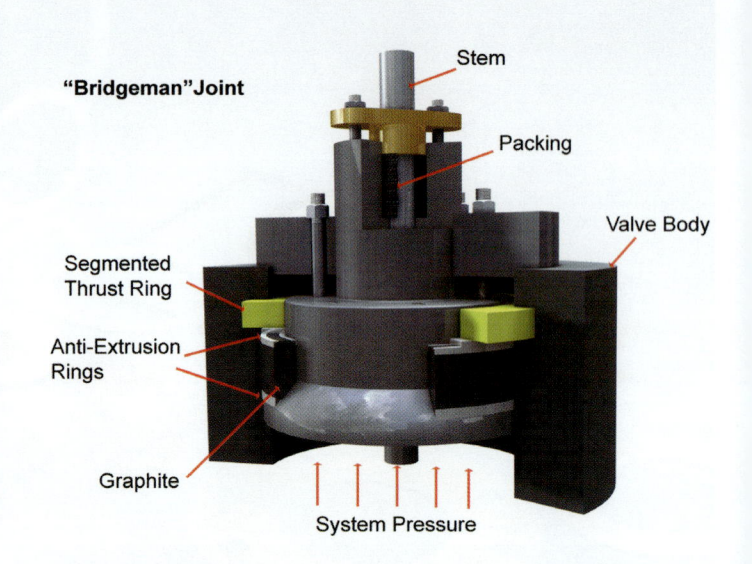

Figure FS2: Pressure Seal Valve bonnet arrangement (James Walker)

Materials such as aramid fibre are used in abrasive services, and PTFE is widely used in sealing aggressive chemical media. Specialised packing grades are even produced having elastomer cores, to improve the sealing of shafts of some pumps or agitators / mixers. This is important where additional resilience is required, often to cope with slight shaft eccentricities, vibration or similar difficult situations.

Figure FS3: There exist a wide variety of Compression Packings (James Walker)

When a packing set is compressed in a gland, there is some stress decay down the set (Figure FS4). Internal losses mean that each ring in turn progressively provides less axial load to the next, and each ring provides less radial force to the stem and gland. Therefore, if a large number of rings are used, the bottom ones perform relatively little sealing function as shown in the diagram alongside.

Many older valves were manufactured with as many as 10 or 12 rings of packing, sometimes with a lantern ring at the centre. However, it is now more common to use five or six rings of packing for economy as well as the efficiency of distributing the gland load throughout the set. A solid bush is typically used to accommodate any surplus depth at the bottom of the packing chamber when repairing old valves.

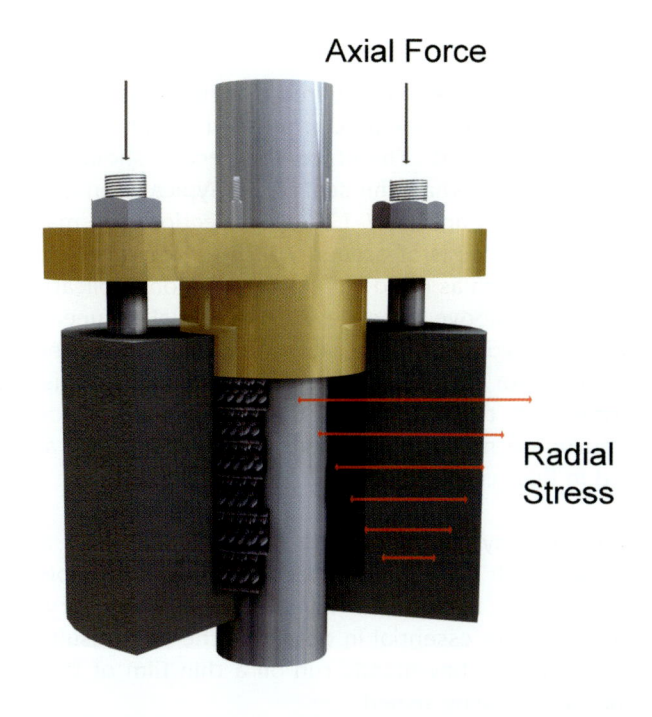

Figure FS4: Stress distribution down a packing set (James Walker)

A good packing set requires the right degree of boundary layer tribology in terms of the lubricity of the packing against the stem, to avoid excessive wear in operation. The addition of too much volatile lubricant, particularly in thermal cycling conditions can lead to loss of packing volume, which reduces compression on the stem or shaft potentially resulting in leakage.

In recent years, the use of expanded graphite in valve glands has become widespread, with many manufacturers using combination sets of packing. This is where rings of moulded expanded graphite are combined with a braided carbon yarn ring at each end. Whilst some people believed that these rings acted as wiper rings to remove graphite particles from the stem, and as anti-extrusion rings, it is now widely understood that much of their purpose is to wipe a boundary layer of lubricant onto the stem to limit and prevent adhesion of graphite particles from the intermediate rings. If un-lubricated graphite rings are used alone in a dry gas application, for example, it is not unusual to see the effects of stick-slip friction on the stem, with graphite particles adhered to it. This can lead to a shortened overall packing life with repeated stem cycling.

Once the boundary layer tribology of the packing has been resolved the sealing longevity is improved, although thermal cycling can also be destructive to the sealing performance. Face centre cubic materials such as austenitic stainless steels have higher coefficients of linear thermal expansion compared to those for body centre cubic materials such as ferritic, martensitic and precipitation hardened stainless steels. Very often the stem or shaft is one type and the packing chamber the other. During thermal cycling the packing chamber volume will vary therefore affecting the compressive loading of the packing on the stem or shaft. In an attempt to combat this effect, the use of live-loading has often been used to provide a degree of cushioning to the packing system. This method usually employs a number of disc springs under the gland nuts in an attempt to offer a more consistent load to the packing set as the overall gland and packing volume changes. The life of the set between adjustments may therefore be improved.

Control valves usually require special attention to the stem or shaft sealing as the number of operating cycles of these components is far greater than with an isolation valve. As low friction and smooth stem operation is mostly required, a PTFE chevron seal is commonly used. It is noted, however, that PTFE also has quite a high expansion coefficient, so severe thermal cycling can also cause problems here, and again the use of live-loading has often been used to try and overcome these effects.

Another alternative that is used is a spring-loaded type of lip seal, usually in PTFE or another engineering polymer. However, with these and PTFE chevrons, care is required to keep foreign matter out of the gland area, as dirt and other such particles can damage these relatively soft materials.

Elastomers

In many valves the deceptively simple rubber O-ring is still widely used, though care is often required in selecting the most appropriate compound. Typically a nitrile, or a hydrogenated nitrile material is generally used for oil applications, whereas ethylene-propylene is often employed on water and low pressure steam duties. Materials such as fluorocarbon and silicone may be required for higher temperature applications and certain chemical media. However, care is required if a mix of chemicals is to be dealt with, as these can often give rise to more complex situations than resistance to a single media alone, and the advice of a specialist should be sought.

The way an O-ring works is quite simple, in that the rubber distributes the pressure applied to it throughout the material. The initial squeeze of the section during fitting produces some internal stress within the seal. This is then increased by the system pressure, such that the reactive stress between the seal and the housing is always a little higher than the system pressure itself, so an effective seal is produced and maintained.

Figure FS5: Elastomer seals (James Walker)

As the rubber is a fluid albeit highly viscous, care should be taken to avoid conditions where it can be extruded through gaps in an assembly under high system pressures. It may be necessary to use a back-up ring on such applications, and this will depend to some extent on the seal section as well as the likely extrusion gap and operating conditions, though typically a back-up ring should be considered at pressures over 100 bar.

Another potential problem with high pressure gas is that of Rapid Gas Decompression (RGD). This is a problem where, over time, the high pressure gas diffuses into the seal material structure. If the system is then de-pressurised too quickly, that gas is unable to diffuse back out of the seal. The gas therefore expands and in doing so damages the internal structure of the material. It can sometimes be seen as blisters on the surface of the seal, internal cracking, or by significant tearing and destruction of the seal. Influencing factors are the rate of decompression, temperature, the seal volume to surface area ratio and the degree of seal restraint in the housing. Special elastomer compounds are produced for enhanced performance and resistance to this effect.

Figure FS6: An O-ring with RGD damage (James Walker)

It should always be remembered that rubber is essentially a fluid in that it cannot be compressed within itself, it can only be displaced. Therefore the recess should have a greater volume than the seal itself. Typically an O-ring is compressed by around 7-17% of its section, and may have a fill of the groove of around 70-75%. However, for some conditions such as rapid gas decompression applications as mentioned above, the levels of initial squeeze and percentage volume fill are often increased to limit the total potential for volume swell. Similarly, to minimise gas permeation, the smallest diameter section that does not compromise mechanical sealing efficiency, should also be used in RGD conditions.

Actuators may have traditional hydraulic type seals as well as rotary shaft lip seals to protect bearings for example. Care in selection, as well as knowledge of boundary layer lubrication are essential in obtaining the best results from such seals, as they mostly run on a thin film of the oil or other media being sealed.

Fluid sealing material manufacturers have a wealth of information available to help valve manufacturers and users with the correct selection, installation and use of their products and should always be consulted.

BVAA 70 years of service to industry

Nearly 50 years ago, BVAA published the very best Valve & Actuator Users' Manual ...

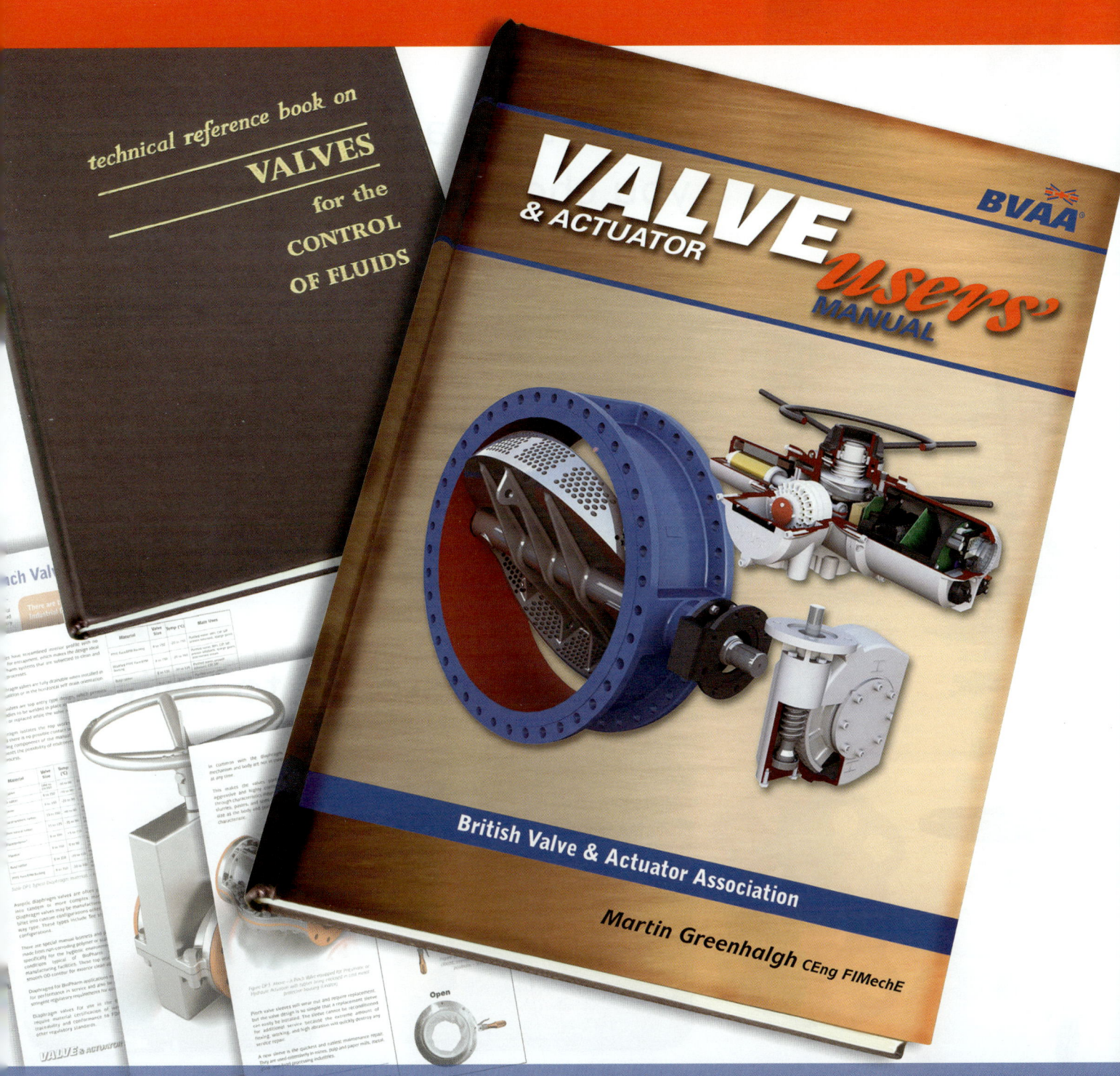

... 50 years later, we still do!
Contact BVAA and order the latest 6th Edition today!
£49.50+p&p, £34.50+p&p for BVAA Members

BVAA

British Valve & Actuator Association
9 Manor Park, Banbury, Oxfordshire OX16 3TB
Tel: 01295 221270 Fax: 01295 268965
Email: enquiry@bvaa.org.uk
www.bvaa.org.uk/usermanual.asp

econosto®
when the valves
you choose are critical
trust Econosto UK to deliver
Valves
Testing
Actuation
Special products
econosto®
Quality Since 1892
Leicester +44 (0)116 272 7300
Aberdeen +44 (0)1224 898 800
Glasgow +44 (0)141 941 1900
www.econosto.uk.com
sales@econosto.uk.com
An ERIKS Company
Econosto UK Ltd | Whiteacres | Whetstone | Leicester | LE8 6ZG | UK

Valve Selection Techniques

Figure VS1: Selecting the right valve and actuator for your application can be critical (Tomoe)

Deciding on the correct choice of valve and actuator for a particular application appears at first sight to be a daunting task. There are literally thousands of suppliers to choose from offering hundreds of different design options.

Every one of these suppliers certainly offer a valve or actuator that is perfect for someone's application. Almost all of them are unlikely to make a valve or actuator that is the ideal choice for your application.

As with most engineering decisions it is generally a question of weighing up different requirements against the different design features available and selecting the best match. There is no one valve or actuator that fits all applications and in most cases there is no perfect solution to any particular application. If there were then the variety of different designs and choice of suppliers would be much more limited.

When trying to select a valve or actuator for an application a matrix type assessment procedure is a useful tool. If 'bottom left' is a simple application and 'top right' is a difficult application, then deciding where your application fits into this pattern is an important choice that must be made. Small valves and actuators are usually bottom left and large ones top right, similarly this applies for low and high fluid pressure, low and high fluid temperature and non-corrosive and corrosive fluid. However there is a point when 'low' actually becomes difficult, for example a valve with 1mm bore, or extreme low pressure vacuum service or low temperatures below -196°C.

Figure VS2: Valve Application Matrix

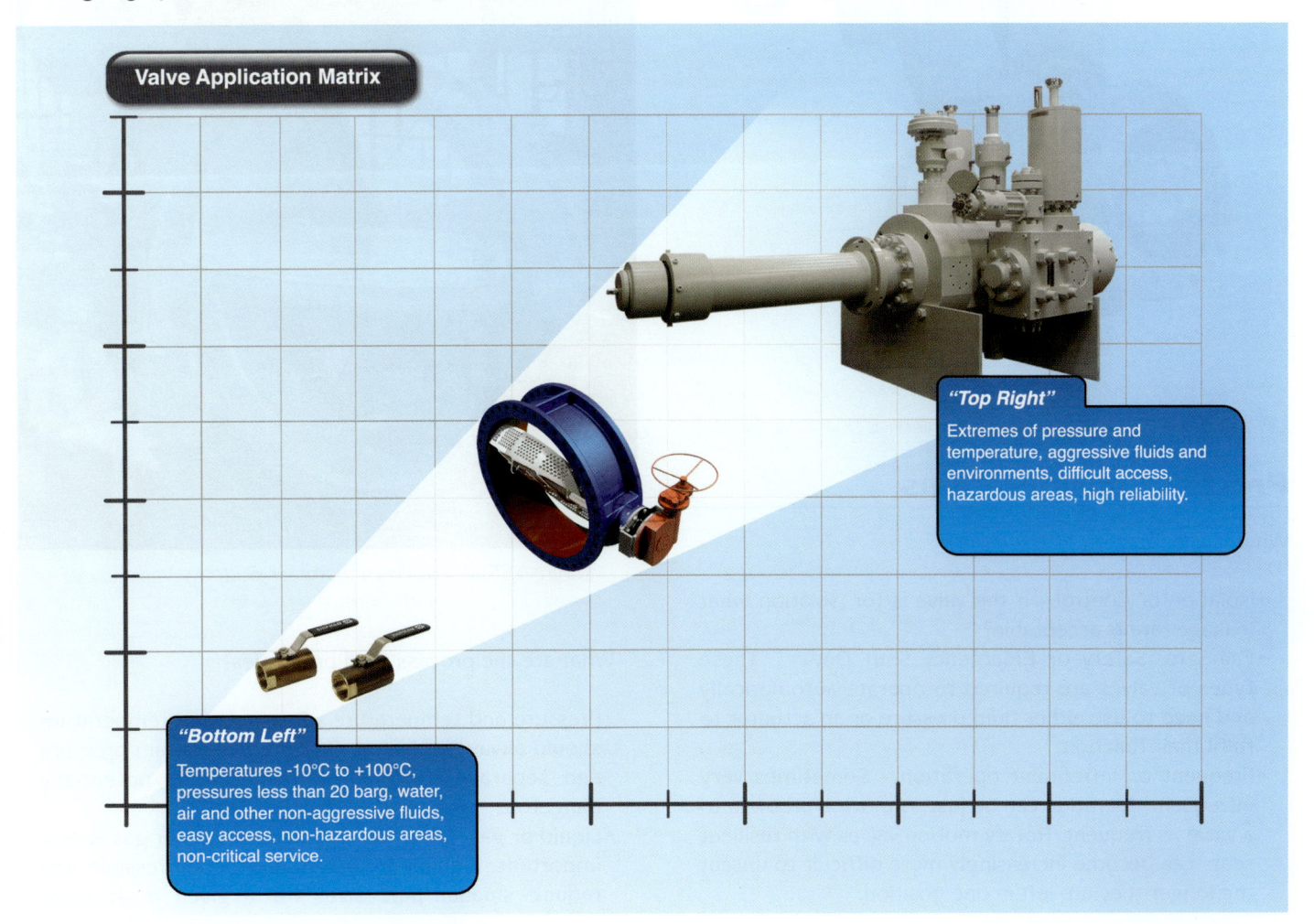

Valve Application Matrix

"Top Right"
Extremes of pressure and temperature, aggressive fluids and environments, difficult access, hazardous areas, high reliability.

"Bottom Left"
Temperatures -10°C to +100°C, pressures less than 20 barg, water, air and other non-aggressive fluids, easy access, non-hazardous areas, non-critical service.

Having made a matrix and decided where the application sits upon it, the valve or actuator selected needs to be compatible with the estimated service difficulty. Far too often people choose a valve or actuator that is designed for a simple application and try to use it for a difficult application. They are then surprised when after a short period of time it fails to live up to expectations. In the valve and actuator world the old adage of 'you get what you pay for' is still largely true.

When analyzing the application, the requirements of the Process (valve function and fluid), External Environment (location and connection to the system), Reliability (maintenance) and Legislation (documentation and testing) should be considered. The requirement is a combination of all of these and where the most difficult process conditions combine with an exacting environment the need for high reliability and an onerous legislative requirement - for example the primary circuit isolation valve of a pressurised water nuclear reactor - the challenge to design a suitable valve is huge and the number of possible options small.

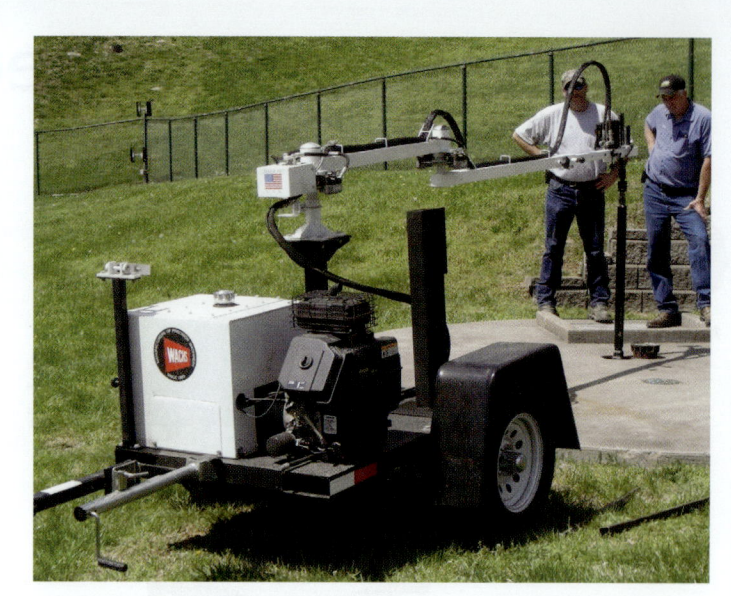

Figure VS4: Infrequent valve operation can lead to seized valves. A valve 'exercising' regimen (opening and closing) can be an effective solution (E H Wachs)

- Full open to closed operating time - Ensuring that the valve functions in the correct time can be very important. A valve closing too quickly can set up pressure surges in the system. A valve not closing quickly enough can lead to downstream process problems.

Figure VS3: A metal-seated, severe service ball valve (Velan)

Process Considerations

What is the function of the valve?

- Isolation or Control - If the valve is for isolation what leakage rate is acceptable?
- Fail-Safe, Safety or Emergency Shut Down - These types of valves are required to operate automatically and need to use either fluid pressure or an actuator to fulfil their function.
- Frequent or infrequent operation - Sometimes very infrequent operation can be just as much a problem to a valve as frequent. Rotary motion valves with resilient seats can become increasingly more difficult to unseat the longer they are left in one position.

Figure VS5: A 2400mm diameter Butterfly Valve under performance test (eTec)

What are the process fluid properties?

- Pressure and temperature - Pressure and temperature should always be linked. Quoting a maximum pressure and separate maximum temperature is potentially dangerous.
- Liquid or gas - Whether the fluid is liquid or gas is very important. Liquids are essentially incompressible and require smaller pipe sizes for a given mass flow.

However the leak of a small quantity of liquid can result in the formation of a large gas cloud to the environment. Therefore liquid service valves can present a greater risk than gas service valves that are a number of pipe sizes larger.
- Flow rate and line size - Normally the line size is given rather than the flow rate. For isolating valves this is generally acceptable but for butterfly valves and check valves dynamic or momentum forces can be important. For control valves full details of flow rates must be given ideally for minimum, normal and maximum flow conditions and for safety valves the maximum conditions must be specified.
- Pressure drop - What is the acceptable pressure drop across the valve and how does that impact on the energy costs of generating the pressure upstream of the valve?
- Leakage to the environment - Failure to contain the process fluid is a major risk for the valve user depending on the nature of the fluid. Fluids which are toxic, flammable or explosive require high integrity containment. Special designs which minimise leakage from the body bonnet joint or from the gland will be necessary.

Figure VS6: 12" cryogenic ball valve undergoing pressure test at -196°C (Adanac)

- Seating Face Leak Tightness - How important is the seat leakage rate? A new valve will be tested to an internationally recognised standard but its ability to continue to provide the same leak tightness will depend on how well the design protects the seating faces and maintains them in the as new condition.
- Corrosive or Abrasive Fluids - Correct selection of all materials for body, bonnet, body bonnet gasket and trim is essential if the effects of corrosion or erosion are not to cause failure of the valve.

- Leakage from the environment - In some processes contamination of the fluid by the environment can be equally as damaging as fluid leaking from the valve.

Environmental Considerations

- Access to the valve and available space or weight restriction - The location where the valve is to be installed can be a key factor in the type of valve best suited to the application. It can make the difference between selecting a gate or a butterfly valve if space or head room is restricted. The ability to lift the valve into position can make weight a major consideration. Access to operate the valve must be considered together with how the valve is to be operated. Valves installed sub-sea or in a radioactive environment have very special design requirements.

Figure VS7: Accessed only from above via a 15ft deep pit, this non-standard 1.5m, 2.2 tonne butterfly valve was a bespoke solution (Flow Group / Leeds Valve).

- Manual or power operation - The requirement to be able to operate every valve is an essential process and safety consideration. Manual operation may not be possible or desirable in which case an actuator must be fitted. There is then a requirement for a power source either electric or fluid and a whole new set of issues relative to the power source come into play.
- Hazardous Area Classification - Valves, actuators and accessories need to be explosion or flame proof when installed in Hazardous Areas.
- Ambient Temperature and Humidity - The corrosive nature of the ambient conditions can cause failure of bolting or seizure of operating mechanisms or in extreme cases failure of bodies or bonnets. High or low ambient temperatures can make operating the valve difficult or hazardous. They can also create problems when the fluid is a liquid. It can freeze and/or expand subjecting the valve components to excess pressure unless appropriate design features are included.

Figure VS8: Manually operated stainless steel Floating Ball Valves on a CO_2 injection facility (Cameron)

Figure VS9: A heavy duty Hydraulic Failsafe Actuator (Rotork) operating a 24 inch Double Flanged Class 1500 Triple Offset Butterfly Valve (Tomoe) on a floating production storage and offloading vessel (FPSO)

Figure VS10: A combination of low air temperatures and wilderness location can result in very innovative flow control solutions (Rotork)

Figure VS11: The ambient temperature range for this electro-hydraulic actuator on the Baku-Tibilisi-Ceyhan crude oil pipeline was -29°C to +40°C (ITT Midland-ACS)

- Connection to the Pipe and to the Power Source - There are many forms of end connection each bringing with it issues for valve design. Pipe joints can be sources of leakage and difficult to maintain if the pipeline temperature varies significantly between a flowing and a non flowing situation, e.g. steam or cryogenic pipe lines. Under these circumstances all welded pipeline joints offer advantages and a valve design must be selected that supports that type of joint. The heat input when welding the valve into the pipeline can affect resilient valve seating. The power source must also be connected to the actuator and the reliability and availability of that power supply must be considered.

- Face-to-Face and End-to-End Dimension - Over the years there has been co-operation between manufacturers and users to standardise the face-to-face and end-to-end dimensions. There are two basic systems either the European EN or the American ASME. Each valve type has its own set of lengths so that replacement of an identical valve type from a different manufacturer is possible. However changing from one valve type to another will usually require changes to pipe work to accommodate the replacement valve type.

Reliability Considerations

- Reliability - The consequences of a particular valve or actuator failing to do its job must be assessed. The consequences can vary from minor inconvenience to catastrophic environmental damage or major loss of life or both. Special consideration must be given to the valve's design relative to the perceived risks due to failure.

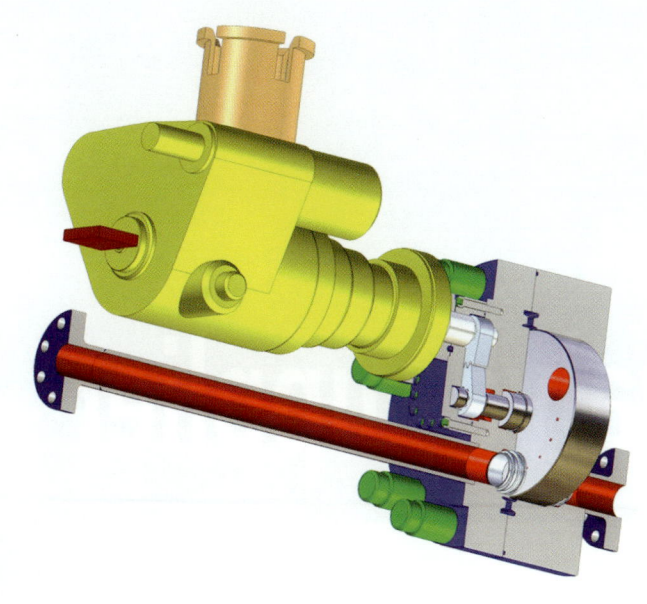

Figure VS12: Rotary Gate Valve for subsea MEG injection on Ormen Lange field had to be resistant to a high level of particulates, hence an innovative design was required (Weir)

Figure VS13: A sub-sea Autonomous Shut Down Valve unit, with an 8 inch Class 600 Ball Valve, automatically isolates pipeline risers on termination of pumping operations or in the event of a line break condition (Paladon)

- Maintenance - The reliability of a valve to perform its intended service is also affected by maintenance. The level of maintenance and the access required to do that maintenance are important aspects of the selection process. For example gate, globe and top entry ball valves all have the ability to be maintained without removing the valve body from the pipeline. If maintenance is not going to be possible for example in a gas pipe line valve buried below ground then high reliability in the original design is essential.
- Testing - New valves are subjected to standard internationally recognised testing. These tests are production tests designed to ensure basic functionality and as new performance. They give little indication of long term reliability or performance. Type testing or special production testing should be considered when there is a high level of perceived risk due to valve failure.
- Quality - ISO 9001 gives the purchaser a basic level of confidence that the manufacturer will produce a consistent product. It also helps to ensure that the necessary material and test certificates can be provided to show compliance with the original valve specification.

Legislation and Standards Considerations

- National Regulations - Different countries have different legislative requirements. These requirements affect the valve and actuator design. Within Europe the Pressure Equipment Directive specifies essential minimum safety requirements that each valve must comply with. Actuator design is affected by other directives. Manufacturers must provide appropriate valve and actuator designs and declare that their products meet all the applicable directives. Products, which comply, are marked with the CE mark. Outside of Europe American codes of practice such as API and ASME are widely used.
- Product Certification - A fire test certificate is an example of the need to provide a design which meets specific certification requirements.

Figure VS14: Valve undergoing fire test (Colson)

Working practice

Working practice may also determine the type of valve selected. Different industries prefer particular valves types or designs to meet their requirements and the conservative nature of the industry means that these traditions are not easily changed even though improved products are sometimes available.

Specification sheets

Transfer of relevant information from user to manufacturer is essential for the manufacturer to offer his most appropriate product for the user's application. A formal specification sheet ensures that all the relevant information is included in a rigorous process. This is particularly important for safety valves and control valves. In the latter case not only has the valve information to be provided but also the information required for the actuator and accessories.

The complete valve supplier

D & D International Valves Ltd
Successfully providing valve solutions since 1982

- Stockist and Distributor
- Gate, Globe and Check valves
- Butterfly valves. Concentric, Double & Triple offset
- Ball valves
- Safety Relief valves
- Actuation
- Valve modification

www.ddvalves.com
sales@ddvalves.com

Telephone +44 1284 700350
Facsimile +44 1284 700650

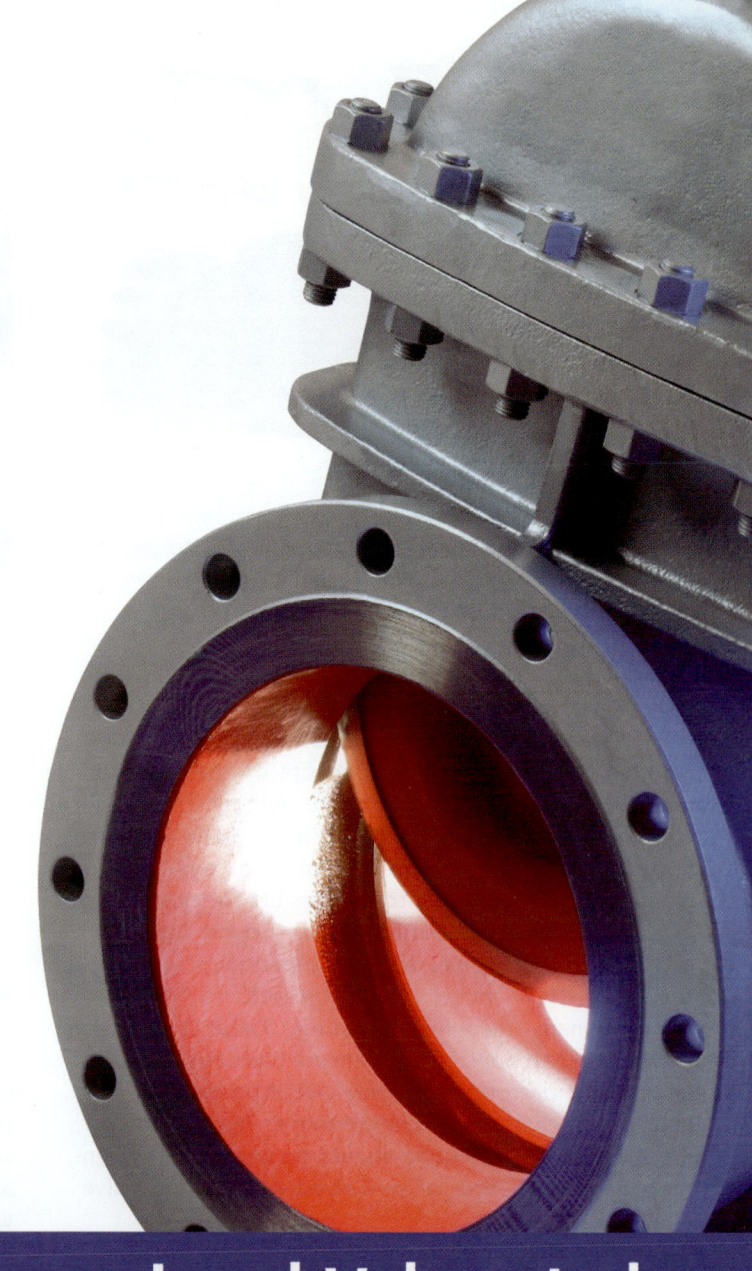

D & D International Valves Ltd

Saxham Business Park, Saxham, Bury St Edmunds, IP28 6RX, England

Linear Valves

Gate, Globe, Diaphragm and Pinch Valves

H Valves Ltd.

Low Pressure PSV

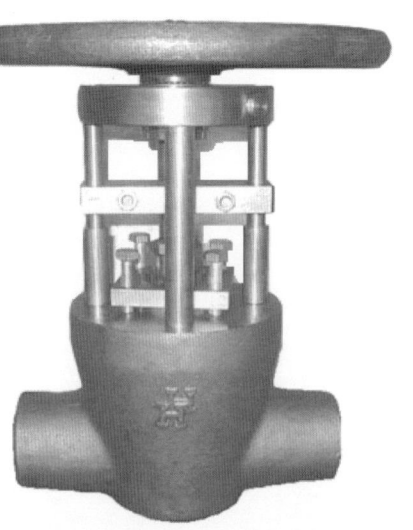

4500 Class PSV

High Pressure
Large Bore
PSV

HH Valves Ltd
Unit 1 Leopold Centre
Smethurst Lane, Pemberton
Wigan, WN5 8EG

High Pressure PSV

CE

LLOYD'S REGISTER QUALITY ASSURANCE · ISO9001

UKAS
QUALITY
MANAGEMENT
001

Tel: +44(0)1942 218111
Fax: +44(0)1942 224800
e-mail: sales@hhvalves.com

For Full Product Range & Stock Availability Visit
www.hhvalves.com

Gate Valves

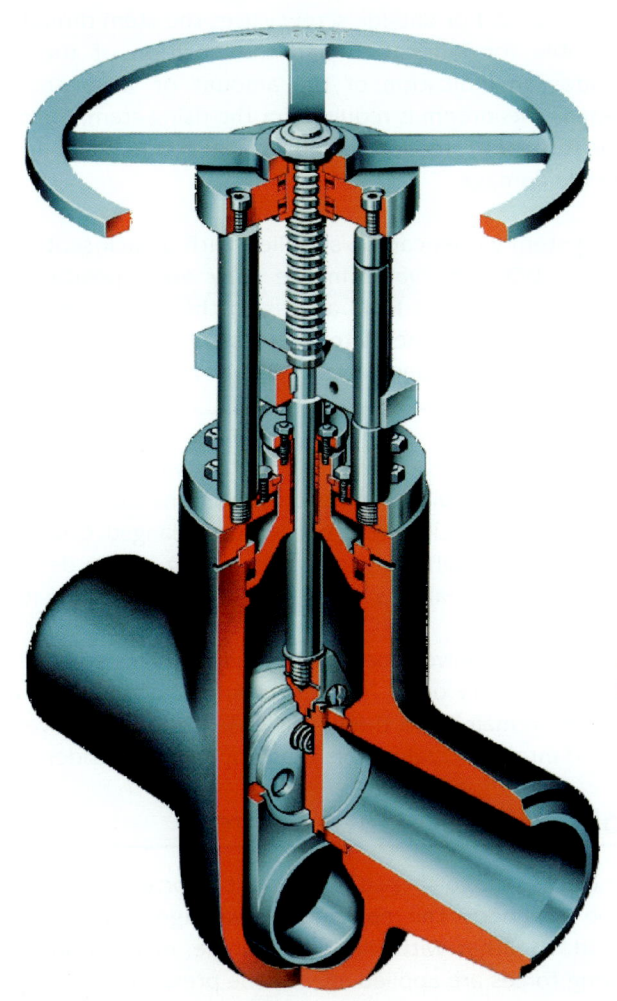

Figure GV1: Parallel Slide Gate Valve with Pressure Seal Bonnet (Tyco Dewrance)

Gate valves are still the most common valve in use today and consist of a body, bonnet, gate and stem. The gate is moved across the body seats by the stem, shutting off the flow of fluid. In general the gate valve will seal against the flow in either direction.

The gate has to move a distance equal to the diameter of the seat so the manual operating time open to close of a gate valve is the longest of all the valve types. Conversely the operating torque is typically the lowest although the need to drive a solid wedge into its seat and subsequently remove it creates a high seating and unseating torque. More details can be found in the chapter on Valve Operating Torques.

The fundamental advantage of the gate valve is that the gate fully retracts into the bonnet and creates minimum pressure drop when the valve is in the full open position. Size and weight are also important factors to take into account when selecting a gate valve. The need to create a space into which the gate can move makes the shell of the gate valve larger than any other valve type. Shell wall thicknesses are a function of size and pressure so large diameter high pressure gate valves can be very heavy in addition to being large in size. For rising stem outside screw design sizes DN250 and greater, the height from pipe centreline to end of stem in the open position is typical 5xD, where D is the seat diameter.

Gate valves are intended to be used as isolating valves. They are used extensively on steam, water, gas, and oil. They are not recommended for regulating purposes, as severe erosion can occur around the leading edge of the gate when approaching closure.

Figure GV2: Typical Wedge Gate Valve (Shipham)

When proposing a gate valve for isolation service the level of required seat leak tightness should be considered. Typically a gate valve will have spent most of its life in the open position. During this period of open service the different types of gate valve will experience differing degrees of fluid damage to the seat faces by abrasive wear or by corrosion or the creation of a deposit on the seat surface. Debris may fall out of the fluid and collect in the bottom of the body in the space below the seats. All of these factors may affect the gate valve's ability to seal on closing. For clean non-corrosive service simple cost effective solid wedge gate valves are a suitable choice, but as temperatures and pressures increase abrasives e.g. sand are present, and the fluid is corrosive, designs become increasingly more complicated and sophisticated.

Figure GV3: Wedge Gate Valve for Nuclear Service (Weir)

Figure GV4: Nuclear Gate Valve (Velan)

The valve may have an inside screw in which the actuating thread of the spindle is contained inside the valve and thus in contact with the line fluid. An option with an inside screw design is a non-rising stem. This arrangement is particularly useful where headroom is limited. The elimination of the vertical movement of the stem also reduces the wear on the gland packing. Inside screw valves are not suitable for dirt-laden or corrosive fluids and for these conditions the outside screw is used. For outside screw valves the stem threads are accessible for lubrication and the position of the stem provides an indication of the amount of valve opening. Adequate headroom is required for the rising stem, for which some form of protection should be arranged to guard against possible damage.

Rising stem valves can be supplied with a back seat in the bonnet. With the valve in the fully open position this feature reduces the fluid access to the underside of the packing and either helps to stop leakage from the packing chamber should it develop or helps to prolong packing life in the first place.

Trim materials can be varied making the valve suitable for a wide range of services, and end connections can be supplied either screwed, socket weld, butt weld, or flanged. Gate valves are divided into different types, depending on the design of the gate and its seating surfaces to suit different applications.

Linear motion valve types are top entry valves so that it is possible to carry out a full refurbishment of the valve with the body remaining in the pipeline. As many gate valves are welded into the pipeline this is an important consideration.

Solid Wedge Gate Valve

The disc is wedge shaped and seats on corresponding faces in the valve body. The mechanical advantage of the actuating thread, together with the wedge angle, ensures adequate seating forces are applied against the pressure. Typically the body seats are separate rings threaded or welded in place, except for valves DN50 or less the seats can be pressed or rolled in. The body seat rings are usually hard faced. Wedge seating surfaces can be integral or hard faced. Other types may have soft seating surfaces or the whole seat constructed with a resilient material such as PTFE. This is to ensure high integrity shut-off with low torques. The soft seat design is generally used for isolating corrosive fluids.

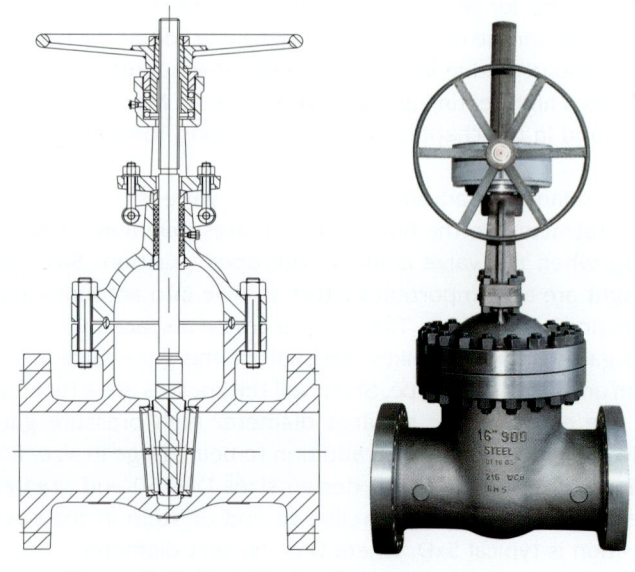

Figure GV5: Flexi Wedge Gate Valve (Tyco Fasani)

The metal seated valve will normally give a tight seal on the downstream seat. When a higher integrity seal is required a double block and bleed design is available, incorporating soft seats which ensure tight shut off both upstream and downstream. Line fluid is then drained from the central chamber of the body, which indicates seat leakage if the fluid continues flowing. This type of valve is specified when there is a need to ensure the integrity of the downstream seat or when the valve is sealing against two separate sources of flow and is required to prevent contamination between these two sources. A variant of the solid wedge gate valve is the sluice valve. As the name implies it is generally used in the water industry and manufactured in cast iron. The wedge is coated with an elastomer, and machining reduced by eliminating conventional body seats and by casting appropriate contours within the body, corresponding to the moulded shape of the wedge gate.

Conduit Gate Valve

The gate is in the form of a plate with parallel faces and incorporates a circular aperture of the same diameter as the valve bore for the open position and a blank area for the closed position. The valve is characterized by its body design, which extends above and below the centre line to provide the necessary cavities for the double-length disc plate. The clearway design has the same pressure drop as an equivalent length of pipe, allows passage of all types of "pigs" and destructive turbulence is eliminated.

The seats are outside the flow stream and in full contact with the gate, in both the fully open and fully closed positions and therefore seat life is greatly extended.

The valve is usually soft seated with floating seats to enable line pressure induced seating forces on both upstream and downstream seats.

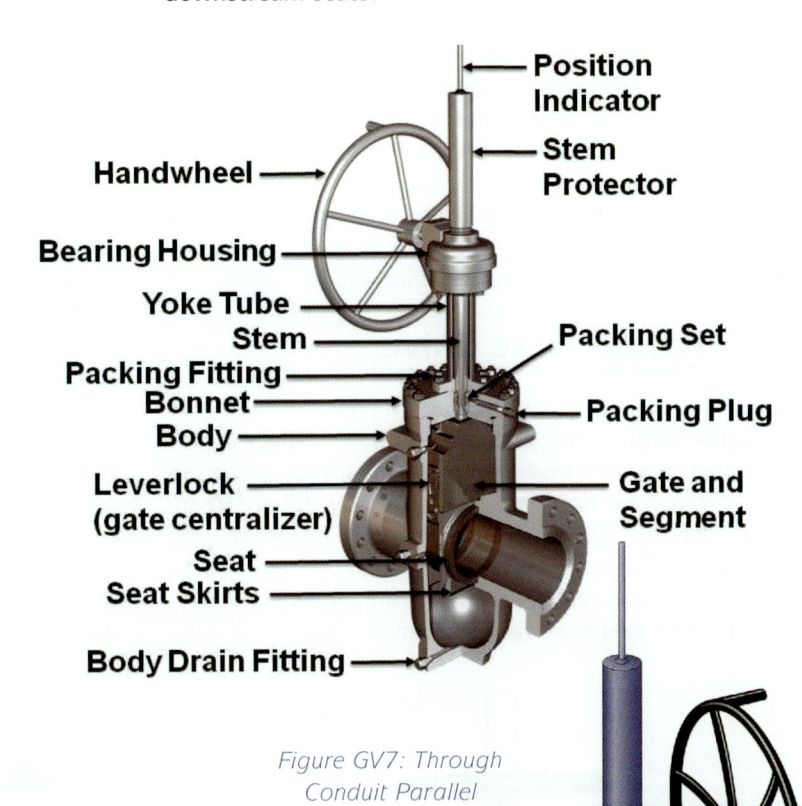

Figure GV6: Subsea, Hydraulic Spring Return, Through Conduit Gate Valve (BEL Valves)

Figure GV7: Through Conduit Parallel Expanding Gate Valve (Cameron)

Flexible Wedge Gate Valve

Developed to overcome jamming at high temperatures of the solid wedge, the shape of the flexible disc is like two wheels on a very short axle, allowing some degree of flexing in the two halves. It is this flexibility that makes the disc tight on both surfaces over a wide range of pressures and temperatures at lower operating forces than the solid wedge gate.

Generally a flexible wedge should be fitted as standard in valves ≥ DN150.

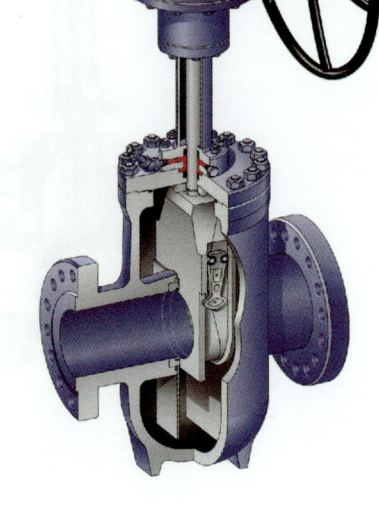

Conduit Parallel Expanding Gate Valve

The parallel expanding gate design is a variation of the conduit gate valve providing good isolation, which is normally unaffected by pressure variations.

The gate-segment assembly consists of gate, segment and gate centralizer. In the closed the position, the segment stops moving when it contacts the body stop. Continued stem movement causes the gate to slide down the segment upper angle, expanding the gate and segment outward against both seats. In the open position, the segment stops moving when it contacts the bonnet stop. Continued stem movement causes the gate to slide up the segment lower angle, again expanding the gate and segment against both seats. Flow is isolated from the valve body. The gate centralizer allows the gate and segment to move freely between the full open or full closed position and only expand at the ends of travel.

Seat sealants can also be injected to effect a seal in an emergency should the seats become damaged. The valve in the closed position forms a tight mechanical seal on both seats simultaneously, which allows the body cavity to be vented preventing a dangerous pressure build due to thermal expansion of liquid trapped between the closed discs. Applications include oil and gas pipelines and storage. It has also been used successfully on high temperature applications such as steam and geothermal.

Parallel Slide Gate Valve

A further variation of the gate valve is the parallel slide gate valve (Figure GV8), where the gate is a pair of discs lightly held against parallel body seats by a spring enclosed between them. Seat tightness is achieved by the upstream pressure forcing the downstream disc against the downstream body seating surface. The effort required to seal a parallel slide valve is therefore much reduced over that required to seal a wedge gate valve.

Figure GV8: Parallel Slide Gate Valve (Peter Smith Valve Co.)

The fact that the sealing load is generated purely by the fluid pressure requires there to be a minimum differential pressure for this type of valve to work satisfactorily. The required minimum differential pressure is approximately inversely proportional to the seat diameter.

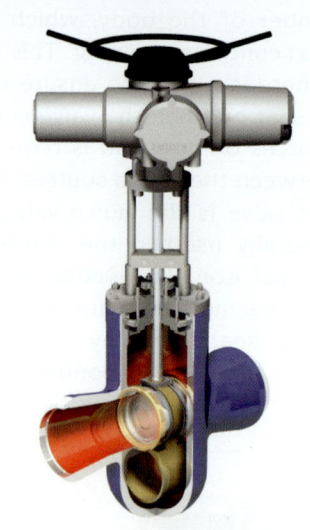

Figure GV9: Venturi Parallel Slide Gate Valve with Pressure Seal Bonnet (Weir)

Each manufacturer will make their recommendations but typically for a DN25 valve 4 bar is the minimum differential required. The distance between the seats is greater than the width across the discs. This allows for expansion and contraction when the valve is subjected to wide variations in temperatures. This makes it ideally suited for steam and high pressure hot water applications.

The parallel slide valve as the name implies has a sliding action as the discs move across the seats. This means that the seating faces must be made of materials that do not suffer from galling wear. Stainless steels are not recommended and usually a cobalt or nickel alloy is the preferred choice. The sliding action also has the effect of wiping the seat surface helping to remove any particles or contaminates which might get between the seating surfaces.

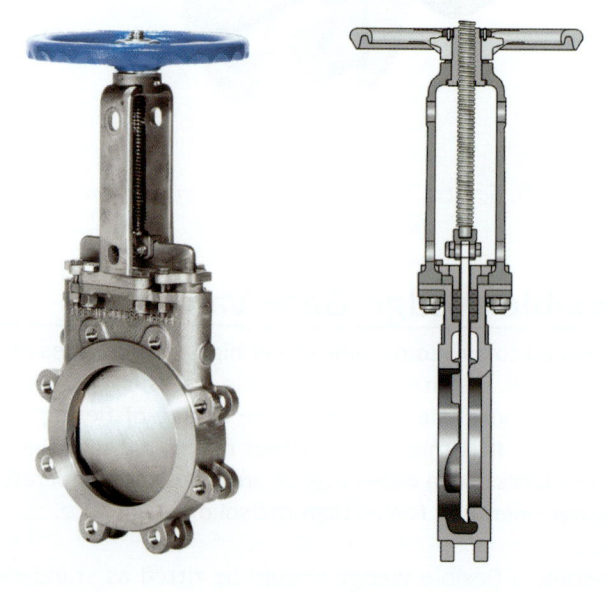

Figure GV10: Stainless Steel Knife Gate Valve, Wafer Lug Type with Metal Seat (Velan)

As with the double disc through conduit valve there is the potential to trap liquid between the discs of a parallel slide gate valve. If this liquid expands due to heating, an excessively high pressure can develop in the body cavity requiring either operating torques greater than the strength torque to open the valve or in extreme cases causing a leak at the body bonnet joint. For valves only required to seal in one direction the provision of a body cavity relief from the inter-gate space to the upstream side of the valve will prevent this problem. Valves that are required to seal in both directions are fitted with an equalising by-pass valve, which combines the equalising function with a normal by-pass system. The manufacturer will recommend when this additional feature should be incorporated.

An option for the parallel slide gate valve is the reduced bore conduit style body. The design is known as a venturi style. The gate incorporates a circular aperture of the same diameter as the seat bore for the open position and the discs for the closed position. The seat bore is typically one pipe size lower than the inlet and outlet body ends. The body reduces the flow area gradually to the seat bore diameter and then expands it back to the full nominal inside diameter of the body end port on the downstream side. The fluid velocity through the seat area is increased by a factor of between 2 and 1.5 depending on valve size. This does lead to slightly higher pressure drops but the conduit design protects the seating faces from the faster flowing fluid and the advantage is a lower weight and lower cost solution.

Knife Gate Valve

As its name implies, the knife gate valve uses a knife-like gate or guillotine to stop the flow of liquids or solids in the line. Unlike conventional gate valves, the knife or slide passes through the body, and a standard gland packing cannot be used. The valves are ideal for use in low pressure (PN16 or ASME Class 150) applications handling viscous media, dry powder, slurries, and sludges, and can be found in mining, sewage, food, cement, brewing, pulp, and the paper industry.

Knife gate valves can be produced with all metal seating faces or with a soft seat, the latter giving a bubble-tight shut-off in both directions. Valves up to DN2500 are available, the larger sizes being of a fabricated design. As the gate moves into position it slides across the seating faces which helps to clean it and ensure effective isolation of these difficult media. Face to face dimensions are the shortest of all the gate valve types.

Penstock Valves

The Penstock is a specific type of gate valve designed for controlling large volumes of water and/or sewage and is usually fabricated from plate material or made from cast iron. The valves are normally rectangular in shape and the preferred width to depth ratio is 2:3 in the vertical position and 4:3 in the horizontal position. Hand wheel or actuated operation of the valve is possible, using a rising stem to avoid it being immersed in the fluid and to allow easy access for lubrication.

Figure GV12: Typical penstock arrangement with electric actuators (Rotork)

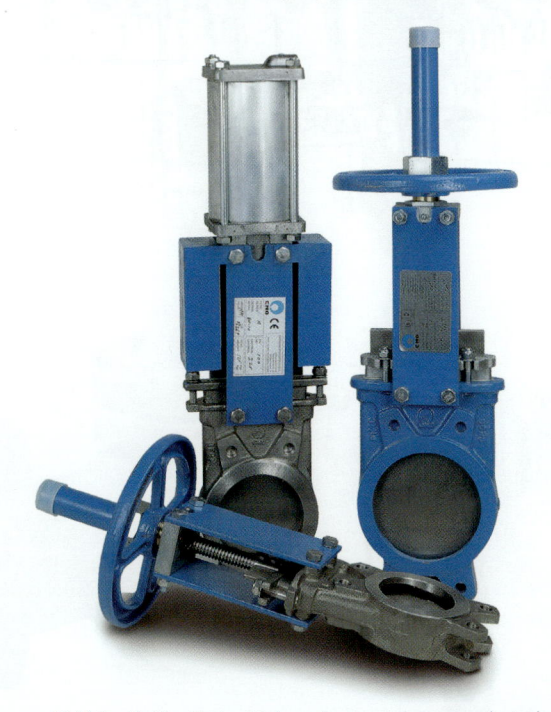

Figure GV11: Knife Gate Valves in cast iron and stainless steel showing hand wheel and actuator operation (Econosto)

kentintrol

KOSO KENT INTROL
– THE ENGINEER'S CHOICE

Engineers from the oil and gas, petrochemical and power industries choose our valves to perform in the toughest environments around the world.

They know that there's no room for compromise, and that's why they've trusted us to help overcome operational conditions for over forty years.

From design to manufacture, service to refurbishment, our highly skilled specialist teams are ready to share their expertise to solve your challenge – however demanding.

KOSO KENT INTROL LIMITED
ARMYTAGE ROAD
BRIGHOUSE
WEST YORKSHIRE
HD6 1QF
UK

TELEPHONE
+44 (0)1484 710311

FACSIMILE
+44 (0)1484 407407

EMAIL
info@kentintrol.com

WEBSITE
WWW.KENTINTROL.COM

FPAL empowered by Achilles

Koso Kent Introl is part of the KOSO Group

KOSO

Globe Valves

Figure GB1: Globe Valve (Shipham)

The globe valve has in theory the best possible isolating characteristic as it is the only valve type where the disc movement is parallel with the direction of flow and the torque applied through the operating mechanism directly clamps the seating faces together.

In the open position flow is through the skirt area under the edge of the disc, which equals the seat bore area after the disc has moved D/4 away from the body seating face. Hence the operating time, open to close, of a globe valve with no pressure in the valve is nominally ¼ of that of the same size gate valve and therefore globe valves are often used where frequent operation is required.

Globe valve bodies can either be straight (Fig. GB3), angled (Fig. GB4) or oblique pattern (Fig. GB5). In all cases flow through the body follows a changing course. This leads to a pressure drop across the valve, greatest for the straight pattern body and least for the angle pattern. This pressure drop is a negative feature when the globe valve is used for isolation, if it is mostly in the open position, but when the valve is mostly closed it is of little consequence. The pressure drop can be a positive feature as it allows the globe valve when fitted with a profiled disc e.g. parabolic to be use for regulating service. When manually throttling, the flow area is proportional to the number of turns of the handwheel but flow will also depend on how the upstream pressure varies. If the system maintains approximately constant upstream pressure flow will be proportional to flow area. The globe valve is the most commonly used valve type in control applications.

Figure GB2: Basic Globe Valve (Weir)

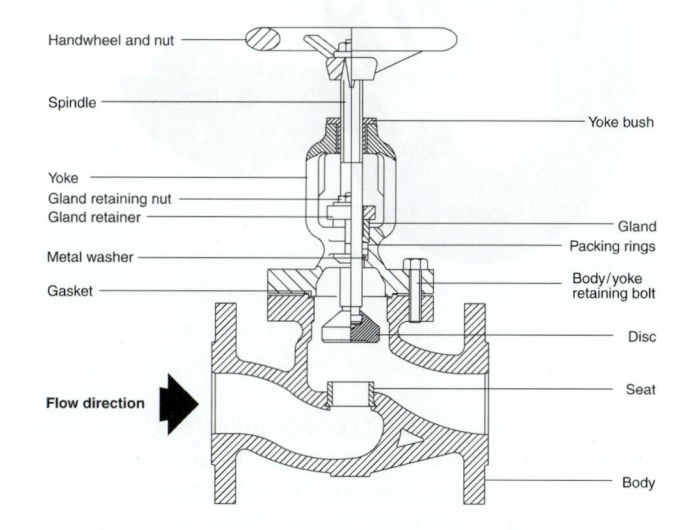

Figure GB3: A typical Globe Valve (diagram prepared by Spirax Sarco)

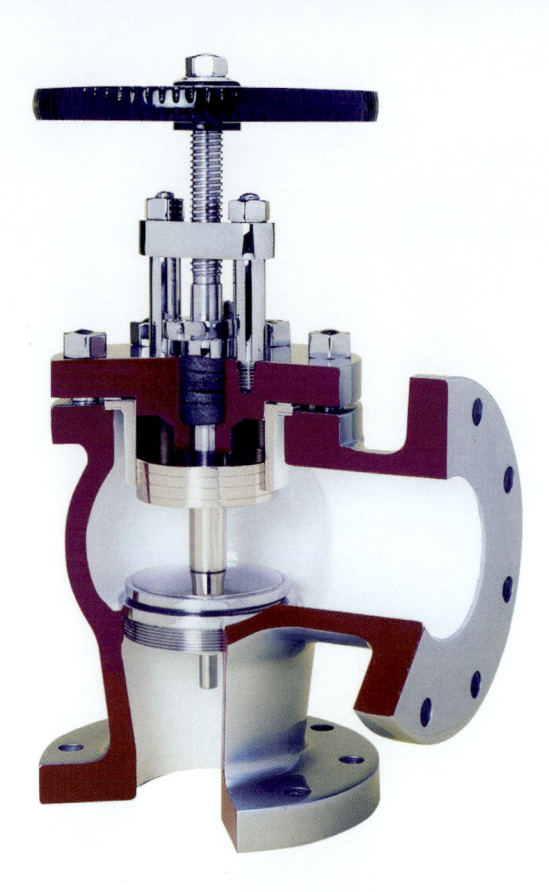

Figure GB4: Angle Pattern Globe Stop and Check Valve for steam isolation (Peter Smith Valve Co.)

Figure GB5: Y-pattern Globe Valve with regulating disc and live loaded gland (Peter Smith Valve Co.)

Figure GB6: Bellows Seal Globe Valve for Hazardous Chemicals (Taylor Shaw/Blackhall)

Flow direction is an important consideration for globe valves. If the flow is from under to over the disc, the pressure acts on the underside of the disc and the operating mechanism must provide sufficient thrust to overcome the pressure load plus providing the additional force necessary to clamp the seating faces together. With the valve closed the bonnet, body/bonnet joint and gland packing are isolated from the upstream pressure and this has therefore been the traditional direction of flow for globe valves. If the flow is from over to under the disc the pressure load helps to clamp the seating surfaces together but the pressure load must be overcome to open the valve. As the seat diameter increases the pressure load increases proportional to D^2. This leads to very high operating torques for high pressure large bore globe valves and a DN150 PN40 valve is the largest valve that can be operated solely by direct manual effort applied to a handwheel. For larger sizes and pressure ratings gearboxes and/or actuators are required.

Designers have sought ways to reduce the required operating torques apart from using the valve with the flow from over to under the disc. The reduced bore valve has a smaller seat diameter than the full bore valve and therefore a lower operating torque. The price to be paid is a significant increase in pressure drop across the valve. A better solution is to provide a balanced design using a cage around the disc. The disc is lengthened and fitted with a sealing ring such that as it moves it acts as a piston inside the cage. This allows a bleed hole to be drilled vertically through the disc. Inlet pressure is then introduced to the top of the disc and the pressure differential across the disc is now balanced significantly reducing the operating torques. The sealing ring introduced into the disc seals the sliding contact surface between the disc and the cage maintaining the leak tightness of the valve in the closed position. As always there is a price to be paid and that is either additional maintenance on the sealing ring which wears as function of operating cycles and fluid cleanliness or an acceptance of a small but increasing leakage past the ring.

The D/4 maximum travel of the globe means that it has a much shorter overall height than the gate valve and it is most suitable valve type for the fitting of a bellows to the operating mechanism seal. The combination of best possible isolation characteristic and a bellows seal makes the globe valve the preferred choice of the chemical industry for the isolation of hazardous chemicals.

Globe valves can be either inside or outside screw. Typically the smaller valves for lower pressures have an inside screw and larger valves for high pressures have an outside screw. The standard globe valve whether inside or outside screw has a rotating and rising stem. This helical motion of the stem leads to above average wear of the packing and potentially causes rotation of the disc across the body seating surface as the two come into contact. By introducing a yoke sleeve into the design the globe valve stem becomes non-rotating and this produces a superior version of the globe valve with reduced gland leakage and better isolation performance over a longer period.

Figure GB7: Non-rotating Stem Globe Valve (Taylor Shaw/Blackhall)

As with the gate valve, rising stem globe valves can be supplied with a back seat in the bonnet. This feature reduces the fluid access to the underside of the packing and either helps to stop leakage from the packing chamber should it develop or helps to prolong packing life in the first place.

Generally, the disc and stem are separate components and connected together in such a manner that the disc is free to swivel independently of the stem. This allows the disc to sit squarely on its seat and minimises frictional contact that might damage the seating surfaces. For metal to metal seating surfaces either the body or disc seating is narrow so that the clamping force produces a high sealing stress. This requires globe valve seating materials to be hard so that there is only limited plastic deformation on contact. Soft seat options are available with most globe valve designs. The design must have a wider seating surface and care must be taken that the operating torque does not produce an axial force which crushes the soft seat material. To minimise the possibility of this happening manufacturers may fit a smaller size of handwheel on soft seat valve.

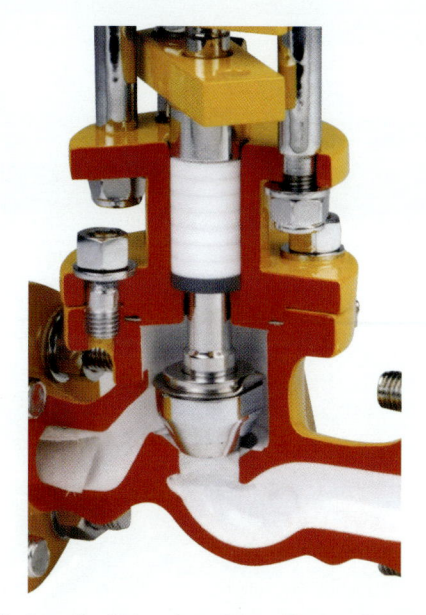

Figure GB8: Detail of gland packing (Taylor Shaw /Blackhall)

Linear motion valve types are top entry valves so that it is possible to carry out a full refurbishment of the valve with the body remaining in the pipeline. The globe valve has only one seat and the seating face is always parallel with the body bonnet flange making this the easiest valve type to repair.

The body style used by the globe isolation valve is also used for the lift type check valve. It is possible to combine the two functions of isolation and check into the same valve and the Globe Stop and Check valve was created. More details are to be found in the chapter on Check Valves.

Figure GB9: Firesafe LNG/LPG Globe Valve (Bestobell Valves)

BVAA VALVE & ACTUATOR USERS' MANUAL - 6th Edition

Saunders®
the science inside

Since P K Saunders invented the original diaphragm valve in 1928, and founded Saunders Valve Co. five years later, Saunders has led the way in providing the highest standards of reliability, engineering and safety. The range has been continually expanded over **75 years** through innovation in both design and new materials technology. As a result the Saunders diaphragm valve has gained a widespread reputation for its versatility and established a presence in diverse process industry sectors.

VISIT OUR WEBSITE
STOP

75 years of science inside!

75

CRANE®

www.cranechempharma.com
© 2010 CRANE ChemPharma Flow Solutions

Diaphragm Valves

Figure DP1: Weir Type Diaphragm Valve with liner (Crane Saunders)

Diaphragm valves were introduced in the 1920s to control compressed air and proved so successful they quickly spread into other areas of fluid handling. There is hardly an industry today which does not make use of this design of valve. The use of a resilient diaphragm as one of the seating faces ensures excellent isolation but does restrict the valves use to pressures and temperatures for which the diaphragm material is suitable. The pressure rating limit is generally PN16/Class 150 but some manufacturers do offer Class 300 rated valves.

There are two basic designs for Industrial Diaphragm Valves:

The Weir Type (Figure DP1) which has a weir formed in the body above which is mounted an elastomeric diaphragm. As the hand wheel is screwed down to close the valve, the diaphragm is pressed against the weir stopping the flow of fluid. The weir and diaphragm form a vena contracta which creates a pressure drop and allows the valve to be used for regulating service as well as isolation.

The Straight Through Type (Figure DP2) in which the body bore is circular and the flow passage transforms to an ellipse of equal area directly under the cone shaped diaphragm closure device. The Straight Through Type offers minimum resistance to flow in the open position and therefore is well suited for slurries. This type is limited to isolation service.

The valve consists of three units — the body, the diaphragm, and the operating bonnet assembly which is isolated from the line fluid by the diaphragm. The body can be manufactured from a wide range of materials including cast iron, SG iron, bronze, gun metal, and stainless steel. It can also be lined with various elastomers, polymers or glass for highly corrosive and/or abrasive applications.

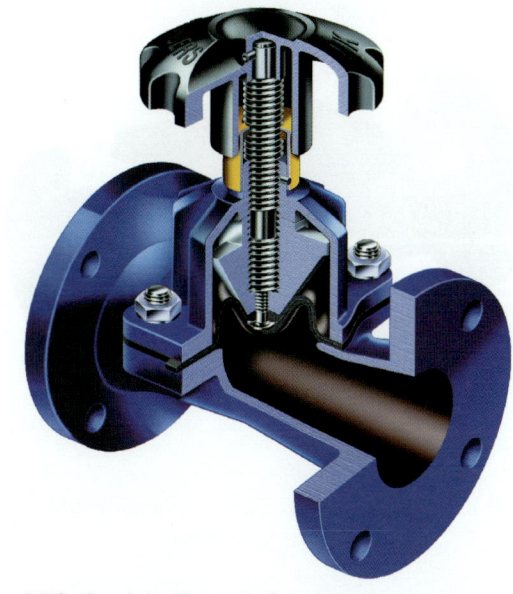

Figure DP2: Straight Through Type Diaphragm Valve without liner (Crane Saunders)

Because the diaphragm isolates the moving parts in the bonnet from the line fluid, the bonnet assembly is manufactured from cast iron materials or plastic coated cast iron for the majority of applications. The design eliminates the need for a seal on the operating mechanism so there is no gland packing or a potential for a gland leak. The valve is suitable for handling aggressive fluids including those containing suspended solids, as well as clean fluid applications. Therefore, the valve is used in a wide variety of process industries including chemical processing, food, oil and gas, water, pulp and paper, minerals processing and power generation, breweries, dairies, petroleum and the gas industry.

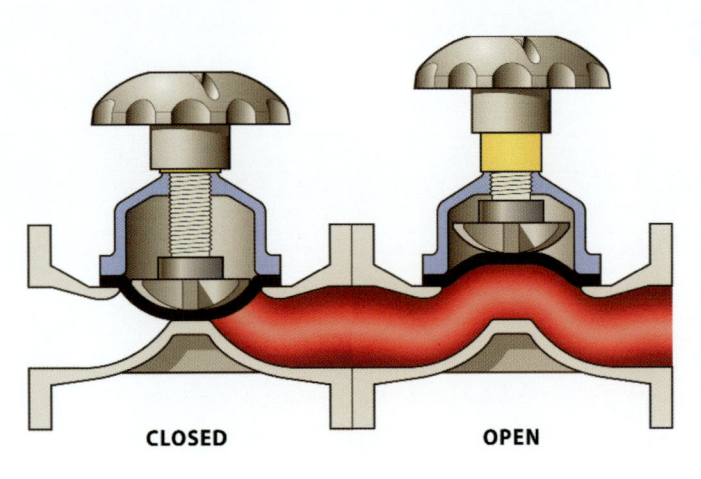

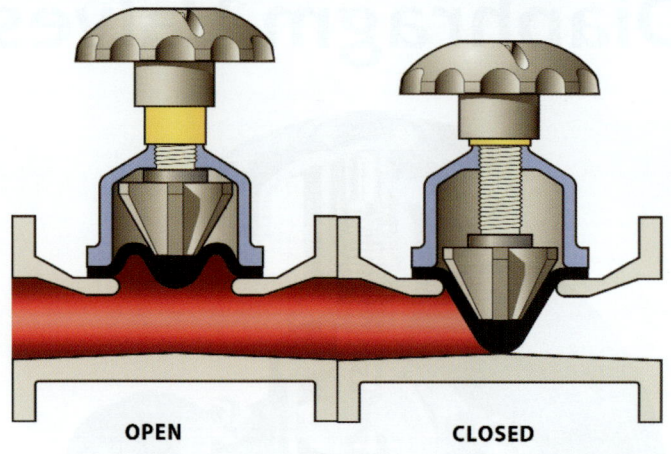

Figure DP3: Weir Type Diaphragm Valve on left and Straight Through Type on right (Crane Saunders)

Diaphragm valves have streamlined interior profile with no cavities or areas for entrapment, which makes the design ideal for use on BioPharm systems that are subjected to 'clean and steam in place' processes. Two way diaphragm valves are fully drainable when installed in the vertical position or in the horizontal self drain orientation.

Diaphragm valves are top entry type design, which permits the valve bodies to be welded in place as the diaphragm can be serviced or replaced while the valve remains in line. The diaphragm isolates the top works from the process media and there is no possible contact between process and the working components of the manual bonnet or actuator. This prevents the possibility of environmental contamination of the process.

Figure DP4: Straight Through Type Diaphragm Valve with liner (Crane Saunders)

Diaphragm materials can be either elastomer or PTFE with an elastomer backing to provide support to the PTFE and also resiliency required to achieve positive closure. Materials have been widely researched over the years and the material shown in Table DP1 is a typical example of the range available to suit the wide variety of fluid temperatures and applications.

The weir diaphragm valve is considered the most hygienic valve concept and has become the primary valve used in Bio-Pharmaceutical applications. Unlike industrial diaphragm valves, which use alloy bodies with rubber or plastic linings, diaphragm valves used in life science applications use steel bodies, which are highly polished to enhance cleanability.

Figure DP5: Straight Through Type Diaphragm Valve, lined, showing top entry detail (Crane Saunders)

Aseptic diaphragm valves are often ported and welded into tandem or more complex manifold assemblies. Diaphragm valves may be manufactured from solid bar or billet into custom configurations other than standard two way type. These types include Tee style and multi-port configurations.

There are special manual bonnets and pneumatic actuators made from non-corroding polymer or stainless steel designed specifically for the hygienic environment and wash down conditions typical of BioPharm development and manufacturing facilities. These top works typically feature smooth OD contour for exterior clean ability

Material	Valve Size	Temp (°C)	Main Uses
Butyl rubber	DN8 to DN350	-30 to 90	Acids and alkalis
Nitrile rubber	8 to 350	-10 to 90	Oils, fats and fuels
Neoprene	8 to 350	-20 to 90	Oils, greases, air and radioactive fluids
Natural/synthetic rubber	15 to 350	-40 to 90	Abrasives, brewing and dilute mineral acids
White natural rubber	15 to 125	-35 to 90	Foods and beverages. White diaphragm
Fluoropolymer*	8 to 350	+5 to 150	Hydrocarbon acids, sulphuric and chlorine applications
Hypalon*	8 to 350	0 to 90	Acid and ozone resistant Sodium Hypochlorite, Chlorine gas
Butyl rubber	8 to 350	-20 to 120	Hot water and intermittent steam services, sugar refining
PTFE Face/EPM Backing	8 to 350	-20 to 160	Acids, solvents, hot water, steam

Table DP1: Typical Diaphragm materials – Industrial Applications

Diaphragms for BioPharm applications must be suitable both for performance in service and also be non-toxic and meet stringent regulatory requirements for extractables/leachables.

Material	Valve Size	Temp (°C)	Main Uses
PTFE Face/EPM Backing	8 to 150	-20 to 150	Purified water, WFI, CIP, SIP, protein solutions, sparge gases
Modified PTFE Face/EPM Backing	8 to 150	-20 to 160	Purified water, WFI, CIP, SIP, protein solutions, sparge gases, intermittent steam
Butyl rubber	8 to 130	-20 to 120	Purified water, protein solutions, CIP, SIP
EPM rubber	8 to 140	-20 to 150	Purified water, protein solutions, CIP, SIP

Table DP2: Typical Diaphragm materials - BioPharm Applications

Diaphragm valves for use in the BioPharm industry require material certification of wetted components, traceability and conformance to FDA, USP, cGMP and other regulatory standards.

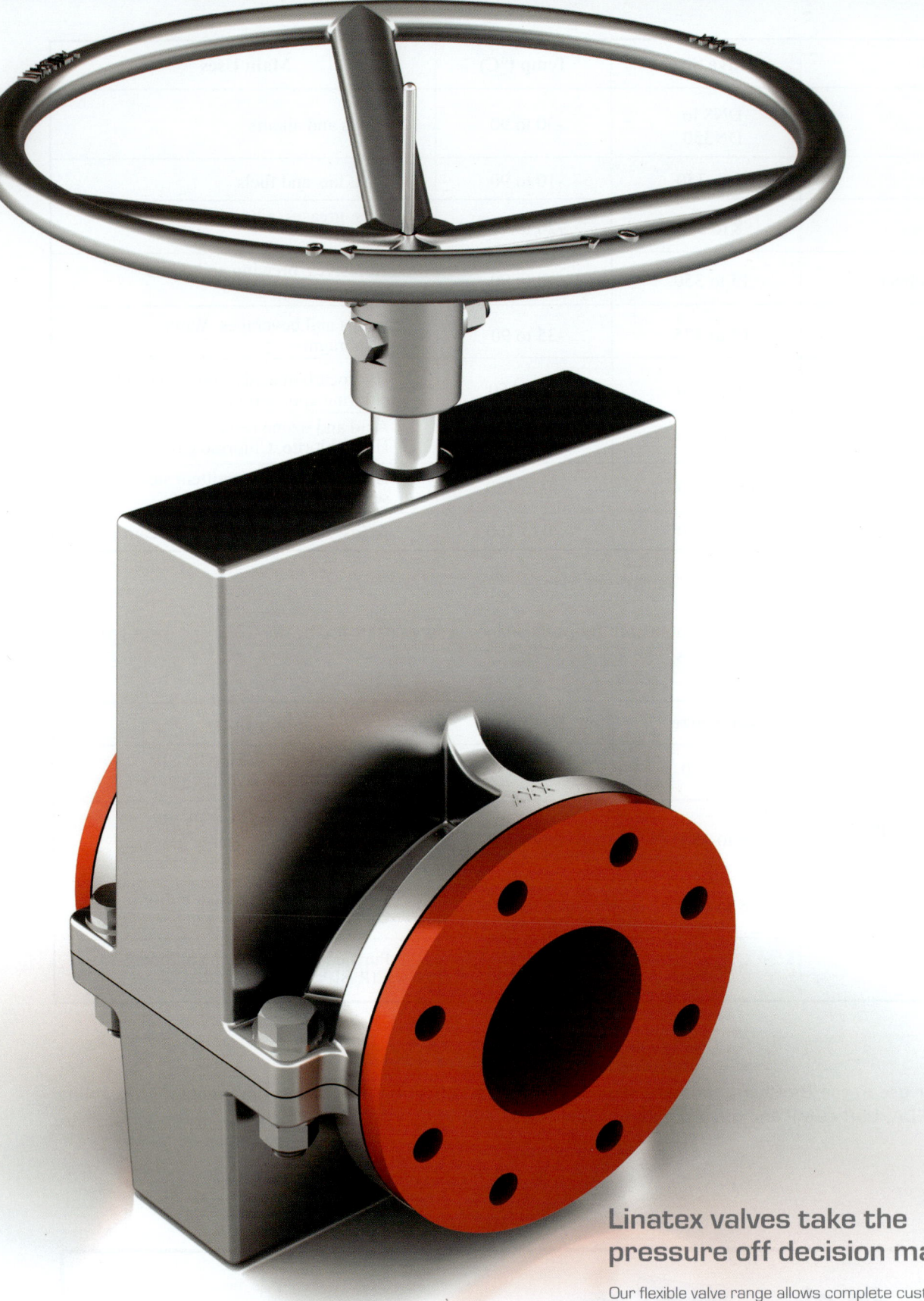

www.linatex.com

Linatex valves take the pressure off decision making

Our flexible valve range allows complete customisation for every processing challenge. Just like all Linatex products, our valves and liners are built to outlast and outperform the rest.

Featuring our unique, replaceable Linatex rubber liners, the range includes: Mechanical Pinch Valves both open and closed bodied, Pneumatic Pinch Valves - 1 or 2 piece, Check Valves both single and double non-return, Control Pinch Valves and Knife Gate Valves.

For more information about our range of products, expert technical advice and extensive global distribution network visit **www.linatex.com**

LINATEX
OUR STRENGTH. YOUR ENDURANCE.

MOULDED PROCESS EQUIPMENT SERVICE HOSE SHEET SCREEN MEDIA

Pinch Valves

Figure DP6: Left – A Hand Wheel Operated Open Body Pinch Valve with full bore, flexible pinch sleeve construction and above – A Pinch Valve equipped for Pneumatic or Hydraulic Actuation with rubber lining enclosed in cast metal protective housing (Linatex)

A forerunner of the diaphragm valve, the pinch valve is the simplest in design of any valve and uses an elastomeric tube or sleeve which can be squeezed at its mid-section similar to the pinch cock used in a laboratory to control the flow of fluid through rubber tube. The industrial version utilizes a bar and spindle with handwheel, and the tube walls are pinched together to produce full closure of the flow path (see Figure DP6).

In common with the diaphragm valves, the operating mechanism and body are not in contact with the working fluid at any time. This makes the valves particularly useful in handling aggressive and highly corrosive fluids, and its straight-through characteristics mean it is suitable for the handling of slurries, pastes, and semi-fluids. The seat bore is the same size as the body end ports giving this type a clearway flow characteristic.

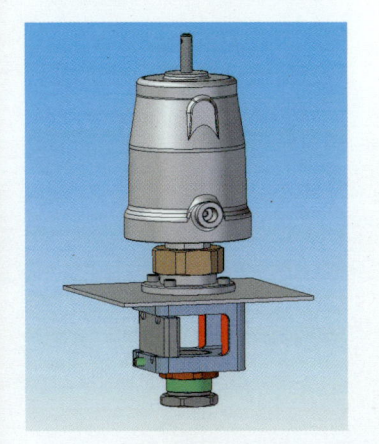

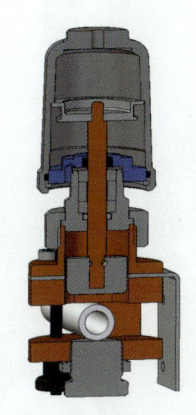

Pinch valve sleeves will wear out and require replacement, but the valve design is so simple that a replacement sleeve can easily be installed. The sleeve cannot be reconditioned for additional service because the extreme amount of flexing, working, and high abrasion will quickly destroy any service repair. A new sleeve is the quickest and easiest maintenance repair. They are used extensively in mines, pulp and paper mills, metal, glass, and food processing industries.

Iris Valve

A variation on the pinch valve is the iris valve. Instead of the flexible elastomeric tube being squeezed the tube of an iris valve is twisted. One end of the tube is held fixed and the operating mechanism rotates the other end at the same time moving it towards the fixed end. When the two ends are in proximity the flow path has been sealed in the same way that the light entering a camera is restricted.

This type of valve is used in low pressure powder and granule applications. It is also an effective way of preventing fish movement. End flanges are made from aluminium or stainless steel and the flexible membrane is nylon or food quality rubber.

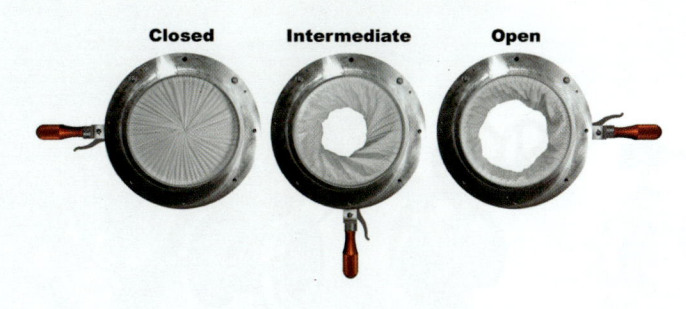

Figure DP7: Single Acting, Spring-to-close Non-intrusive Pinch Valve, with sectioned view showing tube in place and quick-fit tube exchange feature (Schubert & Salzer)

Figure DP8: Iris Valve showing closed, intermediate and open positions – (Vortex Valves)

BVAA VALVE & ACTUATOR USERS' MANUAL - 6th Edition

WHY TRIPLE OFFSET?

Hobbs VALVE
Tomorrow's Valve Today

For this, and many other answers, explaining the **Evolution, Features & Benefits** of **Triple Offset Butterfly Valves** visit us at **www.hobbsvalve.com**

Rotary Valves

Plug, Ball and Butterfly Valves

How do you keep up with valve industry developments?

We all have difficulty getting out sometimes.

Time is a precious resource after all. But if you're a significant user or buyer of valves and actuators, you really do need to keep up with new technology and product developments, and keep an eye out for new suppliers.

BVAA has the Answer!
We bring the exhibition to you!

For many years the BVAA has been organising 'desktop exhibitions' for major users, inside their own premises.

These zero cost, hassle-free events are managed by BVAA and are customised to suit your needs at your convenience.

Solve your supply chain issues over lunch

Designed for rapid set-up and breakdown, 'desktops' typically fit around your lunch period, to minimise downtime. We demonstrate the latest products, provide unrivalled industry advice, and have over 100 leading UK companies to choose from.

"We have had very positive feedback from exhibition attendees. We are already looking forward to doing it all over again"
- Dave Anderson, Score.

Previous hosts include:-
Ministry of Defence, Foster Wheeler, AMEC, MW Kellogg, Stone & Webster, Snamprogetti, British Energy, Score, Aker Kvaerner, KBR, Parsons...

BVAA

British Valve & Actuator Association
9 Manor Park, Banbury, Oxfordshire OX16 3TB
Tel: 01295 221270 Fax: 01295 268965
Email enquiry@bvaa.org.uk
www.bvaa.org.uk/exhibitions.asp

Plug Valves

Figure PV1: Non-lubricated Taper Plug Valve, Sleeved Design (Flowserve Durco)

The plug valve or plug cock is the oldest form of valve comprising a body with a tapered or, less frequently, a parallel seating into which a plug fits. The plug has a through-port and a 90 degree rotation of the plug fully opens or closes the valve. The taper plug is intended to be loaded from the large end such that it is forced into the taper seat to provide isolation. The parallel plug valve has a means of loading the flexible sleeve surrounding the plug thereby compressing it against the plug forming a seal.

Whether parallel or taper type, the operating torque of the plug valve is a function of the end load and of the co-efficient of friction between the plug and the body seat therefore over loading the plug - or the sleeve in the case of the parallel plug type - will make the valve difficult to operate. This fundamental conflict between tight shut-off and ease of operation has been the driver of most developments in the design of plug valves.

As plug diameters and isolation pressures increase the force to operate a plug valve rapidly exceeds the capability of one man and a lever. Plug valves are therefore generally fitted with a gear box or a powered actuator. Of the three rotary motion valve types, plug valves have the highest operating torque. More details can be found in the chapter on Valve Operating Torques.

On large DN valves the weight of the plug can be a significant factor in the loading between the seat and the plug and also the large end presents problems with gland sealing. These problems can be reduced by inserting the plug from the bottom of the valve thereby putting the small end at the top. An adverse feature of the taper plug valve in the open position is that fluid pressure can find its way into the larger end chamber behind the plug. This pressure then acts to force the plug into its seat causing an increased valve operating torque or even seizure.

The plug valve, like other rotary valves, requires a minimum of installation space, is simple to operate, exhibits fast response relative to linear valves, and has a low pressure drop / high C_v as a result of its straight through flow path. The operating shaft only rotates so therefore the gland seal does not have to contend with contamination being dragged into the packing either from the fluid or the atmosphere.

Finally it is easily adapted to multi-port construction. Plug valves can be separated into two basic groups: -

Non-lubricated — which can incorporate mechanical design features to reduce friction between the plug face and body seat or sleeve (see Figure PV1).

Lubricated — which incorporates a design feature where an insoluble lubricant/sealant, specially suited to a particular service, can be injected under pressure between the plug face and body seat. They can be further categorized by the shape of the flow port.

- Round opening — round ports in both plug and body.
- Rectangular opening — rectangular ports in both plug and body.
- Diamond port opening — the opening in the plug is diamond shaped.

The area of the port is either full bore which has an area 100% of the body end port area or Standard port, which has the port area of a full bore valve, one pipe size smaller. The smaller port area of the Standard port increases the pressure drop across the valve for a given flow rate but the resulting smaller plug requires a reduced operating torque. The body of the Standard port valve reduces the flow area gradually to the seat bore diameter and then expands it back to the full nominal inside diameter of the body end port on the downstream side. The design is known as a venturi style.

The port pattern in the plug can be either: -

(a) **Multi-port** — with three or more ports and corresponding pipe connections, used mainly for transfer or diverting services.

(b) **Eccentric plug** — utilizing one half of a plug to offer straight through flow and higher capacity. This option also has the benefit that liquid is not trapped inside the plug port when the valve is in the closed position.

The valves are produced in a variety of materials such as bronze, brass, and carbon and stainless steels, as well as exotic alloys and plastics. There is also a range of fully lined plug valves where all the wetted parts inside the valve are lined with PTFE or formable fluorocarbon for highly corrosive and toxic applications. The use of a sealant or fluorocarbon sleeve or lining generally restricts the use of plug valves to below fluid temperatures of 250°C.

Plug valves allow in line maintenance of the seats or sleeve as they can be accessed for maintenance through the top or bottom of the valve without removing the valve from the line.

Non-lubricated plug valve - These are the simplest form of plug valve and are used extensively in the chemical and petrochemical industries, particularly where lubricants are unacceptable (see Figure PV1). An adjustable gland and a spring loaded plug may be provided to compensate for wear and permit ease of operation.

Figure PV2: Lined Plug Valve (Crane Xomox)

Sleeved design - A sleeve of PTFE is fitted into the valve body and accepts the metal plug, which rotates against the PTFE with a minimum of required torque. PTFE, being highly inert and having a very low coefficient of friction, is ideal for this service. Metal plugs can also be coated with PTFE based anti-seize compounds to further reduce wear and operating torques.

Lined design - To avoid the use of expensive high alloy materials for corrosion resistance, plug valves can be manufactured from relatively inexpensive materials, with the body and plug fully lined on all wetted areas with Teflon. The linings are moulded into dovetail recesses in the body to lock them in place. The lining is generally 3mm thick in all areas to ensure adequate resistance to abrasive materials. If the lining is destroyed, the less corrosion-resistant metallic body will be exposed to the aggressive line fluid. The lining must also be impervious to the fluid thereby prevent blistering or separation of the lining from the body wall. These valves are generally used in conjunction with lined pipe systems which offer a much more economical alternative to high alloy pipe systems, and with greater strength and higher pressures than a totally plastics system is capable of handling. See Figures PV2 and PV3.

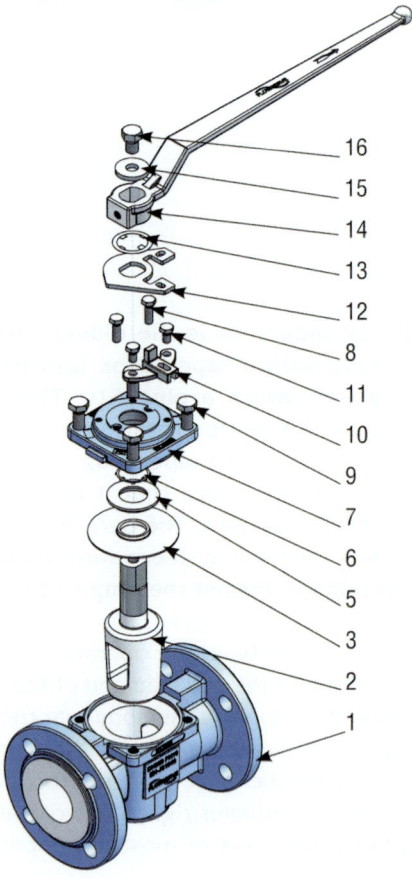

Parts List			
1	Body	9	Top cap bolt
2	Plug	10	Stop
3	Diaphragm	11	Stop fastener
5	Thrust gland	12	Stop collar
6	Grounding spring	13	Stop collar retainer
7	Top cap	14	Wrench
8	Adjuster bolt	15	Washer
		16	Hexagon bolt

Figure PV3: Typical construction of a Lined Plug Valve (Flowserve)

Expanding plug valve - A variant of the non-lubricated plug valve is shown in Figure PV4, where the tapered plug moves across two slip segments with permanently bonded resilient seals, which are pressed against the body of the valve to form the seal. The tapered plug is moved vertically when the valve shaft is rotated either loading or unloading the slip segments. This design reduces wear on the seals because they are always unloaded during the plug rotation and when provided with a secondary metal seat can be made fully fire-safe. Bubble tight sealing on both the upstream and downstream ports is possible, thus providing the opportunity for a 'double block and bleed' feature.

Figure PV4: Expanding Plug Valve (General Cameron)

The eccentric plug valve - The eccentric plug is half of a full plug and offers a full bore port (Figure PV5). This option also has the benefit that liquid is not trapped inside the plug port when the valve is in the closed position. The valve is single seated and provides bidirectional isolation in clean service. The valve is used extensively in the water and waste industry and is available with a variety of resilient plug facings to provide tight shut off without the use of sealing lubricants. In the open position the plug is neither in contact with the body wall nor the seat. As it moves to the closed position there is only contact with one seat face so operating torques are lower than with the standard full plug designs. The valve can be used for slurries and dirty service but correct orientation of the valve in the pipeline must be considered. It should be installed such that in the open position the plug is at the top of the pipeline, i.e. the shaft must be horizontal and in the closed position the flow direction should be onto the outer face of the plug. It is a design that can be adapted for control applications as detailed in the Control Valves chapter.

Figure PV5: Eccentric plug valve (DeZurik)

Lubricated Plug Valves - In this type of plug valve an insoluble lubricant/sealant, specially suited to a particular service, is injected under pressure between the plug faces and the body seat. The lubricant is fed through a non-return valve by means of a pressure screw or grease gun and reaches the seating surfaces via a system of ducts and grooves in the plug and body. The lubricant forms a film between the plug and the body surfaces minimising corrosion, erosion and hence operating torque. Lever operation is possible up to an including DN80 size. The lubricant also assists with leak tightness between plug and body seat. The application of lubricant behind the small end of the plug will counteract any load created by line pressure accessing the chamber above the large end of the plug. Lubricated plug valves have been used extensively in oilfield production, distribution and refinery installations, up to pressures of 400 barg. They are also used in gas production and distribution systems.

Pressure balanced taper plug valves - An alternative solution to the problem of line pressure forcing the plug onto its seat is the pressure balanced (also called dynamic balance) taper plug valve where the line pressure is used to replace the sealant pressure (Figure PV6). The pressure balanced system consists of two holes in the plug connecting chambers at each end of the plug with the line pressure. The line pressure to the small end chamber passes through a ball type non return valve helping to maintain the balancing force which is produced. The need for sealant injection to keep the valve operable is eliminated. A balanced design is also available for dirty service. By drilling the plug its full depth the balancing pressure is fed directly from the large end to the small end. Again a ball check valve is fitted into the balancing flow path to help maintain the pressure above the small end. The balanced design also reduces the operating torque allowing higher pressure designs to be developed while keeping actuator costs down.

Key:

A   Blowout proof shaft with Double D Drive for wrench

B   Weatherseal

C   Graphite packaging rings give normal sealing and firesealing

D   Shaft packing compound injector to renew sealing to atmosphere

E   Thrust washer

F   Plug sealant injector to renew sealing to downstream

G   Pressure balance holes

H   Plug with metal-to-metal seating

J   Plug loading screw

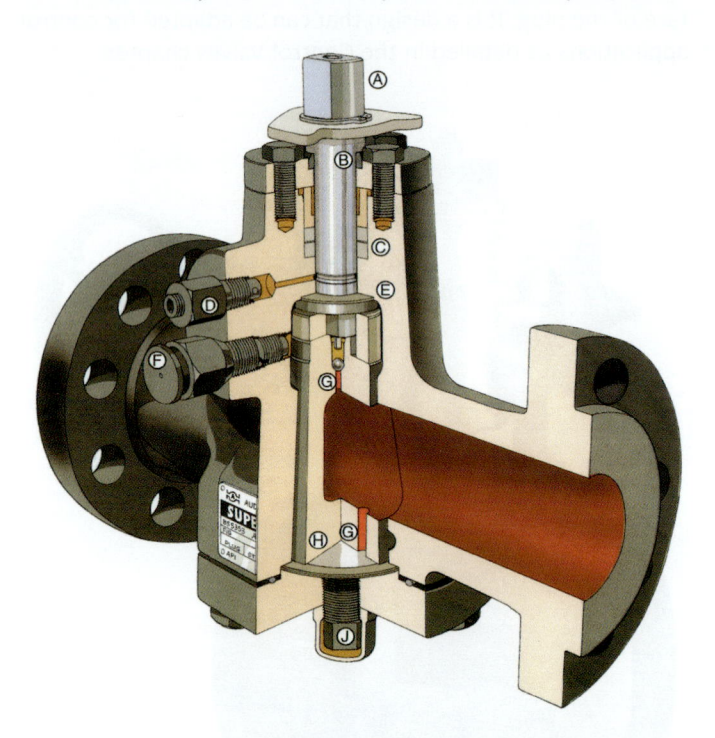

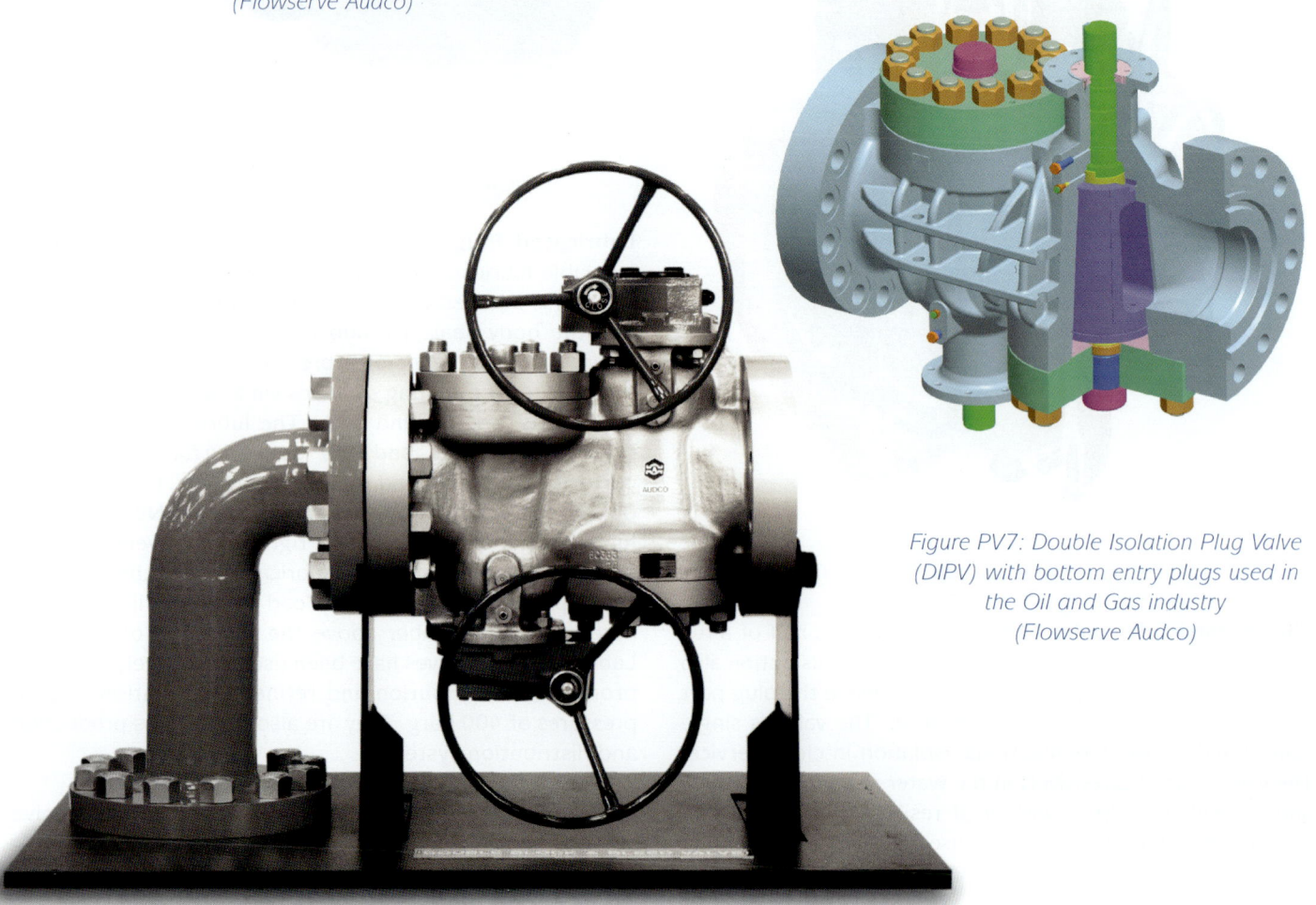

Figure PV6: Pressure Balanced Taper Plug Valve (Flowserve Audco)

Figure PV7: Double Isolation Plug Valve (DIPV) with bottom entry plugs used in the Oil and Gas industry (Flowserve Audco)

What do over 140 world class companies......have in common?

...Superb Backing!

The British Valve & Actuator Association
Professional Support For The Process Flow Control Industry
9 Manor Park, Banbury, Oxon. OX16 3TB (UK) Tel: (0) 1295 221270 Fax: (0) 1295 268965

www.bvaa.org.uk

*Pageland Foundry
South Carolina, USA*

Conbraco Industries was created when two Detroit-based manufacturers of brass valves and fittings merged in 1928 forming Consolidated Brass Company.

Apollo Valves, manufactured by Conbraco Industries, Inc., is a family owned business supplying the commercial and industrial valve markets since 1928.

Apollo's Industrial Valves are designed, engineered, cast, machined, assembled and tested in our state-of-the-art facilities in the Carolinas, USA. At our USA foundries, we cast 81 Bronze, 85 Bronze, Lead Free Bronze, Carbon Steel, Low-Temp Carbon Steel, Stainless Steel, Low-Carbon Stainless Steel, Alloy 20, Hastelloy "C", Nickel, Nickel-Copper Alloys and Titanium.

Our vertical manufacturing integration assures better quality control, better cost control, and the shortest delivery lead times possible for our range of Ball Valves, Automation Products, Safety Relief Valves, Backflow Preventers and Plumbing/ Heating Products.

HEADQUARTERS
Apollo Valves, a division of Conbraco Industries, Inc.
701 Matthews Mint Hill Road
Matthews, NC 28105 **USA**

QUALITY
INTEGRITY
COMMITMENT

Strategically located regional offices in the United Kingdom, Asia-Pacific, Caribbean, Mexico, Central/South America, Canada and the USA

"Apollo" Valves
manufactured in the USA
by CONBRACO Industries

Customer Service +001 (704) 841-6000
www.apollovalves.com

Ball Valves

Figure BV1: One Piece Design Ball Valve with Floating Ball showing the open (left) and closed positions (Tyco Hindle)

The ball valve is the newest member of the valve family. Today so many variations exist and it is so widely used that it is considered by most to be a valve type in its own right. When it was first introduced it was known as the spherical plug valve, which confirms its origins, namely that it is a variation of the plug valve type.

The ball valve has developed from 1950 to the successful valve it is today because of the availability of machines that can accurately produce the spherical polished outer surface of the ball and the availability of inert resilient materials for the seats and seals. More recently the ability to produce metal seats in a variety of low wear corrosion resistant materials has further extended the application range of the valve.

The valve consists of a body, ball, seats, shaft, and gland seals (see Figure BV1). Characteristics of the ball valve are its smooth operation, low pressure loss, high Cv, compact design, and the ability to obtain a fire test certificate. The operating shaft only rotates so therefore the gland seal does not have to contend with contamination being dragged into the packing either from the fluid or the atmosphere.

Ball valves can be supplied manufactured from bar stock, forgings, or castings with screwed, socket weld, butt weld, or flanged ends. Their applications are as wide and varied as industry itself. They range from simple services, such as water, solvents, acids, and natural gas, to more difficult and dangerous services such as gaseous oxygen, hydrogen peroxide, methane, and ethylene.

Materials commonly used in ball valve manufacture are carbon steel grade WCB and LCB for the body and stainless steel grade 316 for the ball and shaft. Alternatively, the body can also be manufactured from stainless alloys for corrosive or low temperature applications. Virgin PTFE is usually used for the seats and gland seals, as it is chemically inert to a wide range of fluids and its coefficient of friction is less than 0.1. However, it tends to lose its rigidity over 100°C which leads to the down rating of the pressure/temperature curve for floating ball designs and a maximum service temperature of 230°C. Nylons, polyether-etherketone (PEEK) and filled PTFE derivatives are used to gain higher rigidity of the seating face, less down rating and an increase in the maximum service temperature to 280°C. For higher maximum service temperature up to 1000°C metal seating is required.

Figure BV2: Tungsten Carbide Coated Ball Valve balls and seats (Hardide)

Floating Ball design

In this design, the ball is held between two elastomeric seats by the compression of the seats against it. The ball is driven through its 90° travel by a shaft, which connects into a slot on top of the ball. This slot allows for some lateral movement of the ball, relative to the shaft, due to the influence of upstream line pressure. In Figure BV3 the movement of the ball is exaggerated. In this way the ball is loaded by line pressure against the downstream seat and this seat should always be regarded as the primary seat in the case of a floating ball design. The upstream seat may also provide a secondary seal if the design incorporates a spring or other form of pre-loading in the seat design. Floating ball designs are generally capable of providing bi-directional shut off. The fact that the seating surface contact pressure is provided by the upstream fluid pressure means that once the valve is moved away from the closed position there is a much reduced operating torque compared with the plug valve.

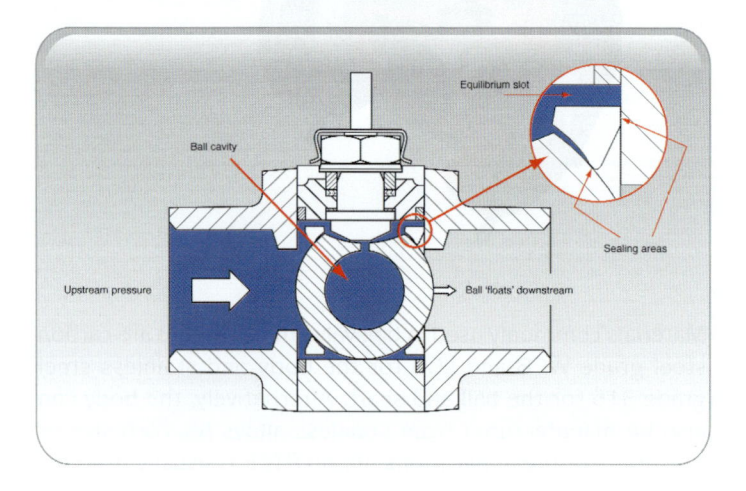

Figure BV3: Schematic of Floating Ball design principle

As the physical size of the ball, and hence its weight, increases, a point is reached beyond which the ball can no longer be adequately supported by the seats alone. Similarly, as line pressure increases, a limit is reached beyond which the loading of the ball on the downstream seat would be too great for the seat to bear without distortion and damage. Exceeding either or both of these parameters requires an alternative design.

Figure BV4: Meta-Seal Trunnion Mounted Ball Valve (Tyco Hindle)

Trunnion Mounted design

In the trunnion mounted design the ball is supported by a trunnion and not the valve seats, thus permitting much higher pressure/temperature ratings (see Figure BV4). In concept, the ball, shaft and trunnion may be regarded as being a one piece construction. The shaft and trunnion are held in bearings preventing the ball from moving to load the downstream seat. In order to facilitate the loading between seat and ball the upstream seat is designed to move forward. A force is generated by the line pressure P_L acting on the area $A_1 - A_2$ pushing the seat tightly on to the ball. Thus, in a trunnion mounted design the primary seat is formed at the upstream side. A spring mechanism is incorporated behind both seats (not shown in Figure BV5 for clarity) to ensure seat loading even in low pressure conditions and to provide a secondary seal at the downstream seat. As with the floating ball design, trunnion mounted design types are capable of bi-directional shut off. Lower operating torque is an important additional benefit and may be as much as one third to a half of that of a floating ball valve, significantly lowering the cost of actuation.

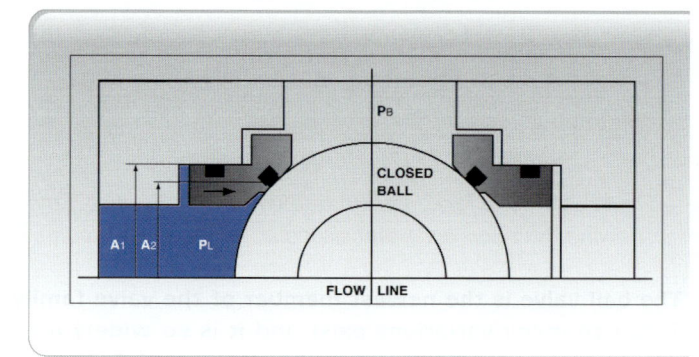

Figure BV5: Schematic of Trunnion Mounted Ball Valve design principle

Rising Stem Ball Valve

The Rising Stem Ball Valve operation is achieved by lifting of the stem to tilt the ball away from seat contact, followed by friction-free rotation of the stem and ball to the open position.

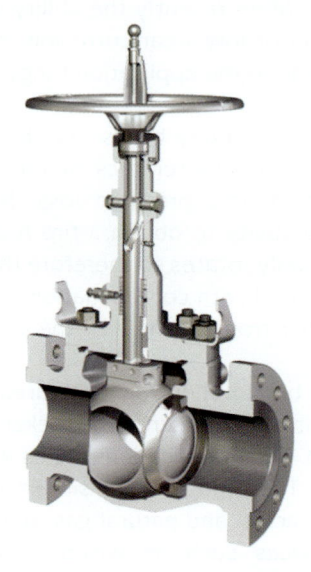

Figure BV6: Rising Stem Ball Valve (Cameron Orbit)

In the full open position, there is unobstructed flow through the valve.	Precision spiral grooves in the stem act against fixed guide pins, causing the stem and core to rotate	Continued turning of the handwheel rotates the core and stem a full 90 degrees without the core touching the seat	Final turns of the handwheel mechanically cam the stem down, pressing the core firmly against the seat.
Side View	Top View		Side View

Figure BV7: Rising Stem Ball Valve action (Cameron Orbit)

Closing is accomplished by turning the stem and ball with ball face retracted, followed by pushing down the stem to tilt the ball against the seat. The stem has parallel cam angles on its lower end that move between inserted pins to accomplish mechanical closure and retraction of the ball. The ball is fitted with pins as a load bearing surface for the stem cam and support pins for the stem are also provided. Ball and support pins are designed to be replaceable, when worn or damaged.

Position of the ball is controlled by threaded steel guide pins that are accessible from the outside of the assembly. The single seat is of a metal body construction with inserted soft seal, or all metal with flexible tube insert.

The bonnet is either an enclosed type with stem packing injection for services up to 260°C or outside screw and yoke (OS&Y) adjustable packing for higher temperature service up to 427°C. Valves are also suitable for low temperature service.

It is a reliable valve with positive shut-off and repeated re-sealing capability. The valve is used in critical, high cycle and high temperature applications. Applications include molecular-sieve dryer isolation.

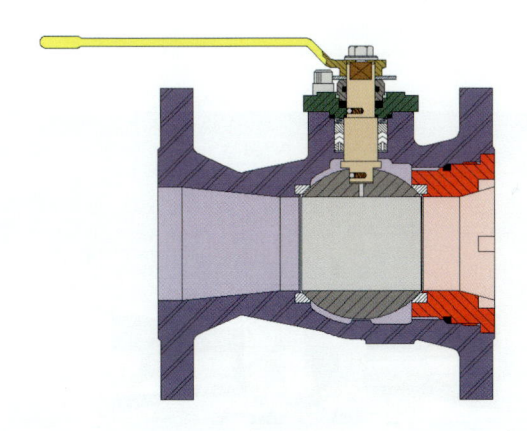

Figure BV8: One piece design schematic (Tyco Hindle) and assembly (Heap & Partners)

Body Styles

There are four common methods of inserting the ball into the body of the valve: -

- The ball is placed into the body through an entry in one of the pipe flanges and secured with an insert that typically forms part of the gasket raised face. This is termed a one piece or end entry valve. The one piece body has inherent strength and minimises leak paths.

- The body is split in one or two places in the same plane as the valve flange, and the body is bolted around the ball. These are termed two or three piece valves (see Figure BV9 and BV10). The three piece design is common amongst small sizes, due to economics of manufacture, and for large sizes - particularly trunnion mounted designs - due to component weight. Two or three piece body valves have the advantage of simplified maintenance.

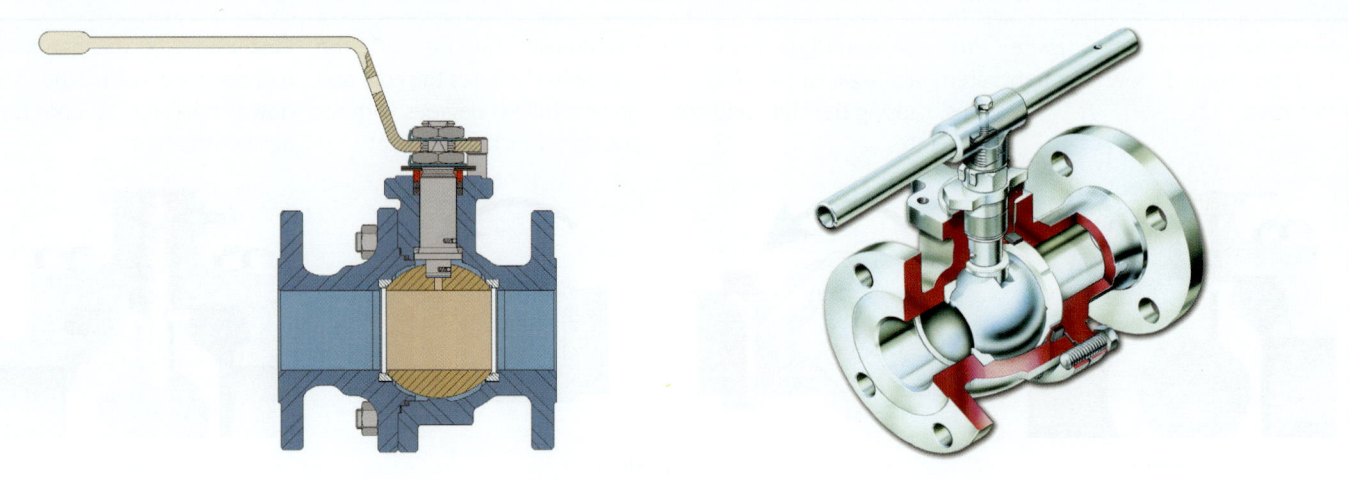

Figure BV9: Two piece design schematic (Tyco Hindle) and sectioned view (Econosto)

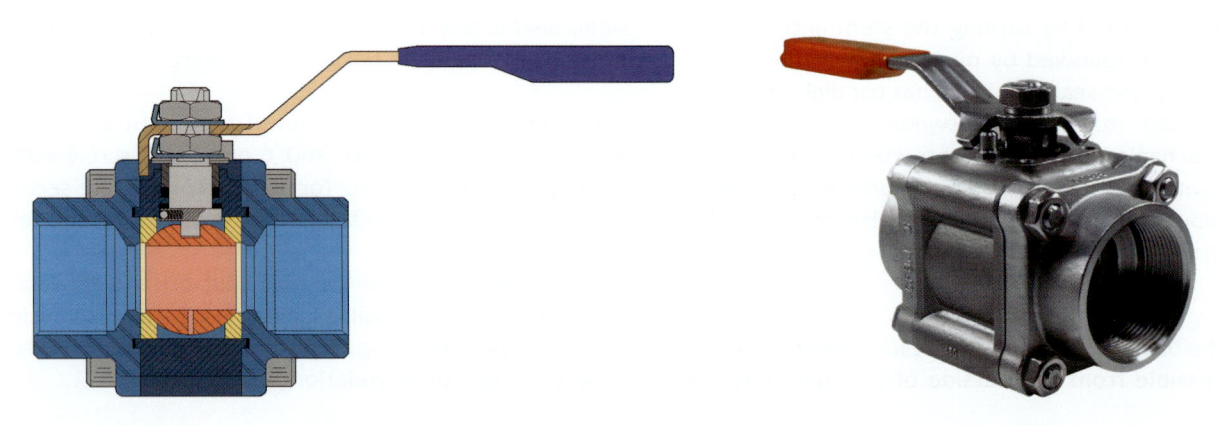

Figure BV10: Schematic of three-piece, screwed end ball valve (Tyco Hindle), with an exterior view of a similar three-piece ball valve (Flowserve Worcester)

- The ball is inserted through a bonnet in the top of the valve. This is termed a 'top entry' valve (see Figure BV11). This design enables the critical parts to be accessed while the valve remains in the pipeline and is therefore used in welded pipe systems.

- The body of the valve is an all welded construction and effectively non-maintainable. Such valves are often to be found in gas transmission pipelines.

Figure BV11: Top entry ball valve, reduced port (Truflo Marine)

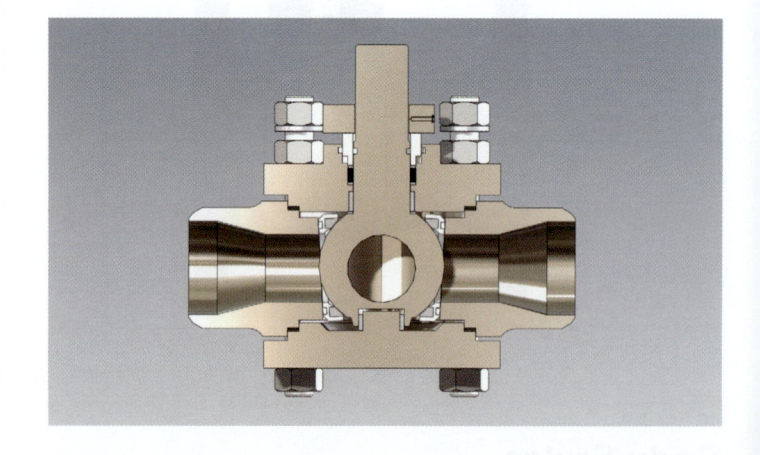

Reduced or Full Port

An important consideration with a ball valve is the port size. The diameter of the port D determines the size of the ball. For small D the ball diameter is approximately 2D and for larger D approximately 1.5D. As ball weight is proportional to d^3, where d is the ball diameter, reduced port valves are therefore significantly lighter and lower cost than their full port brothers. Full port generally means the ball port is the same as the inside diameter of the pipe. Typical reduced port diameters are specified in Table BV15.

Cv is proportional to D^4 so the savings in weight and cost of the reduced port valve may have to be partly paid for by higher pump energy costs.

Anti-Static Design

If the ball of a ball valve is held between seats without any metallic contact between it and other components, the flow of fluid can lead to a build up of static electrical charge on the surface of the ball. To prevent the possibility of such static being discharged in the form of a spark, possibly igniting fluid or causing an explosion, valve design standards used in many process industries require the incorporation of a mechanical means, which ensures electrical continuity between ball, body and downstream pipe work. See Figure BV14.

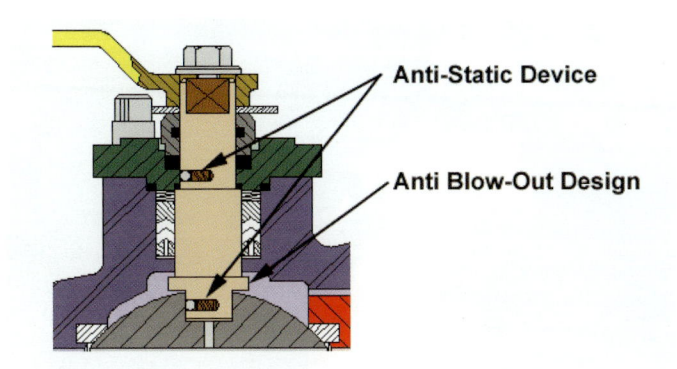

Figure BV14: An anti-blowout design stem, with anti-static devices fitted (Tyco Hindle)

Anti-Blowout Design

In the floating ball design the shaft is not attached to the ball. It is essential to make sure that it cannot be blown out past the packing in the event that the gland bolting is removed or slackened. See Figure BV14.

Figure BV12: A 42 inch, Class 600, all welded steel design Ball Valve for underground service (Schuck Valves)

Port Dimensions in BS 5351		
Valve Size (inch)	Full Port	Reduced Port
½	0.50	0.31
¾	0.75	0.49
1	1.00	0.62
1½	1.50	1.12
2	2.00	1.50
3	3.00	2.25
4	4.00	3.00
6	6.00	4.00
8	8.00	6.00
10	10.00	8.00
12	12.00	10.00
16	16.00	12.00

Table BV15: Reduced and Full Port Dimensions

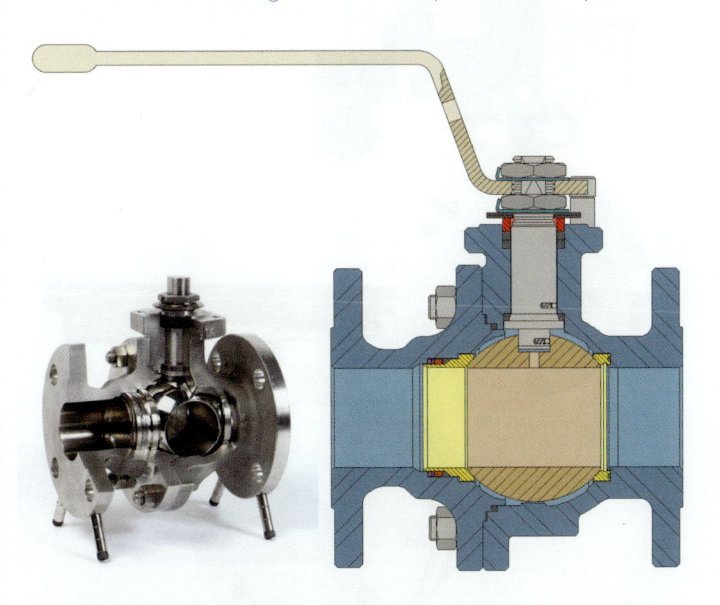

Figure BV13: Full port, metal seated ball valve (Hindle). Note: A reduced port design is shown in Figure BV1

Fire Testing

Since most ball valve designs rely upon soft seats for their performance, there is justifiable concern over the effect of conditions which would jeopardise the integrity of those seats. If a soft seated ball valve is exposed to a fire which is intense enough to damage or destroy the seats, certain industries, notably the oil, chemical and petrochemical industries, demand that a valve should continue to be operable.

Ball valve standards include details of the fire tests to which representative valves are subjected before a given ball valve design can be certified as 'fire tested'.

Figure BV16: Fire test on a 3" Class 150 Cyrogenic Ball Valve (Apollo / Adanac)

- **Cavity pressure relief** - The body cavity of a ball valve is full of line fluid when the valve is in the closed position. In a floating ball design, if the ball is not being loaded onto the downstream seat by differential pressure then both seats may be providing the same level of isolation and neither providing an obvious escape route for the pressure increase of the fluid inside the ball. If the fluid is incompressible i.e. a liquid the potential increase in pressure due to an increase in fluid temperature may be extremely high and may even render the valve inoperable. For such applications, manufacturers offer various alternative designs to ensure over-pressure relief, usually to the upstream side. All such modifications make the valves unidirectional and this should be clearly indicated on the valve body. One method for providing cavity over-pressure relief is illustrated in Figure BV17.

  Cavity relief in a trunnion mounted design is part of the design. Any build up in pressure P_B (See Figure BV5) in the cavity is automatically relieved by the downstream seat moving away from the ball.

- **Lined ball valve** - Similar to the lined plug valve it is manufactured for greater corrosion resistance. All the wetted areas have a fluorocarbon lining inside a metal body. The ball can also be fully lined or manufactured from solid ceramic, and offers much lower torques than the lined plug valve. As with the lined plug valve, if the lining is destroyed, the less corrosion-resistant metallic body will be exposed to the aggressive line fluid.

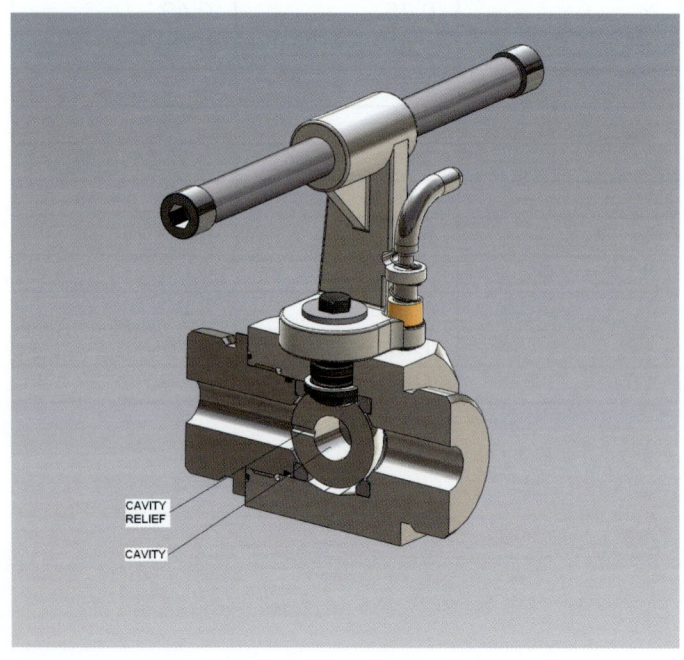

Figure BV17: High Pressure Full Bore Floating Ball Valve with Cavity Relief and Locking Device (Truflo Marine)

Figure BV18: Lined Ball Valves can provide greater corrosion resistance (Tyco Neotecha)

- **Low Temperature applications** - Ball valves are used for low temperature and cryogenic applications including LPG, LNG, liquid nitrogen and liquid oxygen applications. Valves are usually offered in low temperature carbon steels for operation down to -46°C and austenitic stainless steels to -196°C. PTFE seals contract at very low temperatures and this can cause very high operating torque. An extended bonnet is usually provided to isolate the gland seals from the cryogenic temperature zone so as to avoid ice formation around the gland and maintain the integrity of the seal and the operation of the valve (see Figure BV19).

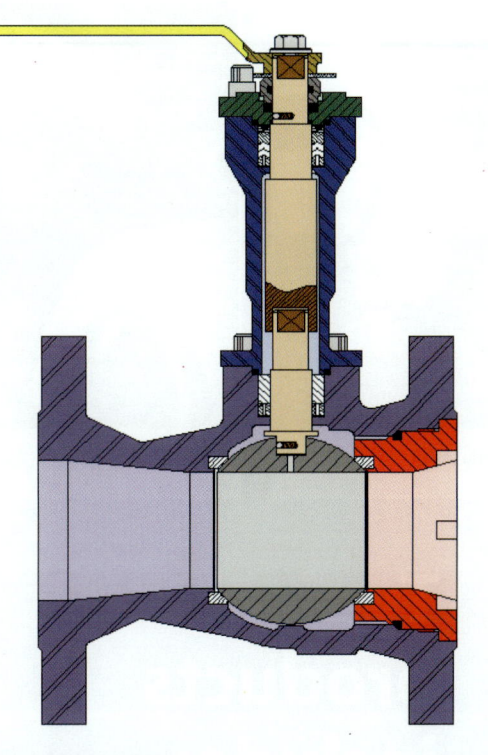

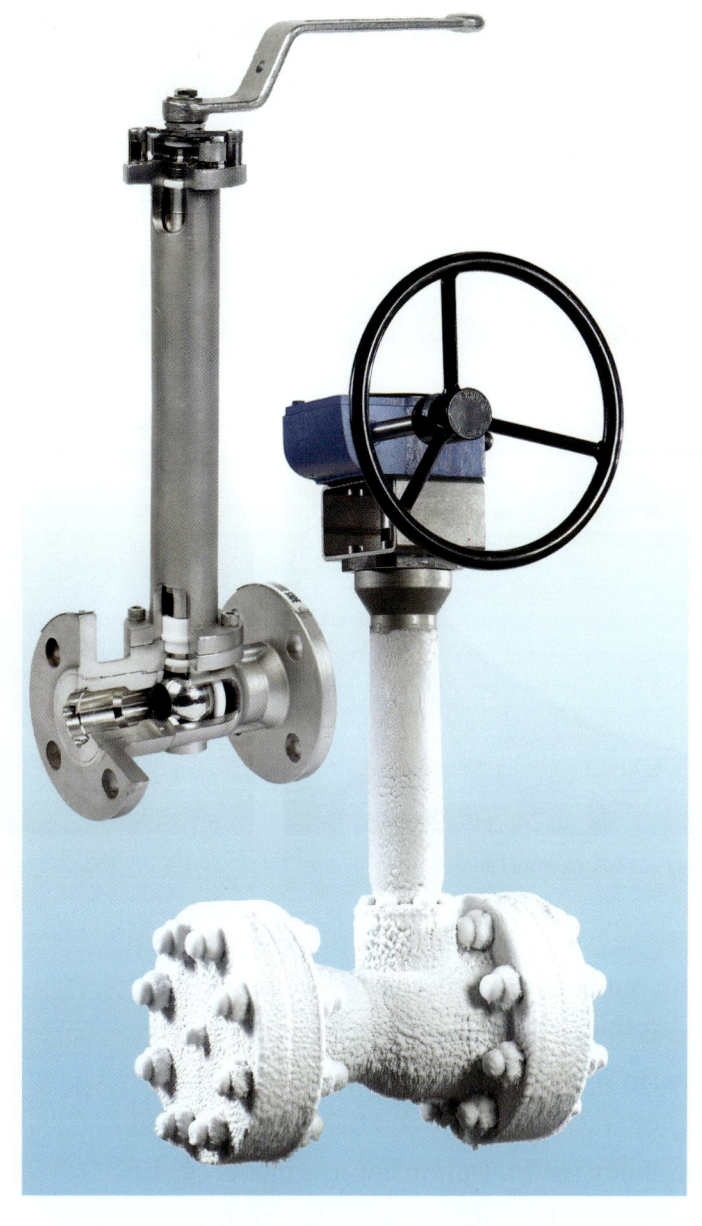

Figure BV19: Extended stem ball valves for cryogenic applications (Tyco Hindle)

Ball Valves as Control Valves

Soft-seated ball valves can be used for less demanding throttling control valves, but partial flow conditions should be avoided as high velocity flow can impinge against a localised area of the ball and seats, thus creating a rapid deterioration of the seat material. However, these particular limitations can largely be overcome by the use of metal seats, hard coatings, and modifications to the ball, making such types of ball valves suitable for control application as detailed in the Control Valves chapter.

BVAA 70 years of service to industry

eTec Engineering Services Ltd provides a bespoke valve design, manufacturing and refurbishment service for the Power and Water Supply Industries.

eTec are one of the few companies that offer valve design and manufacture in the UK.

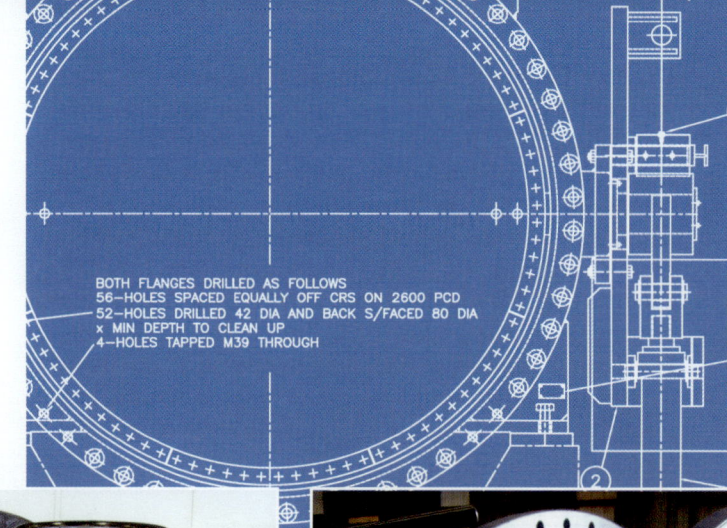

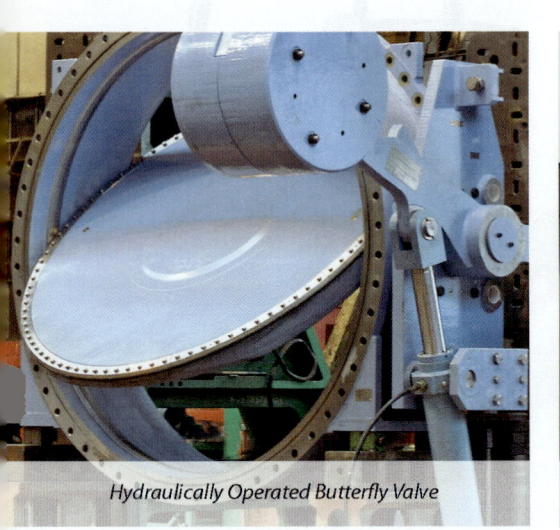

Hydraulically Operated Butterfly Valve

Ebonite Lined Butterfly Valve

Electrically Operated Butterfly Valve

With our engineering expertise we provide a comprehensive service for all types of valves used in Hydro and Thermal Power Stations together with Butterfly and Discharge valves for the Utility sector. Design improvements can be incorporated to existing valves to improve operation and safety requirements including: seal / bearing replacement, alternative actuation systems and locking arrangements.

eTec also offer Onsite Condition Assessment, Technical Support and a Installation Service. Our engineers have historical links with Boving and Kvearner Boving to provide a re-engineering service for valve parts.

www.etec.uk.com

Typical products include:

- **Butterfly valves**
  up to 4000 mm diameter designed to 35 bar

- **Spherical valves**
  up to 3000 mm diameter designed to 70 bar

- **Reflux valve** (butterfly type)
  300 mm to 1200 mm diameter designed to 16 bar

- **Energy dissipating valve**
  300 mm diameter and above designed to 30 bar

- **Four way valves**
  up to 1600 mm diameter designed to 6 bar

T 01709 544111
F 01709 480529
Email: sales@etec.uk.com

Butterfly Valves

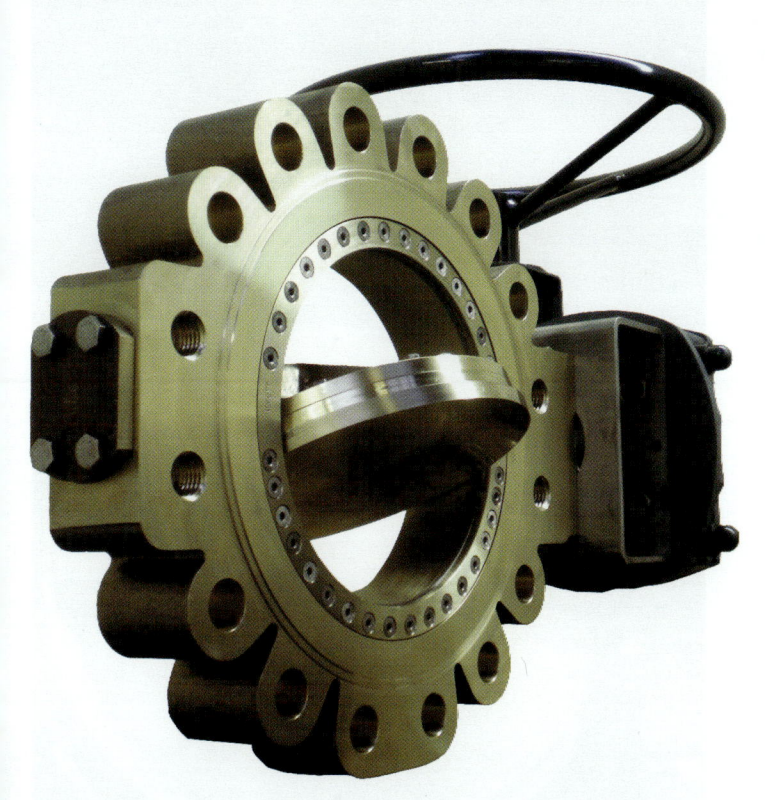

Figure BF1: Triple Offset Butterfly Valve, Class 300, 16 inch Nickel Aluminium Bronze, lugged and through drilled (Hobbs Valve)

There are three butterfly valve types, namely Concentric, Double Offset and Triple Offset.

The butterfly valve consists of a disc rotating on a shaft at right angles within a pipe section body.

They are generally designed for insertion between two pipe flanges using through bolts, thus saving substantial space and weight. This design is known as wafer type (see Figure BF2). The wafer lug type allows end of pipe installation without the use of a second flange. Offset designs are also manufactured with double flanges in both ball and gate valve face-to-face dimensions to facilitate the replacement of these valve types by the butterfly valve (see Figure BF3).

Butterfly valves are available in sizes from DN25 to very large sizes, see Figure BF4. An undesirable feature of the butterfly valve is that in all positions the disc is in the fluid and so it is not a valve type that can be used in a pigging application. In small sizes the disc occupies a significant proportion of the flow area resulting a low C_v, where as in large sizes and high pressures the requirement to carry the pressure load in the closed position results in very thick section and heavy discs, which creates operating challenges for the shaft support bearings.

Operating torques are a combination of friction torques from the contact between the body and disc, torques due to aerodynamic loads on the disc (lower for offset designs) and torques due to unbalanced momentum forces in the case of

Figure BF2: Left – Lever Operated, Wafer Type, Lined Butterfly Valve and right – similar but of Lugged Design (Ebro Armaturen)

offset designs. Full use should be made of manufacturers' expertise when making an assessment of the maximum operating torque required.

Aerodynamic forces on the disc result from the fluid flow across it as it moves from the closed to the fully open position. These forces can be significant in both gas and liquid flow and peak at around the 70 degree open position. The disc also creates a pressure drop, which is a minimum in the full open edge-on position. Recent developments in computer aided fluid dynamics (CFD) have facilitated the optimisation of disc profiles to reduce pressure drop and increase C_v values.

BVAA VALVE & ACTUATOR USERS' MANUAL - 6th Edition

The attachment of the shaft to the disc is an important consideration when assessing butterfly valve designs.

Some designs have an integral shaft and disc, which means that the disc is highly unlikely to separate from the shaft and attempt to follow the fluid down the pipeline. For cost reasons and also when it is necessary to have the shaft and disc in different materials, the disc is a separate component from the shaft. In these designs the shaft must be securely attached to the disc. The presence of pins, screws, etc. can lead to problems in corrosive environments and subsequent detachment of the disc. It is also important to ensure the weak point of the shaft is outside the pressure envelope and that it has a shoulder or similar feature, which prevents it being ejected through the bonnet thereby creating a major leak and hazard. A one piece shaft design ensures that bearing loads are minimised. API 609 is an example of a butterfly product standard that addresses shaft and disc attachment design issues.

Figure BF4: A 138 inch Butterfly Valve (Curtiss Wright)

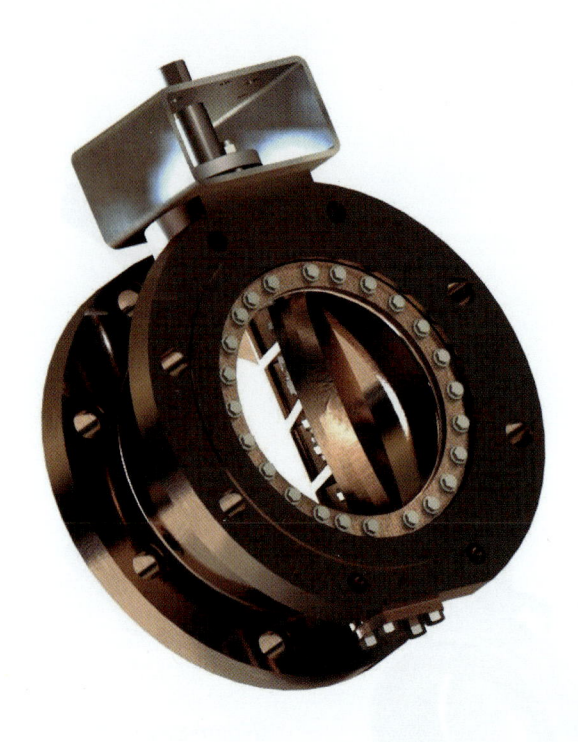

Figure BF3: Double Flange, Short Pattern Butterfly Valve (Hobbs Valve)

General Purpose Butterfly Valve (Concentric design)

In its simplest form the axis of the shaft passes through the centre line of the valve body and the plane of the body seating surface is at 90° to the pipe axis and parallel with and aligned with the shaft axis. Potentially the disc can rotate 360° about the shaft axis (See Figure BF5).

The concentric butterfly valve requires a resilient material either on the inside of the body or on the perimeter of the disc to form the seal as the two components come together in the closed position. Once again there is a conflict for the designer in that the amount of interference between the

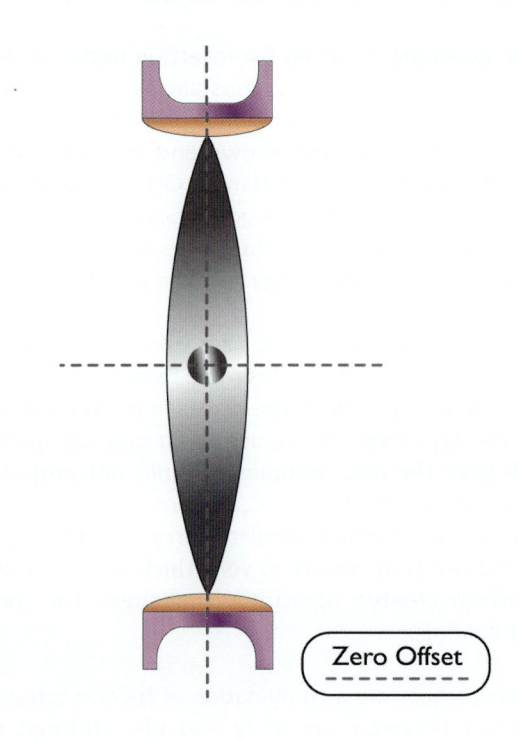

Figure BF5: Zero offset (concentric) designs (Hobbs Valve)

Figure BF6: Wafer Type Virgin PTFE Lined Butterfly Valve with Split Body (Tyco Hindle / Keystone)

two determines how tight the isolation is but also how high the operating torque is and the wear on the resilient seating face. The wear from the disc rotating across the resilient seat material eventually leads to seat leakage.

The shaft must also pass through the body seating face requiring consideration of the shaft seal to prevent fluid leakage along the shaft. Sealing is bi-directional using chemically inert, corrosion resistant materials such as Buna 'N', EPDM, nitrile, or PTFE. Having resilient seating the valve has good isolating characteristics in small size low pressure applications but achieving acceptable leak tightness becomes increasingly difficult as size and pressure class increase. It is also limited to applications within the temperature range of the seating face material.

The concentric design is commonly used in the power generation, brewing, water and food industries, where the pipe line fluid is clean. The body material is normally cast iron for general applications, but bodies in other materials such as carbon steel, stainless steel, aluminium bronze, and aluminium are manufactured. Discs are, again, usually in cast iron, but can be obtained in carbon steel, stainless steel, or aluminium bronze

A fully lined version is also available, with the body seat and disc coated with PTFE. The body is split in a horizontal plane along the pipe axis and assembled with a single piece liner, minimizing cavities and providing an excellent valve for hygienic service (see Figure BF6).

BVAA 70 years of service to industry

Winn

One name embodies design innovation and reliability

Winn Hi-Seal valves set the benchmark for dependability in critical service and isolation applications. That's the result of delivering 150 years of pioneering design and outstanding reliability to the industry. Winn Hi-Seal double offset valves provide superior performance in a wide variety of applications, and are available in soft-, fire safe and metal seated combinations. Get in touch to take advantage of our innovation and experience.

Success flows through us.

To learn more, please visit www.tyco.com/flowcontrol

Copyright © 2010 Tyco Flow Control. All rights reserved.

tyco

High Performance Butterfly Valves

Figure BF7: Wafer Type High Performance Butterfly Valve, Gearbox Operated (Ebro Armaturen)

The butterfly valve's low weight and compact design make it potentially very attractive in saving cost in many isolating applications. However the concentric design proved incapable of meeting expectations in terms of leakage rate.

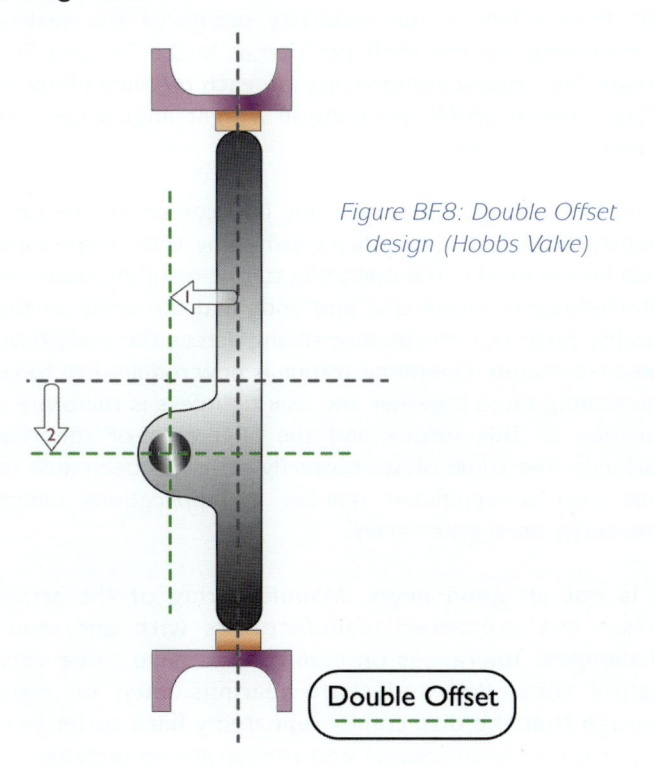

Figure BF8: Double Offset design (Hobbs Valve)

High Performance Butterfly Valve / Double Offset

To try to improve the seat leakage performance designers first developed the High Performance Butterfly Valve (double off set design). The shaft axis is now offset from both the plane of the seat (offset 1) and the axis of the body (offset 2). See Figure BF8.

This design provides two improvements in seat tightness. Firstly the shaft is no longer passing through the body seating face and secondly the offset design leads to a cam action between the seating faces on closure. This allows body disc interferences to be increased and thereby improving seat tightness without affecting the valves operability unduly. The valve still requires either a resilient seal in the body or on the perimeter of the disc to achieve bubble tight isolation but now the disc is only in contact with the body during the first 10% of opening and the last 10% of closing. Pressure-assisted seats are available such that the process pressure from either direction increases the contact load between seat and disc. This increases the valve's ability to seal as the process pressure increases. The double offset design utilises relatively little soft seating material, a thin seat of PTFE being common. This component, often in conjunction with a metallic back up ring, provides the soft seal at a specifically pre-determined position. In the event of the valve being subjected to fire conditions and the PTFE seat being destroyed, a secondary metallic seat is provided enabling the design to obtain a fire test certificate, see Figure BF9.

Figure BF9: High Performance Butterfly Valve being Fire Tested (Tyco Hindle/Winn)

Triple Offset

A significant breakthrough for the butterfly valve occurred when it was realised that by producing offset conical seating faces on the disc and body, bubble tight metal-to-metal seating was possible. Seating is bi-directional but in one flow direction the pressure is helping to close the valve (preferred as seating torque is lowest) and in the other it is trying to open it.

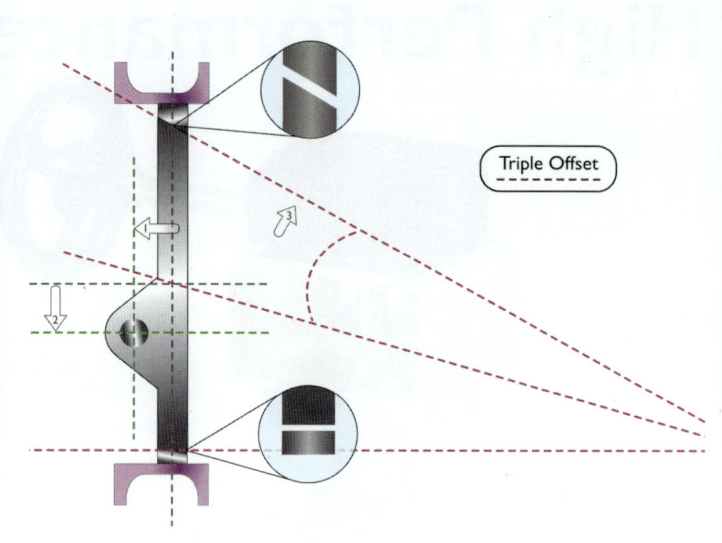

Figure BF10: Conical Geometry of Seat and Disc in Triple Offset Butterfly Valves (Hobbs Valve)

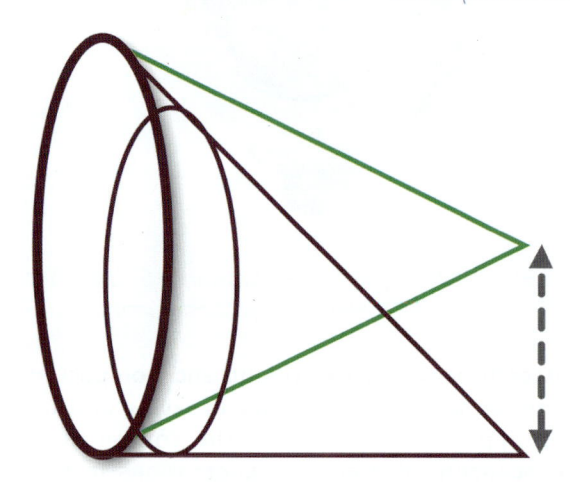

The third offset is the geometry design of the sealing components not the shaft position as in the 1st and 2nd offset. The sealing components are each machined into an offset conical profile resulting in a right angled cone as shown in Figure BF10.

Now the seating faces only come into contact at the final moment of closure in exactly the same way as the globe valve (See Figure BF11). The operating torque resulting from the interference between disc and body and the wear on the seating faces due to the disc sliding across the body have been eliminated. Operating torque is now required to force the seating faces together and leak tightness is therefore a function of this torque and the alignment of the two surfaces. The triple offset butterfly valve has been able to take over a significant number of applications which previously used gate valves.

It is not all good news. Manufacturing of the triple offset has presented manufacturers with enormous challenges. Tolerances on components have to be very tightly controlled and shaft bearings must be rigid enough that the disc comes repeatedly back to its seat in the body. As pressures and temperatures increase so

Figure BF11: Triple Offset Seating Detail (Hobbs Valve)

do these difficulties and there have been occasions when performance expectations have not been achieved. However as the performance envelope is pushed ever further the applications that were extreme last year are now achieved with increasing reliability.

The body material is normally carbon steel or austenitic stainless steel for general applications, but bodies in all types of materials including nickel alloy can be supplied.

It is a design that can be adapted for control applications, as detailed in the chapter Control Valves.

Figure BF12: High Performance Triple Offset Butterfly Valve featuring torque sealing (Velan)

Figure BF14: Double Block & Bleed Butterfly Valve assembly, flanged design, featuring two gearbox operated Triple Offset Butterfly Valves (Curtiss Wright)

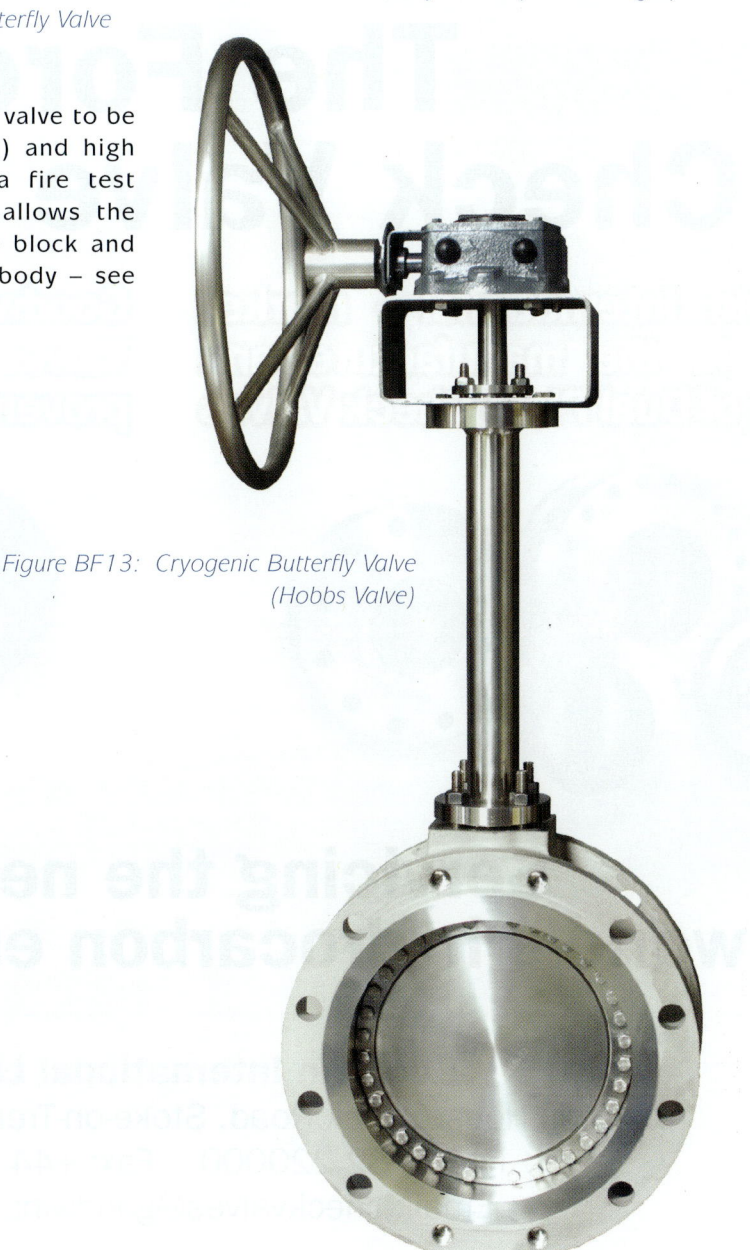

The triple offset design allows the butterfly valve to be used both in low (cryogenic, Figure BF13) and high temperature applications and to obtain a fire test approval. The bubble tight isolation also allows the triple offset design to be used for double block and bleed by mounting two valves in a single body – see Figure BF14.

Figure BF13: Cryogenic Butterfly Valve (Hobbs Valve)

GOODWIN
INTERNATIONAL

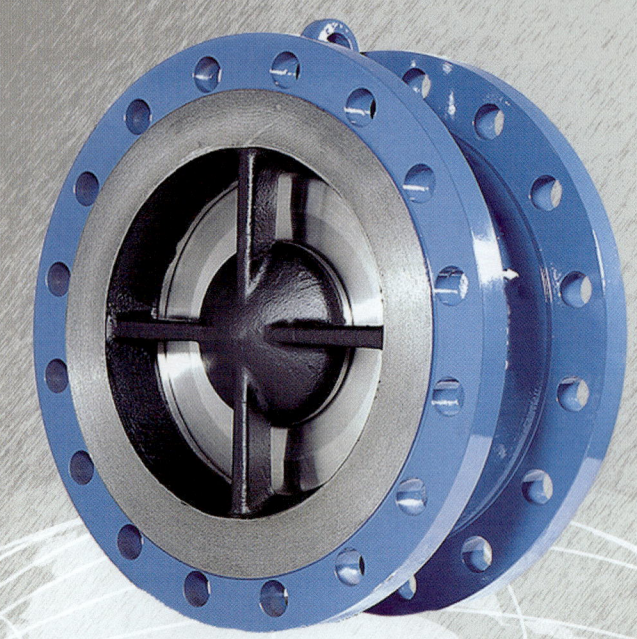

The Force
in Check Valve Solutions

Goodwin International - market leader in the manufacture and design of Dual Plate Check Valves

Goodwin Non-Slam Nozzle Check Valves founded on 70 years of proven design and technology

Servicing the needs of
the world's hydrocarbon energy industries

Goodwin International Limited
Goodwin House, Leek Road, Stoke-on-Trent, England ST1 3NR
Tel: +44 1782 220000   Fax: +44 1782 208060
Email: checkvalves@goodwingroup.com

www.checkvalves.co.uk

Check Valves

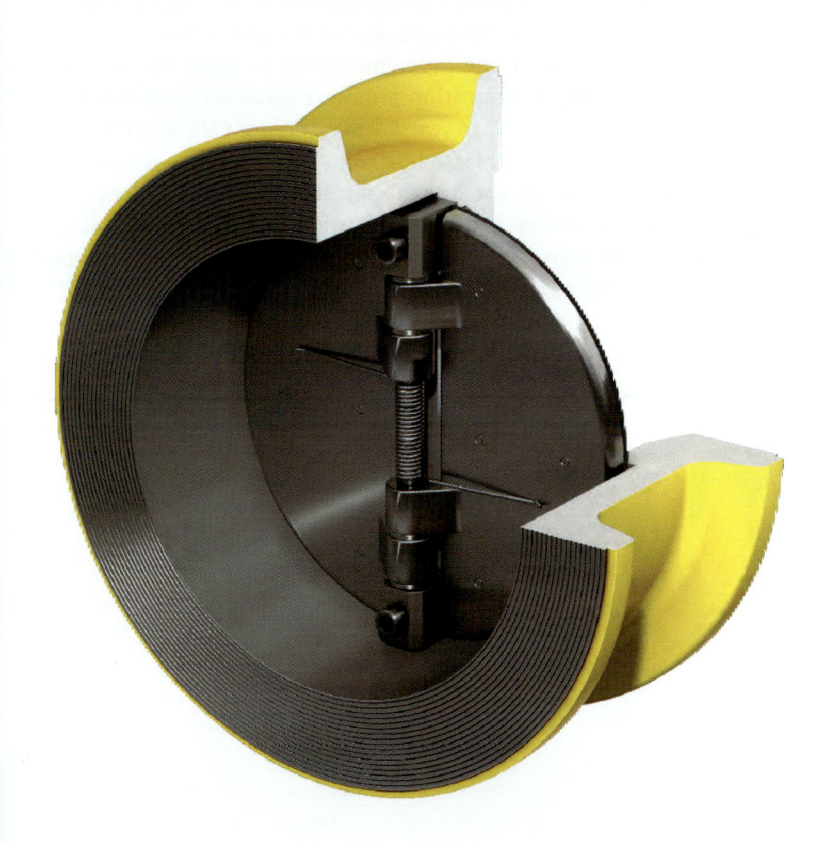

Figure CV1: A Dual Plate Wafer Check Valve (Cameron Wheatley)

The check valve can be likened to an iceberg. There is a lot more below the surface than first meets the eye.

The European Standard definition of a check valve is a valve which automatically opens by flow in a defined direction and which automatically closes to prevent flow in the reverse direction. Another common name for the check valve is the non-return valve.

This straightforward definition has resulted in many alternative designs because each application demands different features and close examination of the options is essential for correct selection. It is definitely the case that one size does not fit all.

The check valve fulfils primarily a safety function protecting equipment, product and people. For example pumps or compressors are used to move fluid against a pressure head. If the equipment stops, the pressure head would cause reverse flow that could be damaging. Check valves also prevent backflow from storage tanks, which would result in different fluids mixing and causing contamination.

Selection Considerations

Check valves can be one of three different valve types, namely globe with the disc moving axially, butterfly with a rotating disc on a hinge pin, or diaphragm with a membrane closing around the seat when the flow reverses.

Factors that should be considered in selecting a suitable valve are pressure class, operating temperature, C_v value, leakage capability, corrosion resistance, flow turn down ratio i.e. minimum flow/maximum flow, liquid or gas, pipework orientation, size limitations, weight, speed of operation, maintainability. Finally, as always, cost is a consideration and the final selection is usually a compromise between many competing considerations.

The function of the check valve is to allow flow in one direction and prevent flow in the reverse. This has to be the starting point for considering which valve to select. In the normal flow direction the C_v value is important, the higher the value the less energy will be used to push the fluid through the valve. In the reverse direction the amount of leakage that is acceptable is an essential criterion.

C_v **value** is a function of design. The C_v quoted by manufacturers is normally the full open C_v. The valve achieves its full open position as a result of the momentum force from the moving fluid pushing the disc into the open position. Typically the ratio of normal liquid to gas flow rates is 1:20 whereas liquid to gas density ratios are between 50 and 100:1. Hence liquids, even though they are travelling slower, achieve the required momentum force more easily than gases.

Designs which have internal components in the flow path result in turbulence and thus pressure loss resulting in a lower C_v but careful streamlining of the internal components such as the nozzle check valve (Figure CV2), will increase the C_v.

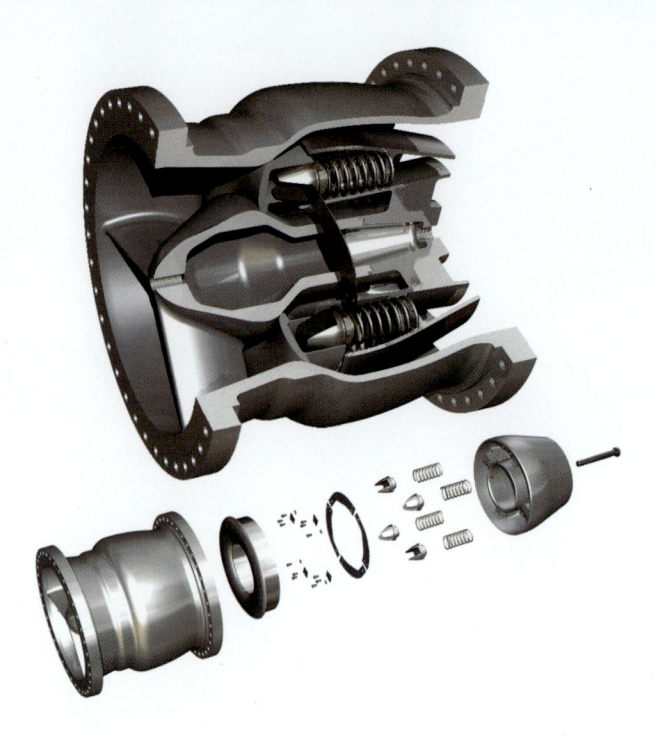

Figure CV2: DRV-B Nozzle Check Valve (Cameron)

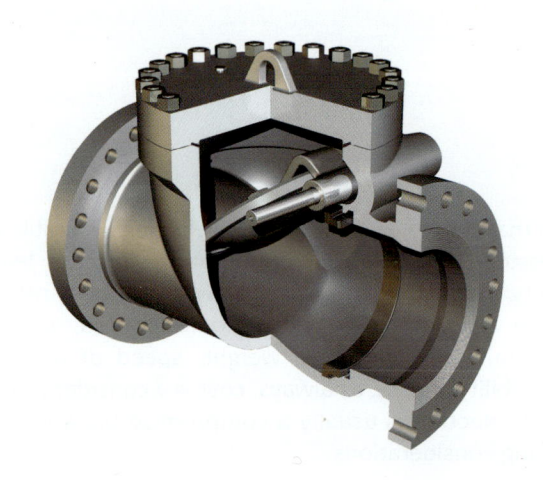

Figure CV3: 16in 600 Swing Check Valve (Tom Wheatley)

The swing check valve (Figure CV3), where the disc lifts out of the flow leaving a full bore orifice, should theoretically have the highest C_v but in practice it is not achieved as the disc does not always fully open. The weight of the disc and its orientation is important in understanding what flow rate is necessary to get the disc fully open. The disc of the swing check valve in a horizontal pipeline has to move from the vertical to a near horizontal position. In the vertical position the disc sees the momentum force acting normal to the surface but as the disc rotates the force becomes increasingly parallel to the surface and the component acting in the normal direction reduces. This makes it difficult to achieve the full open position.

It is not sufficient to consider solely the quantum of the C_v. Consideration must also be given to matching the flow velocity such that the valve is not oversized and the disc(s) will achieve their full open position. Also the minimum flow rate in the case of a high turn down ratio must be considered so that disc chatter is not an issue.

The swing check valve is from the butterfly family i.e. a rotating disc. Other types from the same family are the tilting disc and the dual plate wafer check valve. The tilting disc has a horizontal hinge pin but because the pin is much closer to the centre line of the pipe the forces to open the disc in a horizontal pipe application are more balanced and therefore smaller. The dual plate wafer valve (Figure CV4 and CV5) has the hinge pin vertical in a horizontal pipe and the disc weight is supported by the hinge pin so opening requires little momentum force. This makes these two types more suitable for higher turn down ratios as they will achieve the full open position at lower flow rates.

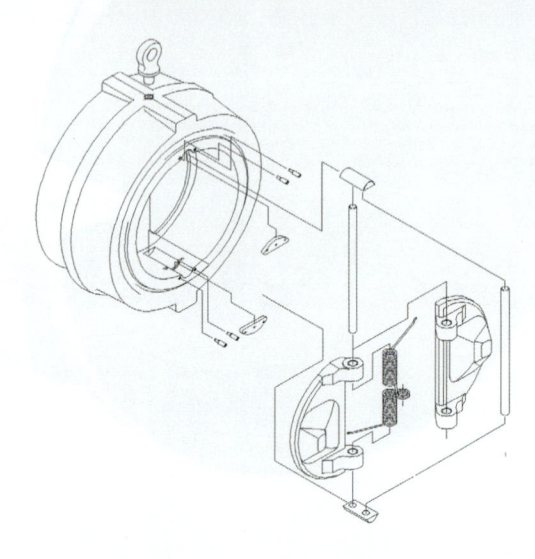

Figure CV4: Dual Plate Wafer Check Valve (Goodwin)

Figure CV5: Dual Plate Wafer Check Valve, partially open (Abacus)

The nozzle and diaphragm types of check valve also require low momentum forces to open either because of their balanced design or because the diaphragm weighs very little. The lift check type (Figure CV6) is a higher momentum force design but unlike the swing type the disc always remains normal to the flow direction.

Additional fluid velocity can be created by reducing the seat bore below the nominal pipe size. In the nozzle check valve this venturi effect can be achieved without much effect on the C_v.

110

BVAA VALVE & ACTUATOR USERS' MANUAL - 6th Edition

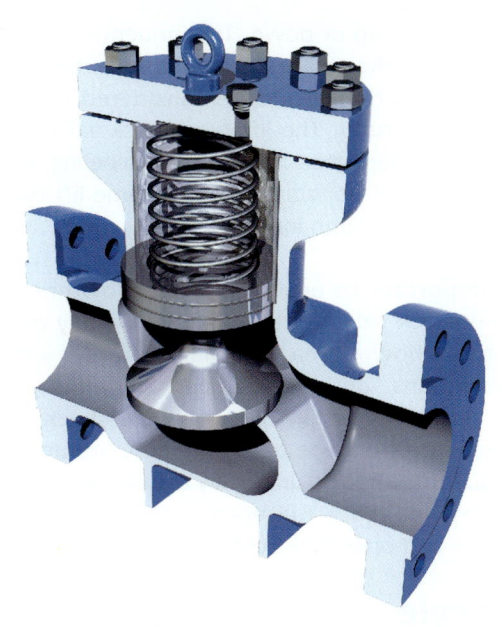

Figure CV6: Lift Check Valve with Piston and Spring (Tom Wheatley)

Seat Leakage in the reverse flow direction is the norm with check valves. The European Standard EN12266-1 specifies seven liquid seat leakage rates, which for a DN150 valve ranges from no visible leakage over 30 seconds to 300mm^3/s Typically the check valve acceptance criteria for metal-to-metal seating is the highest rate i.e. 300mm^3/s using a test pressure of 1.1 x the maximum allowable pressure at 20°C applied in the reverse direction. This is obviously significantly higher than the leakage rate of an isolating valve but it is less than 0.002% of the nominal flow at 2m/s in a DN150 pipe.

actual differential pressure could be significantly lower than the design differential used for the factory acceptance test. For example, a Class 600 valve in ASTM A350 LF2 will be factory tested at a differential pressure of 1.1 x 102.1 barg.

The Class 600 rating may be necessary to match the pump maximum delivery head but the actual reverse flow upstream pressure is the working pressure of the vessel downstream of the valve, which may only be 40 barg. The required minimum clamping force to achieve the as new leakage criteria vary for each design and manufacturers will make recommendations when asked. Generally the required minimum force can be well below the force used during factory testing. However it is always worth asking for advice, if the actual reverse flow differential pressure is going to be less than 25% of the nominal 20°C pressure rating of the valve.

The use of resilient seating e.g. rubbers or plastics can result in valves that meet the no visible leakage criteria both for liquids and gases but these materials are not an option at either high temperature (>260°C) or at low temperature (<-200°C). They may also not be acceptable, if the design has to be fire tested.

The extremes of temperature can lead to problems with differential expansion of the materials in any valve. This is particularly true for check valves, which rely solely on the line pressure to provide the clamping force. Valves selected for cryogenic service (Figure CV8) or for high temperature applications should have been type tested at these extremes to ensure they will perform as anticipated.

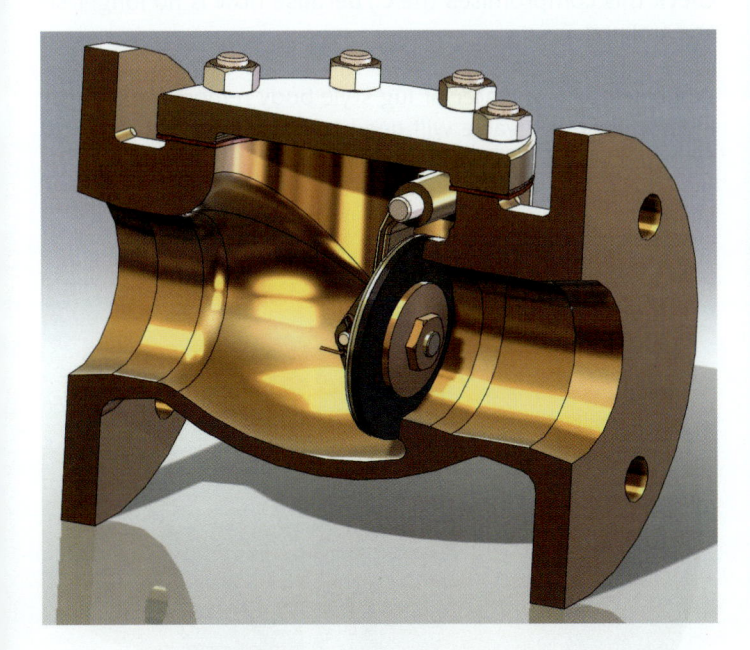

Figure CV7: Nickel Aluminium Bronze PN 16 DN65 Swing Check Valve with a soft seat (Truflo)

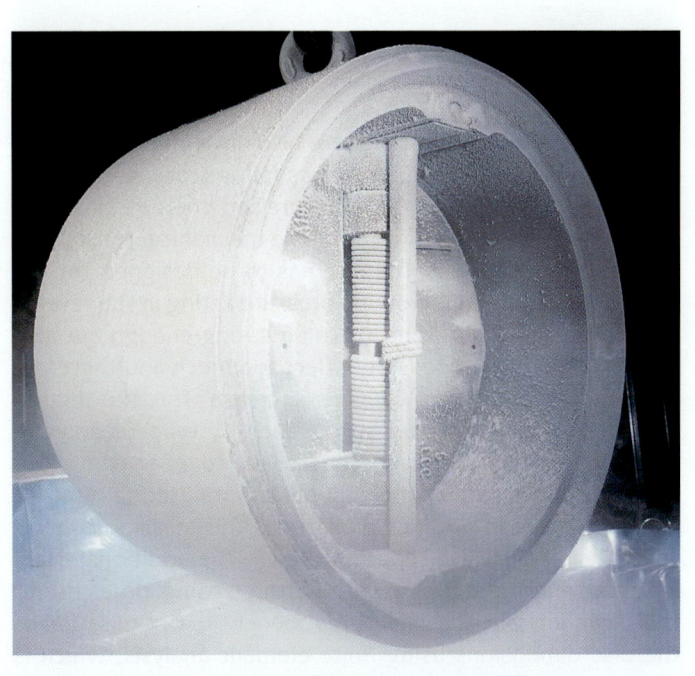

Figure CV8: A 42in ASME Class 600 Dual Plate Wafer Check Valve after Cryogenic test with test plates removed (Goodwin)

The achievable leakage rate is a function of the flatness of the mating components, the materials of the components and the ability of the design to bring the components together. The only force clamping the components together is the differential pressure acting across the valve in the reverse flow direction. It must be remembered that the

The actual factory test results are typically 100mm^3/s liquid leakage for a well built DN150 valve. This level of seat leakage will deteriorate over time due to damage from debris in the fluid or mechanical operation of the valve. If the check valve rarely closes and the fluid is clean, the leakage rate is unlikely to deteriorate and it will take many years before it becomes noticeable. However, if the valve is closing frequently and in a dirty or corrosive environment, consideration must be given to the rate of increase of seat leakage, as this may quickly reach an unacceptable level. In that case the valve selection may need to be changed to one that is more easily maintained or one with a design that minimises seat damage such as seat hard facing.

The butterfly family of metal-to-metal check valves, i.e. swing, tilting disc and dual plate wafer, achieve the API598 rate of 3cc/inch of bore over 1 minute. Assuming 1 inch equals DN25 this is identical to the $2mm^3$/DN/second rate specified in EN12266-1. The globe family of nozzle and lift type check valves can achieve better seat leakage rates because the designs allow the disc to be axially aligned to the seat and the two surfaces remain parallel during closure. These designs may be more appropriate for gases, which are more difficult to seal. The diaphragm check valve with its resilient seat always achieves the no visible leakage criterion.

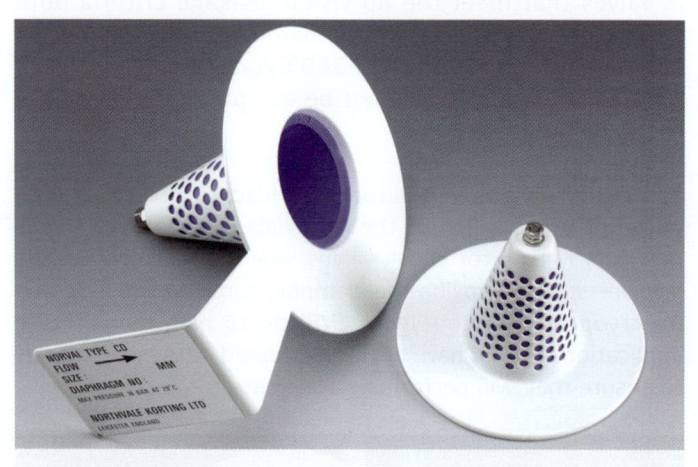

Figure CV9: Diaphragm Check Valve (Northvale Korting)

Pressure class and temperature also have an effect on the choice. The issue of temperature as far as seat materials has already been mentioned. Temperature and pressure class also affect the internals as a result of the mechanical forces that have to be considered. The disc has to be designed for the valve's maximum 20°C working pressure acting in the reverse flow direction. The disc needs to be strong enough but also stiff enough that it does not deflect too much and therefore leak excessively. Disc weights increase for the higher pressures, which for the swing and lift type means the required momentum forces necessary to open the valve increase. In the case of the dual plate wafer, tilting disc and nozzle types the additional material thicknesses take up space inside the valve reducing the C_v. The additional mass increases inertia and slows down the closing time. Under these circumstances there is great benefit to optimisation of the disc dimensions using finite element analysis, which is now readily available through many 3-D design packages. The two requirements of stiffness along the seat contact line and structural strength in the centre can be optimised to the users benefit lower weight, better performance and lower cost. The diaphragm valve is no longer an option for pressure classes exceeding Class 300.

Pipework orientation is another important consideration. All designs are suitable for horizontal pipelines but not all are suitable for vertical pipelines. If the pipeline is vertical then flow can be either up or down. The influence of gravity and flow direction on the ability of the disc to either, open fully or to close correctly need to be considered in discussion with the manufacturer. The lift type check valve in a globe valve body is not designed for a vertical pipeline as the disc movement becomes horizontal but the same lift type works well in a straight through inline body.

It is also important to ensure that the flow entering and leaving the check valve is stable and uniformly distributed across the flow area. Manufacturers will make recommendations on the number of pipe diameters upstream and downstream between the check valve and either other valves or pipe bends. If these are not followed then the flow may not be uniformly distributed and the disc or discs may not reach full open, which will increase the pressure loss and operating cost of the plant.

Body Style

The butterfly types of tilting disc and dual plate wafer check valves very often have a wafer, double flanged or lug style (Figure CV10), which fastens between the upstream and downstream pipe flanges. It is also possible to obtain a version of the swing check in a wafer body called a hinged flap valve. The wafer, double flanged or lug body style offers significant advantages in terms of weight and cost savings. It also allows check valves of very large diameters to be manufactured. It must be remembered with this type of body the discs will extend into the downstream pipework when fully open. Reduced body lengths are also possible in nozzle check and vertical inline lift check valves. In the case of nozzle check this compromises the C_v because flow is no longer so streamlined.

When selecting a wafer or lug style body remember that for maintenance the valve will need to be completely removed from the pipework.

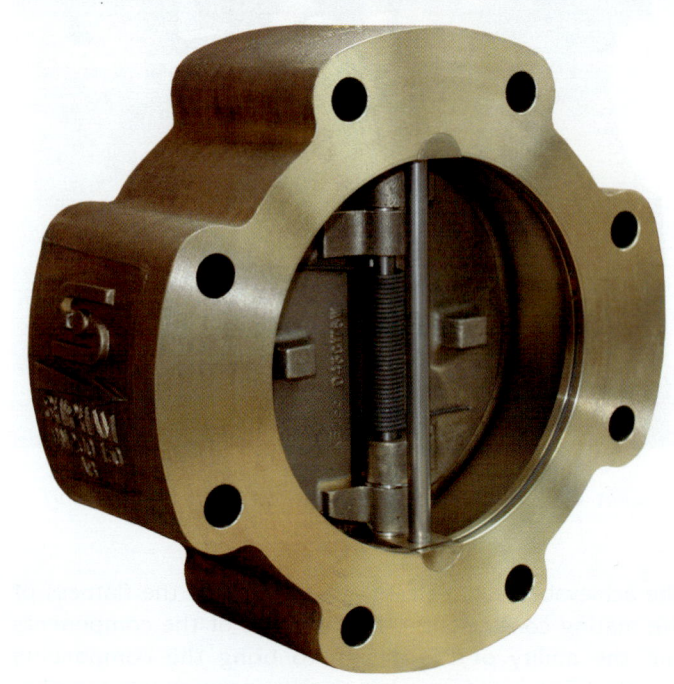

Figure CV10: A lug style Dual Plate Wafer Check Valve (Shipham)

Globe Stop and Check Valves

The body style used by the globe isolation valve is also used for the lift type check valve. It only took a little innovation to realize that it was possible to combine the two functions of isolation and check into the same design and the Globe Stop and Check valve was created. It is also known as a screw down non-return or SDNR. In the globe stop and check design the valve disc is not connected to the valve stem. With the valve in the open position the fluid momentum force opens the valve. Reverse flow closes the valve as a lift check valve. By turning the handwheel and moving the valve to the closed position it becomes an isolating valve and its check function is negated. The advantage of combining the two functions into one is a reduction in the pressure loss as well as cost. This valve type is commonly used on the outlet of steam boilers connected to a common header.

Check Valve Problems

Disc chatter is the manifestation of unstable check valve behaviour. It results when low momentum forces allow the disc to fall back towards the seat. At that point additional opening forces associated with an increasing differential pressure act to open the disc. This behaviour of closing and opening at low flow velocities, particularly for gases, can become unstable leading to the disc coming into contact first with the seat and then the disc end of travel stop. The phenomenon is known as disc chatter and it can cause rapid failure of the valve and great annoyance to those close by.

Water hammer is a potentially extremely dangerous event. It is normally associated with liquids and is most severe in liquids with no entrained gas as these are essentially incompressible. It results from the rapid deceleration of a liquid column associated with a rapidly closing valve. A check valve is intended to close rapidly and consideration must therefore be given to the potential of water hammer in those cases where reverse flow has started before the valve has fully closed.

When the liquid column is rapidly decelerated, the momentum change results in an increase in the static pressure at the surface of the disc. This pressure wave then reflects back through the column of fluid travelling away from the disc. The magnitude of the pressure pulse can be significant and in extreme cases has been known to move pipes from their anchors or burst open flange joints or rupture the pipe. If the system is closed at the opposite end the wave will oscillate backwards and forwards gradually decaying away due to viscosity effects. If the time for the wave to travel the liquid column length is short compared with the valve closing time then the calculation can ignore fluid compressibility and pipe elasticity. Speed of sound in the fluid is an important property in calculating the magnitude of the pressure pulse and the time for the pulse to reach the end of the fluid column. Joukowsky was the first to show that for an instantaneously closing valve the pressure increase is

$$\Delta p = \rho * a * V_r$$

where Δp is the rise in pressure in Pa, ρ is the fluid density kg/m³, a is the speed of sound in the fluid m/s and V_r is the maximum reverse flow velocity m/s.

This equation gives a worst case scenario. It should be used to determine whether there is a need for further investigation. If there appears to be a problem then detailed discussion with the manufacturer will help to ensure a valve is selected that minimises the potential for water hammer.

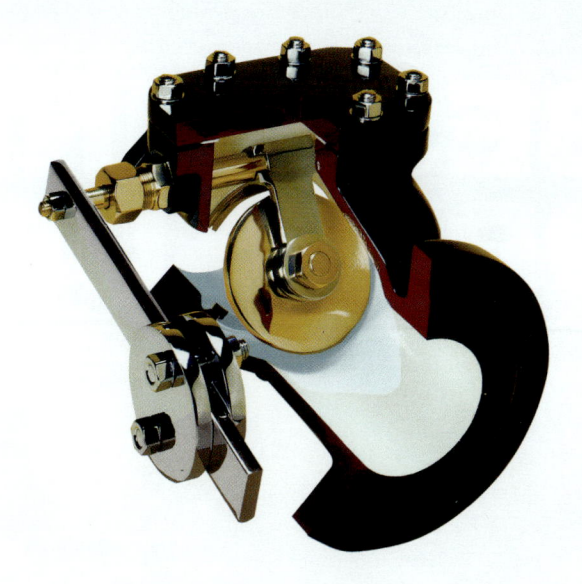

Figure CV11: Cast iron Swing Check Valve with lever and weight (Peter Smith Valve Co.)

Assisted Opening or Closing

Most check valves are supplied with some form of assisted opening or closing. Remembering the Joukowsky equation that water hammer is proportional to reverse flow velocity it is clear that if there is no reverse flow velocity then there is no water hammer. Manufacturers therefore fit their designs with closure assist devices such that as the flow reduces the disc is already moving towards the seat. Ideally if, when normal flow has decayed to zero, the disc has returned to the seat then reverse flow cannot commence. The valve most likely to suffer from water hammer is one where firstly either gravity or reverse flow causes disc movement and secondly, which has the longest distance to travel to reach the closed position. On this latter point the swing check valve disc has to move πD/4, the dual plate wafer disc πD/8 and the nozzle or lift check disc D/4, where D is the pipe diameter. When distance to close is combined with disc inertia a valve's dynamic response can be determined. Valves with the highest dynamic response are least susceptible to water hammer.

Springs are the most common energy source for assisted closure but external weights can be added to swing check valves attached to the hinge pin. As with most engineering solutions to one problem there is a consequence to be paid for in a negative effect somewhere else. In this case it is an increase in the momentum force necessary to fully open the valve. Typically it means that flow rates have to go up or, more likely, valves are not fully open and pressure losses and operating energy needs are higher.

Occasionally in gas applications with large turn down ratios, assisted opening is provided to swing check valves to ensure that they do not suffer from disc chatter.

DRESSER Consolidated

BEST UNDER PRESSURE.

Designed For Dependability

Dresser Consolidated® is the leader in pressure and safety relief valves with the trusted expertise to provide and service reliable flow safety systems in critical applications. Through superior engineering, high quality materials and knowledgeable service, we better support our customers by streamlining processes, increasing efficiencies and reducing costs.

Unique and Versatile Pilot Valve Design

The Type 3900 valve is a highly versatile modular pilot-operated safety relief valve design providing outstanding features and benefits, including:

- Easy maintenance
- High reliability
- Efficient operation and performance
- Separate non-flowing pilot sub-assembly design
- Pop action or modulating action functionality

© 2010 Dresser, Inc. All rights reserved.

www.dresser.com

Safety Valves

Effective communication between user and manufacturer is fundamental in ensuring that the Safety Valve is able to perform its vital role of protecting equipment and personnel.

The user has knowledge of his plant and the process conditions liable to result in over pressurisation and the manufacturer has knowledge of the safety valve's capabilities and limitations. Information needs to be exchanged in a clear and accurate manner so that the correct valve is supplied and installed. To facilitate this, the international standards organisations have created detailed definitions of all the necessary terms used.

The potential risk to life resulting from the failure of a safety valve means that in most places in the world their use is covered by national regulations. In the USA this means the ASME Codes and API Recommended Procedures and in the EU it means the Pressure Equipment Directive 97/23/EC. The ASME regulatory systems are also widely used in other parts of the world. There are small differences in the definitions used and also differences in the maximum allowable accumulation pressure.

The purpose of this section is to raise awareness of safety valve fundamentals. For those directly responsible for the specification, installation and operation of safety valves it is strongly recommended that they undertake a detailed course in safety valves covering all aspects of this type of valve.

Other terms used for Safety Valves also originate from their function i.e. Relief Valve, Pressure Relief Valve and Safety Relief Valve. These terms are defined in the ASME code as well as that for Safety Valve. In Europe only the term Safety Valve is used and the definition covers that which ASME divides into three. The inclusion of the term Relief Valve in the ASME Code recognises that for a liquid, over pressurisation can be prevented by a small release of the fluid because it is essentially incompressible.

Design Basics

The Safety Valve is part of the globe valve family. Operating forces acting to close the valve directly oppose the fluid flow direction. There are three operating states that need to be explained when trying to understand the operation of a Safety Valve.

- The first is the state of equilibrium operation, Figure SV1. In this state the forces acting to close are in equilibrium with those acting to open and the seat and disc surfaces are just in contact. The assumption is that the contact surfaces are perfect and therefore the valve is at the point when flow is about to start. This is the basis of the EN/ISO definition of Set Pressure. In practice it is not possible to achieve zero flow and flow will commence as the equilibrium pressure is approached. This is recognised within the ASME Code, which bases the definition of set pressure on the visible evidence of fluid downstream of the seat.

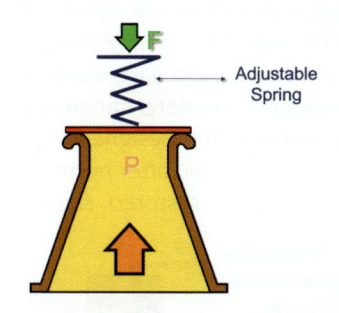

Figure SV1: Equilibrium State
(Tyco Anderson Greenwood Crosby)

- The second operating state is full open. This is the position the safety valve must achieve in order to pass its rated capacity at its specified overpressure. It is an often misunderstood concept that the rated capacity is achieved at the Set Pressure. For the valve to lift the inlet pressure must exceed the equilibrium pressure. Different designs of safety valve have different relationships between lift and inlet pressure, hence between flow and pressure. Some achieve a rapid increase in flow for small increases in inlet pressure, others achieve a more gradual increase.

- The third operating state is the fully closed or nil leakage state. The pressure at which this state is achieved depends on whether the inlet pressure is increasing or falling. Starting from zero pressure the valve will theoretically remain in the fully closed state until the equilibrium state is reached. However after the valve has opened and flow has occurred it is no longer possible to achieve the fully closed position until the pressure has gone significantly below the equilibrium state pressure. This re-seat pressure varies for each valve design and in general the more rapid the flow increase was, when achieving the fully open condition, the lower the reseat pressure is. Once the valve has re-seated it should remain closed if the pressure is once again below the equilibrium state.

The mass flow capacity of the safety valve is a function of flow area and inlet absolute pressure. The ruling flow area is either that of the inlet seat or nozzle bore, if the lift of the safety valve exceeds D/4, or the skirt area under the disc, if the lift is less than D/4. D is the inside diameter of the seat or nozzle outlet. When the fluid is a gas the velocity reaches Mach 1 at the ruling orifice cross section and the flow chokes. The flow in volume terms is then a function solely of flow area but in mass terms it is additionally proportional to the absolute inlet pressure assuming the laws of a perfect gas.

The capacity of each design of safety valve is not something that can be accurately calculated using the theoretical laws of physics as the complications of the turbulent flow path inside the valve make this impossible. Therefore all capacities are based on actual achieved flow tests carried out at a third party facility e.g. the American National Board. These tests give the manufacturer a constant that can then be applied to the theoretical equation of a simple nozzle in choked flow thereby providing a uniform approach to the calculation of capacity for valves from different manufacturers.

Definitions and Practical Relevance

- **Safety Valve (EN/ISO)** - A valve which automatically, without the assistance of any energy other than that of the fluid concerned, discharges a quantity of the fluid so as to prevent a predetermined safe pressure being exceeded, and which is designed to re-close and prevent further flow of the fluid after normal pressure conditions of service have been restored, see Figure SV2.

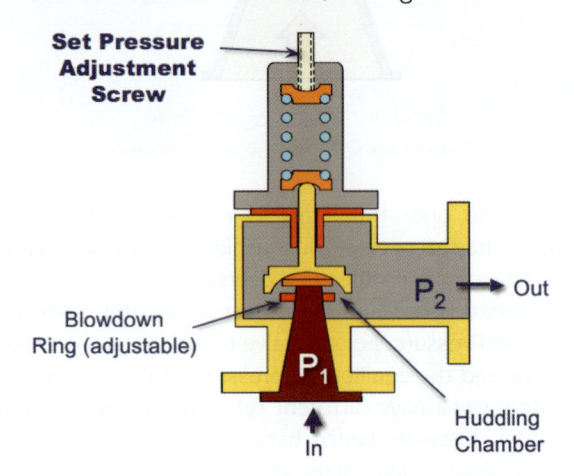

Figure SV2: Unbalanced Direct Spring Safety Valve (Tyco Anderson Greenwood Crosby)

- **Safety Valve (ASME)** - An automatic pressure relieving device actuated by the static pressure upstream of the valve and characterised by a rapid full opening or pop action. It is used for steam, gas or vapour service.

- **Relief Valve (ASME)** - An automatic pressure relieving device actuated by the static pressure upstream of the valve, which opens in proportion to the increase in pressure over the opening pressure. It is used primarily for liquid service. See Figures SV3 and SV4.

- **Safety Relief Valve (ASME)** - An automatic pressure relieving device actuated by the static pressure upstream

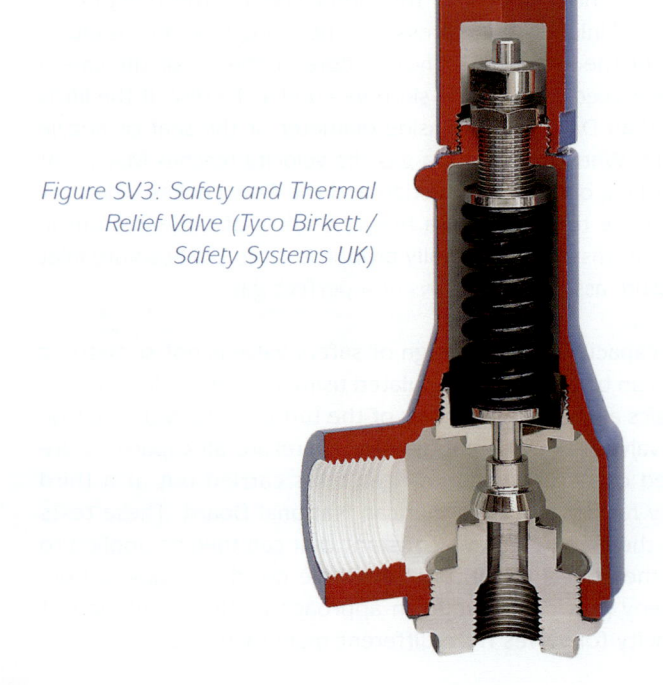

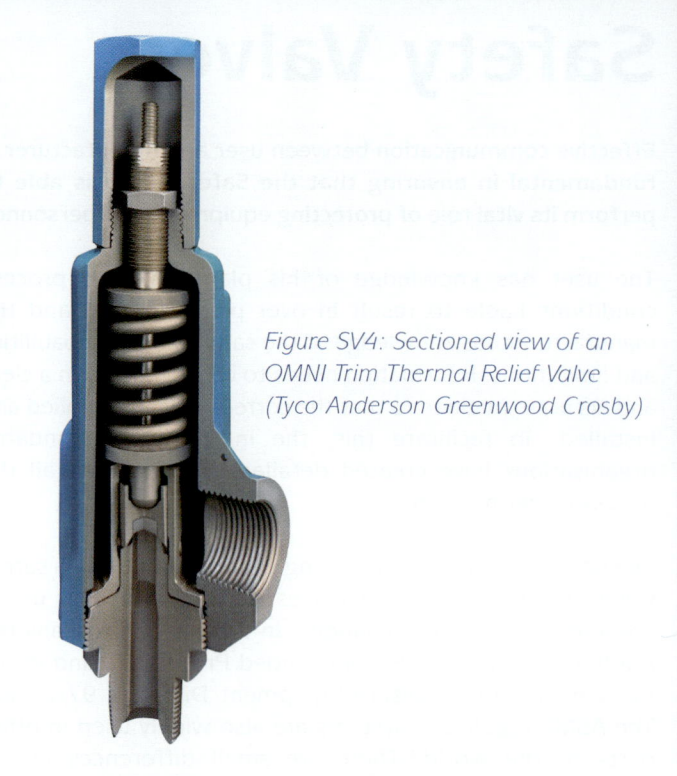

Figure SV3: Safety and Thermal Relief Valve (Tyco Birkett / Safety Systems UK)

Figure SV4: Sectioned view of an OMNI Trim Thermal Relief Valve (Tyco Anderson Greenwood Crosby)

of the valve characterised by rapid opening or pop action, or by opening in proportion to the increase in pressure over the opening pressure, depending on the application and may be used either for liquid or compressible fluid.

- **Set Pressure (EN/ISO)** - Pre-determined pressure at which a safety valve under operating conditions commences to open. Note: It is a gauge pressure measured at the valve inlet at which the pressure forces tending to open the valve for the specific service conditions are in equilibrium with the forces retaining the valve on its seat.

- **Set Pressure (ASME)** - The inlet pressure at which the valve is adjusted to open under service conditions. For a valve on gas, vapour or steam service, the set pressure is the inlet pressure at which the valve Pops under service conditions. For a valve in liquid service, the set pressure is the inlet pressure at which the valve Starts to Open under service conditions.

- **Overpressure** - A pressure increase over the set pressure of a safety valve, usually expressed as a percentage of the set pressure.

- **Accumulation** - The pressure increase over the maximum allowable working pressure (MAWP or PS) of the vessel during discharge through the safety valve, expressed as a percentage of that pressure or in pressure units.

- **Blowdown** - The difference between the actual set pressure of a safety valve and the actual reseating pressure, expressed as a percentage of the set pressure or in pressure units.

- **Back Pressure** - The static pressure existing at the outlet of a safety device due to pressure in the discharge system. See Figure SV5.

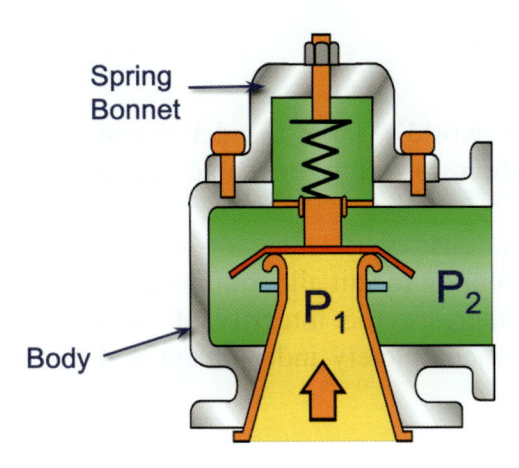

Figure SV5: Back Pressure (Tyco Anderson Greenwood Crosby)

- Cold Differential Set Pressure - The inlet static pressure at which the safety valve is adjusted to open on the test stand, probably using air. This pressure includes the corrections for service conditions of back pressure or temperature or both. Note: Manufacturers of safety valves may use steam as a test medium, thus a later re-test on air (at ambient temperature) is likely to give an apparent different set pressure.

The Set Pressure is stamped on every safety valve and appears on every safety valve data sheet. It is an important parameter but it must be remembered that the Safety Valve's function is to prevent a predetermined safe pressure being exceeded not simply to lift exactly at the nominal Set Pressure.

The Set Pressure whether in EN/ISO or ASME is defined based on operating conditions. In practice the valve has to be set in the factory or at the valve repair shop. There are many variables that have to be considered in evaluating a true set pressure, the accuracy of the pressure gauge, the human judgement as to what is understood by equilibrium (EN/ISO) or Pops or Starts to Open (ASME), the factory test fluid and factory test temperature relative to actual service fluid and temperature. For example a valve destined for steam service may be set in the factory using air at ambient temperature. In service the temperature of the steam or fluid will increase the seat bore area so to compensate, manufacturers apply the Cold Differential Set Pressure. All these practical considerations mean that for the majority of valves it is acceptable for the actual Set Pressure to vary from the nominal figure stamped on the valve and specified on the data sheet within a tolerance of ±3%.

While it is permissible to accept a tolerance on the Set Pressure, the maximum Accumulation Pressure specified in the applicable design code must not be exceeded. For applications covered by the European Pressure Equipment Directive 110% is the maximum. For an ASME single valve installation 110% is specified, 116% for a multi valve installations and 121% when the source of overpressure is a fire.

Actual Set Pressure is also important as it is used to define the Blowdown of the Safety Valve, which is usually quoted by manufacturers as a percentage of Set Pressure. Blowdown can be very important in applications involving fluids with significant amounts of stored energy or high value or which are dangerous. The greater the Blowdown percentage the more fluid is dumped from the system before the valve re-closes. For example a safety valve fitted to a steam boiler will continue to see a pressure greater than the Blowdown pressure until sufficient saturated water has evaporated to allow the saturation temperature to fall to that corresponding to the Blowdown pressure. The use on a steam boiler of a Safety Valve with 10% Blowdown as compared with one with 5% Blowdown can result in a large additional energy release each time the valve lifts.

Back Pressure acts on top of the disc potentially increasing the Set Pressure. The Back Pressure can exist with the valve closed and be either constant or variable. It can also be created when the flow from the safety valve is not able to escape quickly enough through the exhaust pipe work. It is important to fully explain all possible Back Pressure situations to the manufacturer so that he can take these into account in his sizing calculations.

Types of Safety Valve

Direct Spring Safety Valve is the most common type, see Figures SV6 and 7. A spring is pre-compressed and this force acts downwards on the disc holding it against the seat to maintain a seal. As the fluid pressure increases the spring force is exceeded and flow commences. Typically designs make use of the flow to provide additional upward force such that the disc quickly lifts the full amount of D/4. This force either comes from momentum as a result of turning the flow downwards or by creating a chamber using a blowdown ring.

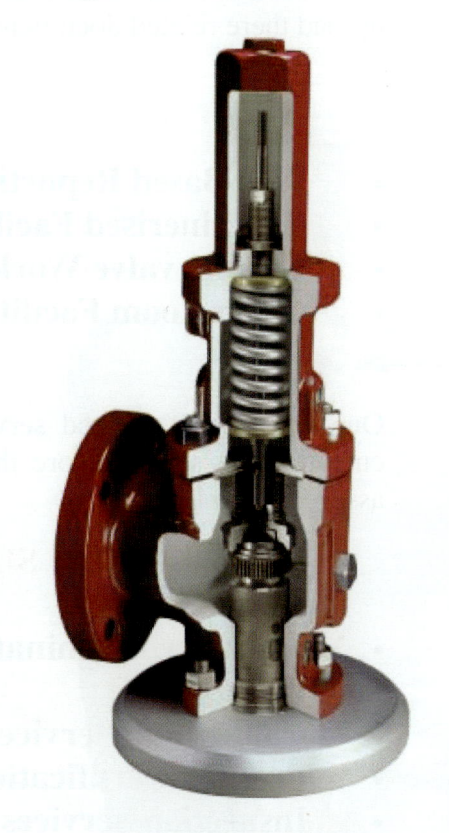

Farris SV6: Full Nozzle, Direct Spring Safety Valve (Curtiss Wright/Farris)

SGS

Valve Services

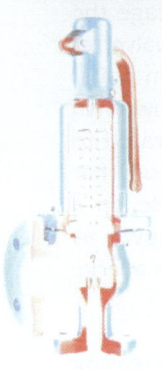

SGS United Kingdom Limited is part of the world's largest independent verification, certification and testing organisation, offering traceable valve services throughout the UK.

Offered from either fully equipped workshop, customers sites, or remote mounted mobile workshop facilities we tailor a solution to suit all requirements. With no manufacturing, insurance or trading interests, the integrity of SGS places it in a commanding position to provide a completely independent and impartial service. A service you can **TRUST**

Valve Overhaul, Repair and Certification of:

- **Relief / Safety**
- **Control**
- **Isolation**
- **Regulating**
- **Pilot Operated**

We have the capability and experience to offer solutions to our valued clients from both basic level turnaround of work package to full management and manning of shutdowns / outages / changeover both on land and off-shore.

SGS offers our custom designed and tailored Web Based Reporting package "**SGSREPORT**" which gives live web based access to both current and historical certification along with defect tracking / analysis, flagging of re-certification due dates. It also gives the opportunity for our customers to upload there related documents to the valves web based history.

SGS: The Total Package

- **Web Based Reporting**
- **Containerised Facilities**
- **Custom valve WorkShops**
- **Clean Room Facilities**

Our portfolio of offered services gives SGS the unparalleled ability to allow customers to service more than one need with a single traceable and Quality assured supplier.

SGS: Further Services

- **Statutory Examinations (PSSR / LOLER / LEV)**
- **PED / TPED**
- **Calibration Services**
- **Product Certification**
- **Inspection Services**

SGS United Kingdom Ltd   Unit 7-8 Manby Road, Immingham, NE Lincolnshire DN40 2LH   **Tel** +44 (0)1469 570248   **Fax** +44 (0)1469 557511
Email Sales: **jon.simpson@sgs.com**

SGS IS THE WORLD'S LEADING VERIFICATION, TESTING, AND CERTIFICATION COMPANY

MEMBER OF BVAA

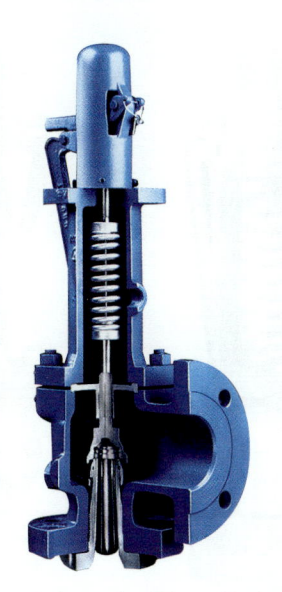

Figure SV7: Sectioned view of a Direct Spring Safety Valve (Tyco Anderson Greenwood Crosby)

The blowdown ring can be moved up or down towards the underside of the disc, see Figure SV8. When it is in its highest position the gap between the disc and the outer edge of the blowdown ring causes a flow restriction and pressure builds up in the chamber. This pressure acting on the area of the chamber provides an additional upward force on the disc. It must be remembered that these additional upward forces are still present when the inlet pressure starts to return to the normal operating pressure. This means that the spring force can only close the valve when the inlet pressure has fallen below the set pressure. The blowdown percentage varies between 5% and 20% depending on the different design and whether the fluid is a liquid or a gas and each manufacturer's published data should be consulted.

Figure SV8: Adjustment of Blowdown Ring in Bellows Sealed Safety Valve

Weight Loaded Safety Valve uses weights on the end of a lever to provide the down force to keep the disc on its seat, see Figure SV9. The limitations of size make this type of valve only suitable for small nozzle sizes and/or low set pressures.

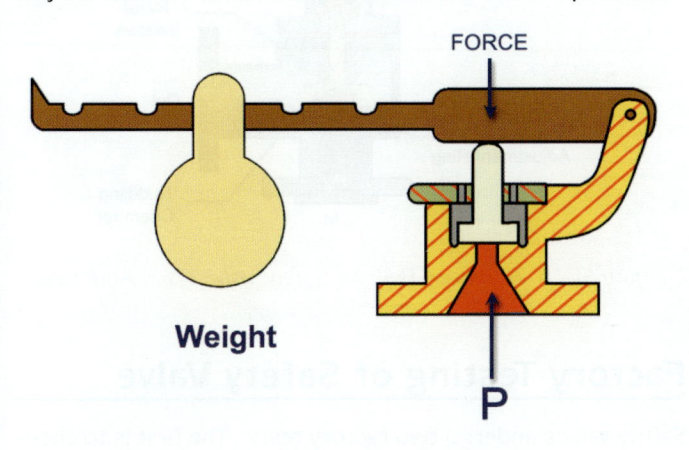

Figure SV9: Weight Loaded Safety Valve (Tyco Anderson Greenwood Crosby)

Torsion Bar Safety Valve uses the torsion generated in a steel bar when it is twisted to provide the down force on the seat. This type of design is used on high pressure applications where the spring size would otherwise be very large.

Pilot Operated Safety Valve uses the inlet pressure to keep the disc on its seat, see Figures SV10 and SV11. The inlet pressure is supplied through a small direct spring safety valve onto the top of the disc. By making this area greater than the seat or nozzle area there is a positive pressure acting downwards on the disc at normal working pressure. When the set pressure is reached the small pilot valve relieves the pressure above the disc to the exhaust thereby unloading the top of the main disc and causing the main valve to open. The amount of fluid that the pilot valve has to relieve is small and therefore the design does not need to incorporate features to provide additional lift forces so it reseats quickly thereby giving the main valve a small blowdown characteristic. One issue to be concerned about with a pilot valve is dirt in the fluid restricting the flow through the small orifices and pipe work leading to the pilot valve. In dirty service a filter should be fitted. The fact that the closing force is provided by the inlet pressure means that size and weight of a pilot valve is typically much less than that of a direct spring safety valve particularly for larger bore and/or high pressure applications. There are different types of pilots used such as snap acting and modulating both of which can be the no-flowing type of design.

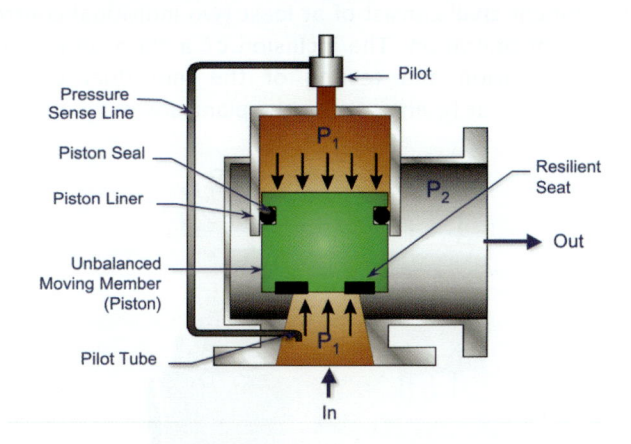

Figure SV10: Pilot Operated Safety Valve (Tyco Anderson Greenwood Crosby)

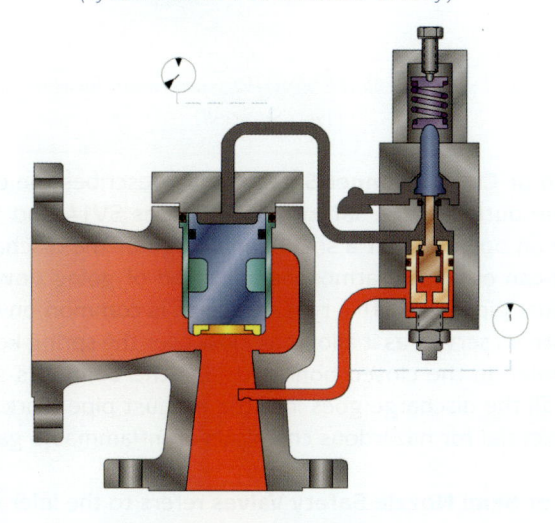

Figure SV11: Cross sectional view of a Pilot Operated Safety Valve - with Snap Action Non Flowing Pilot with Main Valve open and discharging (Tyco Anderson Greenwood Crosby)

Figure SV12: Pilot Operated Safety Valve - with Snap Acting Non Flowing Pilot (Tyco Anderson Greenwood Crosby)

Controlled Safety Pressure Relief System (CSPRS) consists of a main valve and control units, see Figure SV13. On reaching the set pressure the controlling forces on the main valve are by means of the control unit automatically applied, released or so reduced allowing the main valve to operate. This type of design is also known as Supplementary Loaded design. Within the Control Unit there are three systems which allows for there always being two in operation. The standard requires that the control unit arrangement shall consist of at least two individual control systems in operation. The inclusion of a third allows for regular isolation and testing of the individual control systems without having to shut the plant down.

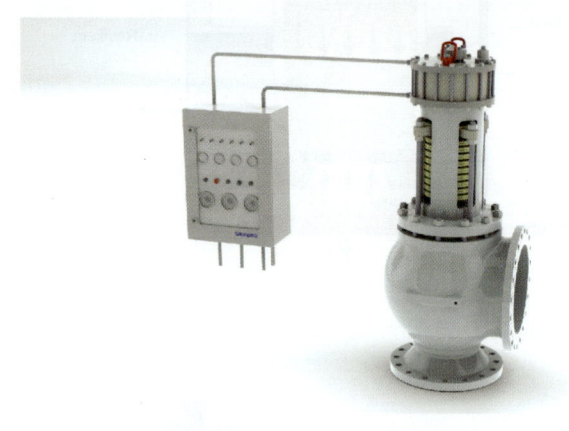

Figure SV13: Controlled Safety Pressure Relief System, CSPRS (Tyco Sempell)

Open or Closed Bonnet Safety Valve describes the design of the outlet of the safety valve, Figures SV14 and SV15. With an open bonnet a small percentage of the discharging fluid can escape to atmosphere instead of going down the exhaust pipe work. This type of design is common on valves for steam service as it allows air flow over the spring keeping it cooler. In the closed bonnet design the bonnet is sealed and all the discharge goes into the exhaust pipe work. This is essential for hazardous chemicals or inflammable gases.

Full or Semi Nozzle Safety Valves refers to the inlet of the safety valve. The semi nozzle design exposes part of the body below the seat to the inlet pressure whereas in the full nozzle design only the nozzle is exposed to the fluid inlet pressure.

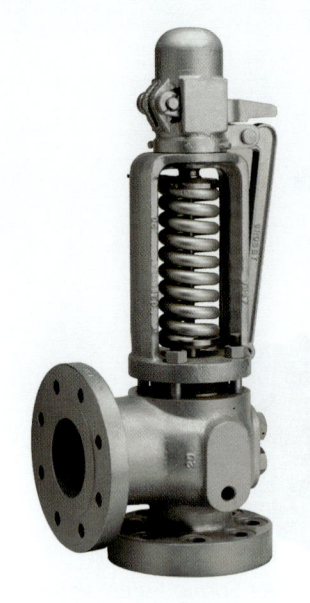

Figure SV14: Open Bonnet Direct Spring Safety Valve (Tyco Anderson Greenwood Crosby)

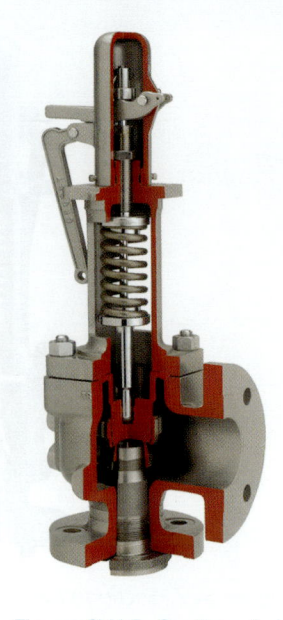

Figure SV15: Sectioned view of an Open Bonnet Direct Spring Safety Valve (Tyco Anderson Greenwood Crosby)

The body of both design types does need to be capable of withstanding the bolt loads required to attach the valve to the pressure equipment.

Balanced or Unbalanced Safety Valve specifies the way the safety valve responds to back pressure, see Figure SV16 and 17. In an unbalanced design the back pressure acts to keep the valve closed (constant back pressure) or restricts the opening during discharge (built-up back pressure). By providing a piston with an equal area to the seat attached to the stem above the disc the effects of back pressure can be neutralised. Any back pressure that exists acts both upwards on the piston and downwards on the disc and the two cancel each other out. A safety valve fitted with a bellows stem seal is usually a balanced design as the bellows outside diameter is made the same as the nozzle inside diameter.

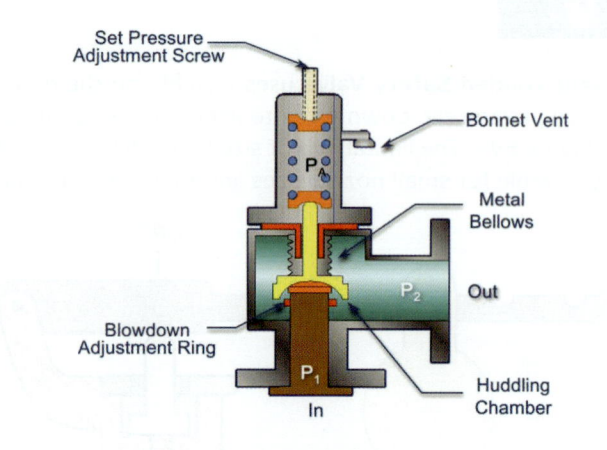

Figure SV16: Balanced Bellows Safety Valve (Tyco Anderson Greenwood Crosby)

Factory Testing of Safety Valve

Safety valves undergo two factory tests. The first is to check that the set pressure is correct. The factory set pressure may include a compensation for the test fluid relative to the service fluid and for the test temperature relative to the service temperature. It is important to use the correct definition of set pressure when assessing the valve's factory performance. The normal tolerance of actual set pressure relative to data sheet set pressure is ±3%.

Seat leakage testing is usually done in accordance with API 527. This standard specifies procedures for testing using either air, steam or water. In all cases leakage testing is done with the inlet pressure at 90% of the set pressure except where the set pressure is 50 psi or less in which case the test pressure is set pressure less 5 psi. The allowable leakage rate is a function of seat bore, set pressure and test fluid. Surprisingly, at first glance, it seems strange that the allowable leakage rate for small nozzle sizes is greater than that allowed for larger nozzles. This recognizes the difficulties manufacturers face in achieving good sealing performance on small diameter nozzles as the forces available to clamp the seat to the disc are small.

Valves that have discharged in service may experience damage to the seat or disc surfaces due to the high velocities (Mach 1) experienced during discharge. Also if the fluid vaporises on discharge or liquid is carried over in the gas stream the high speed impact of these liquid droplets can be severe. In particular if the fluid is saturated steam the disc surface can be scored to a depth of a 2 or 3 mm and the phenomenon is known as wire drawing. The Pilot Operated Safety Valve and CSPRS provide a positive clamping force between seat and disc even at pressures right up to the set pressure. These designs are therefore less likely to leak and the margin between working pressure and set pressure can be reduced, see Figure SV18. Seat and disc surface and flatness is still important with these designs, if good leak tightness is to be achieved. A further option to improve leak tightness is to fit a resilient seal in the disc. Operating temperature and the corrosive nature of the fluid will restrict the application of this option.

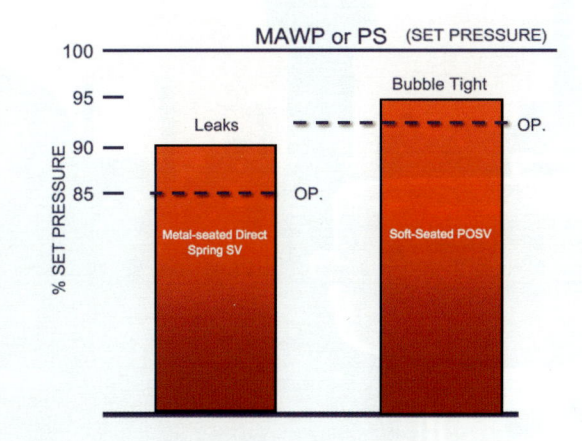

Figure SV18: Maximum Allowable Working Pressure (MAWP) or Maximum Allowable Pressure (PS)

Disc Chatter or Flutter Chatter is the abnormal rapid reciprocating motion of the disc coming into contact with the seat as the valve first lifts, discharges and then the disc returns to the seat for a short period. Flutter is when the disc 'hunts' between open/closed with no seat contact. The end result is vibration, excess wear, leakage, and loss of product. Very often the cause is an oversized safety valve.

Discharge at a pressure less than the Set Pressure can occur due to vibration or excessive fluid temperature or an unexpected reduction in the back pressure below the design back pressure. It can also result from stresses from the discharge piping acting on the safety valve body and affecting the alignment between seat and disc. It is important that the design of discharge piping takes into account thermal expansion or contraction of the pressure equipment and does not apply stresses to the body when the valve is at the working temperature and pressure.

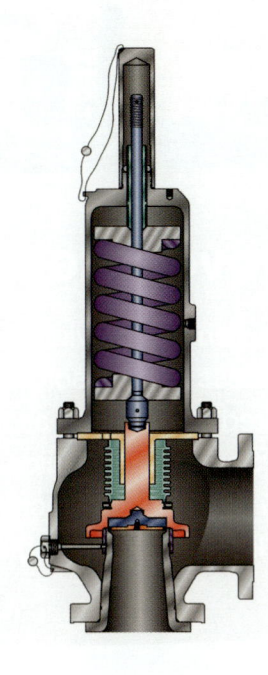

Figure SV17: Cross section of a Balanced Bellows Safety Valve (Tyco Anderson Greenwood Crosby)

Safety Valve Problems and Possible Causes

Seat leakage is the most common safety valve problem. For Direct Spring Safety Valves it is important to ensure an adequate margin between the working pressure and the set pressure. This is normally 10% i.e. working pressure is 90% of set pressure. This ensures that there is an adequate force acting downwards on the disc to clamp it to the seat and prevent gross leakage. Remember new valves with metal-to-metal seating tested on air in accordance with API 527 will have leakage rates of between 20 and 100 bubbles per minute. If the actual in service margin is reduced below the 10% then increased leakage rates will be the result. The surface finish and particularly the flatness of seats and discs will also have a major impact on leakage.

DRESSER Masoneilan

PUTTING YOU IN CONTROL.

Field Proven Reliability

Dresser Masoneilan has provided world-class automated process control solutions for over one hundred years, such as the highly-reliable 41005 Series Cage-Guided globe valve and the 78400 Series LincolnLog® high-pressure liquid letdown angle valve.

Advanced Digital Technology

The Dresser Masoneilan 12400, SVI® II ESD, SVI® II AP and FVP™ instruments and ValVue®, ValVue® ESD and OVD® software provide high performance digital process control and safety with value-added benefits, such as:

- compatibility with existing analog systems,
- on-line diagnostics,
- partial stroke testing,
- improved process yield.

© 2010 Dresser, Inc. All rights reserved.

www.dresser.com

Pressure Regulators & Control Valves

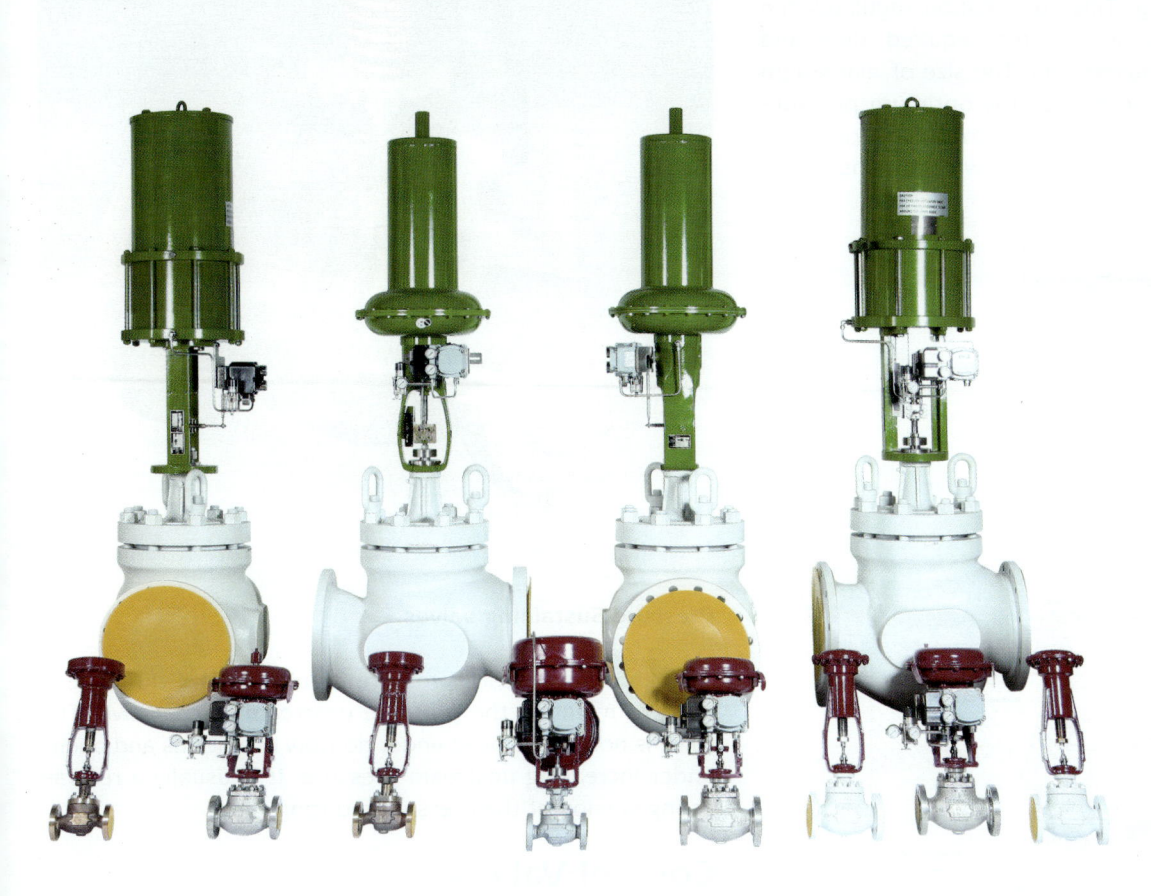

Figure PRCV1: Selection of Globe Control Valves (Dresser)

Introduction

Pressure regulators, control valves and control valve actuators are important items of equipment used in process plants to control fluids. They are used as part of the final control element to achieve process variables such as pressure, flow, level, temperature, etc. This chapter describes the basic principles behind these items, provides an introduction to the different types available and details how control valves are selected for a specific application. For full and comprehensive information it is recommended that the manufacturers of these products are contacted and they will provide the specialist information and advice required.

Pressure Regulators

General

Sometimes known as self operating regulators, these valves are used to control the pressure in a fluid system without the requirement for an external power source to activate them. When valves are used to control fluids and are equipped with power actuators they are classified as control valves. Pressure regulators date back to the late 19th century and were used for regulating the steam pressure in boiler systems. They are still used today for the same function of regulating the pressure within air, steam, industrial gases, water, oil and other fluid systems.

Pressure regulators can also be used for temperature and differential pressure control purposes. They provide an economic solution to the final control element application for relatively simple fluid control requirements. They are used for applications where deviations of the controlled variable from the adjusted set point are acceptable and the set point remains constant for a long period of time, often for the full useful life of the device. These valves normally fall into two categories, direct acting and pilot operated.

Direct Acting Pressure Regulator Valves

Pressure regulators are used for controlling the downstream pressure and are installed where it is required to reduce from one level of pressure to another. This is whilst providing the required fluid flow to satisfy a downstream demand, irrespective of fluctuations in the inlet pressure or flow demand. The valve is automatic in operation and requires no external power source. One of the most commonly used pressure regulators is the pneumatic filter regulator or airset as it is also known, which is used on a control valve and pneumatic actuator assembly. This device provides a constant air pressure to the positioner, plus it protects the positioner and actuator from being over pressurised.

The direct acting pressure regulator typically illustrated in figure PRCV2 is based on the downstream pressure within the valve acting on a diaphragm which is opposed by an adjustable compressed helical spring. This combination regulates the position of the trim to achieve the required flow and downstream pressure requirements. The size of diaphragm and spring are calculated to provide the optimum pressure control for the required range of flows. The pressure regulator is normally open under no flow conditions.

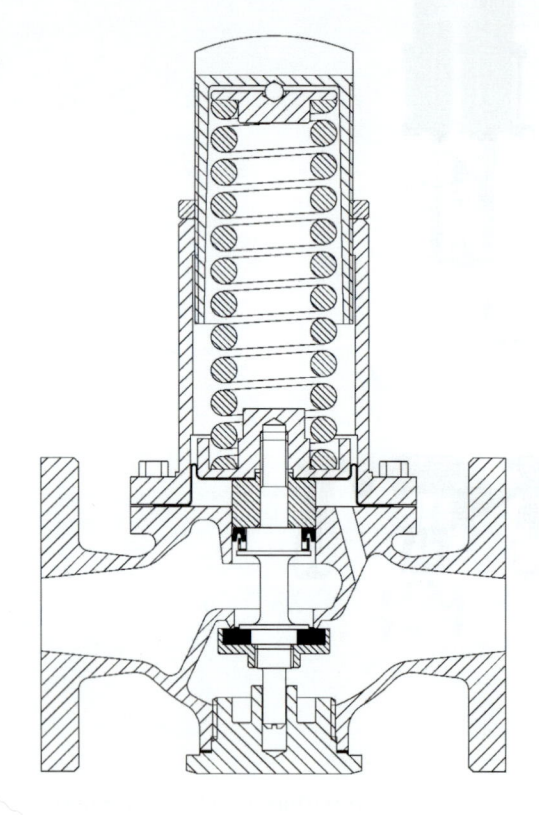

Figure PRCV2: Direct Acting Pressure Regulator (Bailey Birkett / Safety Systems UK)

Pilot Operated Pressure Regulating Valves

Pilot operated valves are available for both upstream and downstream pressure regulator applications. The main valve is either assisted or completely controlled by the operation of a pilot valve, which could be a small direct acting pressure reducing regulator. The precise method of operation is dependent on the particular design of valve, but essentially the pilot valve regulates the opening of the main valve to maintain the flow at the desired pressure value.

Pilot operated pressure regulators provide improved accuracy of pressure control compared to direct acting pressure regulator designs and are usually smaller than direct acting valves for the same duty. The pilot valve can be integral with the main valve or a separate unit suitable for remote pressure sensing. This design can be used for remote on/off control as part of a complex system and as a temperature control device by using the appropriate type of pilot valve. Due to the complexity of design pilot operated pressure regulators require regular maintenance and clean working conditions. This is often achieved by using a strainer or filter directly upstream of the unit. Figure PRCV3 illustrates a pilot operated pressure regulator of the diaphragm and piston design.

Figure PRCV3: Pilot Operated Valve (Auld Valves)

Pressure Sustaining Valves

Also known as a back pressure regulator valve this type is used to maintain the pressure upstream of the valve. The valve is normally closed under no flow conditions and opens under increasing upstream pressure. It is usually a reverse acting version of the pressure reducing valve.

Control Valves

General

A control valve is described as a power operated device which varies the fluid resistance in a process control system. It comprises of a valve connected to a power operated actuator that is capable of changing the position of a plug or disc within the valve in response to a signal from a control system. The self acting valve previously described is generally acknowledged to be the forerunner of the modern control valve. It soon became evident that these pressure regulators were limited to small valve sizes, relatively low pressures and lower temperatures due to the capability of the diaphragm. Based on this was the development of the process control valve and actuator combination. Figure PRCV4 illustrates a typical globe control valve assembly.

Control Valve Terminology

Control Valve terminology is included in the main chapter on Terminology in this manual and particular note should be made of the definitions of Control Valve, body, bonnet and valve trim. Definitions in IEC 60534-1 vary slightly in words but not in meaning from those given in our Terminology chapter because this text is suitable for all valve types whereas the IEC text uses words that only apply to control valves. For example in control valves the operating element is always an actuator so the IEC definition of bonnet states that it is the portion of the valve which closes an opening in the body and through which passes the stem connecting the closure member to the actuator.

Control Valve Standards

Control valves use some of the basic standards for design and pressure temperature ratings listed in Appendix 1.

Standard IEC 60534 Industrial Process Control Valves details those aspects of terminology, selection and testing, which are specific to control valves and these are shown below. Also listed is the ANSI/FCI standard widely used for seat leakage testing.

Part 1 – Control valve terminology
Part 2.1 – Flow capacity – Sizing equations for fluid flow under installed conditions
Part 2.3 – Flow capacity – Test procedures
Part 2.4 – Flow capacity – Inherent flow characteristics and rangeability
Part 2.5 – Flow capacity – Sizing equations for fluid flow through multistage control valves with interstage recovery
Part 3.1 – Dimensions – Face to face dimensions for flanged two way globe type straight pattern and centre to face dimensions for flanged two way globe type angle control valves
Part 3.2 – Dimensions – Face to face dimensions for rotary control valves except butterfly valves
Part 3.3 – Dimensions – End to end dimensions for butt weld two way globe type straight pattern control valves
Part 4 – Inspection and routine testing
Part 5 – Marking
Part 6.1 – Mounting details for attachment of positioners to control valves – section 1 – Positioner mounting on linear actuators
Part 6.2 – Mounting details for attachment of positioners to control valves – section 1 – Positioner mounting on rotary actuators
Part 7 – Control valve data sheet
Part 8.1 – Noise considerations – section 1 – Laboratory measurement of noise generated by aerodynamic flow through control valves
Part 8.2 – Noise considerations – section 2 – Laboratory measurement of noise generated by hydrodynamic flow through control valves
Part 8.3 – Noise considerations – section 3 – Control valve aerodynamic noise prediction method
Part 8.4 – Noise considerations – section 4 – Prediction of noise generated by hydrodynamic flow

ANSI/FCI 70-2 2006 Control Valve Leakage for determining test method and allowable control valve leakage rates for the different classes.

Figure PRCV4: Typical Globe Control Valve Assembly (Valve Solutions)

Control Valve Basics

Control valves are used as the final control element in a closed loop system and these systems can be used to control parameters including pressure, temperature, flow, level etc. The sizing and selection of these control valves is critical for the efficient performance and control of the process. These valves control the flow by absorbing pressure from the line fluid and the variation of pressure absorption and flow is regulated by varying the relative position of the control valve plug to the seat. The control valve is required to handle process conditions from maximum to minimum and can on some occasions also be required to provide 'tight shut off.'

The motive power for these control valves results from an actuator mounted directly on to the control valve assembly. Early developments of the control valve were based on a globe style single seated design fitted with a spring opposed pneumatic diaphragm actuator. The controlling signal was fed either direct or via a positioner in to the actuator to control the position of the trim. Depending upon the control requirements these systems can become very complex and this has also led to the development of pneumatic cylinder, electric, hydraulic and electro-hydraulic types of actuators. Positioning accuracy, speed of movement and reliability has improved significantly over the years. Devices are now available that can provide position feedback and the control valves can be fitted with instruments that will provide diagnostics to assist in the performance and maintenance of these critical items of process control equipment.

SCHUBERT & SALZER
CONTROL SYSTEMS

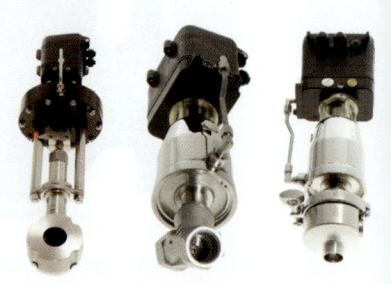

KREATIVE INTELLIGENZ

The sliding gate principle - an idea that is spreading

As long ago as 1983, Schubert & Salzer Control Systems started looking for new approaches to reduce the size of conventional valves (some of which were bulky) and, at the same time, to increase precision. We succeeded with the new principle of the sliding gate valve. The rudiments of this principle can be seen in Leonardo da Vinci's drawings. Today, customers in the chemical industry, steelworks, food and pharmaceuticals companies can use the advantages of the sliding gate valve because we allowed ourselves the freedom to place the priority on a new development.

Digital positioner - made by Schubert & Salzer

The new products developed by Schubert & Salzer also include a digital positioner that not only offers high levels of operational reliability but also the ability to adapt the characteristic curves freely.

Sterile Control Valves

In Line with our philosophy of continuous development we designed and patented a rolling diaphragm system which is fitted to all our Aseptic control valves

---

Schubert & Salzer UK Limited Tel: +44 (0) 1952 462021 **Fax:** +44 (0) 1952 463275
email: info@schubert-salzer.co.uk   www.schubert-salzer.co.uk

Innovative measurement & control solutions for flowing media

Control Valve Types

Control valves are generally divided in to two generic design types, the first being linear rising stem designs and the second rotary stem designs. The globe style control valve illustrated in figure PRCV5 is the basis of the most common design available and comprises of a single seated valve fitted with a pneumatic spring opposed diaphragm actuator.

Figure PRCV5: Automatic process Globe Control Valve, spring opposed, diaphragm actuated (Koso Kent Introl)

The single seated globe control valve is manufactured in a wide range of sizes, valve ratings and material combinations. These valves are provided with either a bolted bonnet, which would be standard for the majority of applications, normalising bonnet for higher temperatures, extension bonnet for lower temperatures, cryogenic bonnet for extremely low temperatures or bellows seal bonnet for critical applications (see Figure PRCV6). They can be designed with a wide range of stem seal designs including PTFE chevron, graphite for higher temperatures and low emission for critical applications.

The trim designs that are manufactured include the low flow type which is normally seat guided and the contoured trim type which uses a heavy top guide. Cage guided trims are used for the larger or more severe applications and enable the plug to be balanced to reduce both the dynamic and actuator forces. The cage trim designs available include the ported cage for standard and high capacity duties, single stage multi hole anti cavitation for liquid applications and the single stage multi hole low noise design for gas/vapour fluids. These cage trims can be extended to multiple cage designs for both liquid and gas/vapour applications. For more severe fluid control applications multi stage multi turn disc stack trims are manufactured, which are capable of handling very high pressure drops. These trim designs are also referred to as 'velocity control trim designs'. Figure PRCV7 illustrates an example of a multi stage stack trim design used for handling high pressure drop liquid applications without the onset of cavitation and its associated problems of premature damage and noise. Figure PRCV8 shows a sectioned Globe Control Valve, with 9 stage/13 turn trim for high pressure drop liquid applications.

Figure PRCV6: Bellows Seal Globe Control Valve (Weir)

Figure PRCV7: Multi Stage Multi Turn Disc Stack Trim design (Valve Solutions)

BVAA **VALVE & ACTUATOR USERS' MANUAL** - 6th Edition

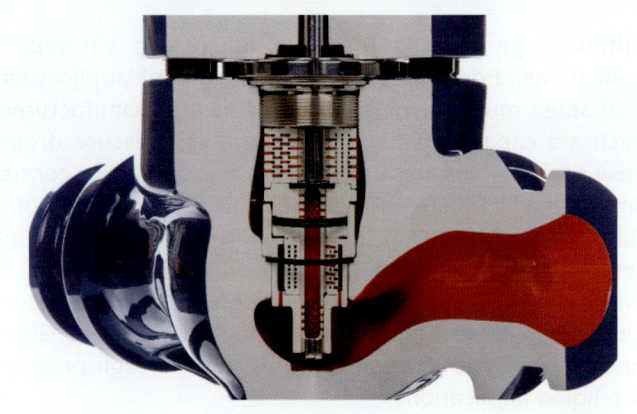

Figure PRCV8: Globe Control Valve, with hubbed ends, 9 stage/13 turn trim for high pressure drop liquid applications (Koso Kent Introl)

Based on the single seated globe design, control valves can be manufactured with the above combinations, but fitted in to an angle pattern body configuration as illustrated in figures PRCV9, 10 and 11.

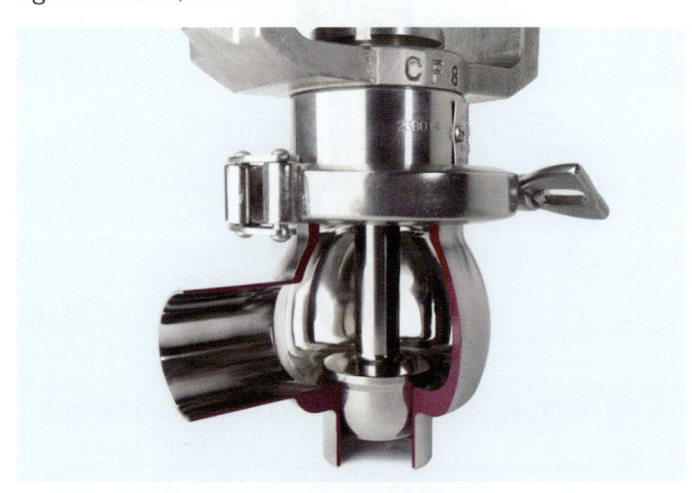

Figure PRCV9: An Angle Style Control valve - this particular model is designed for sterile applications (Spirax Sarco).

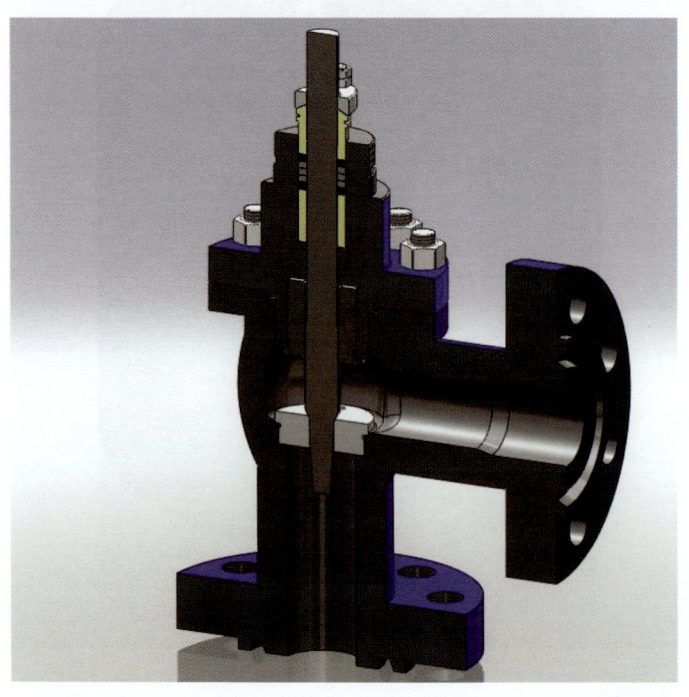

Figure PRCV10: A schematic of an Angle Valve assembly (Valve Solutions)

Figure PRCV11: An Angle Valve assembly with low-noise anti-cavitation features for high pressure drop applications (Dresser)

Double seated globe control valves are used for low to medium duty applications which require a high capacity. They are normally manufactured in sizes up to 300mm nominal diameter and for ratings up to and including ANSI class 600. These valves are normally fitted with a bolted bonnet with the same variations of bonnet designs and stem seal designs as detailed for single seat designs. This design of trim is inherently balanced but limited to applications where the leakage is a maximum of FCI 70-2 class III. The valves can be manufactured with anti cavitation or low noise trim designs but these are limited compared to single seated designs. Figure PRCV12 illustrates a typical double seated control valve design.

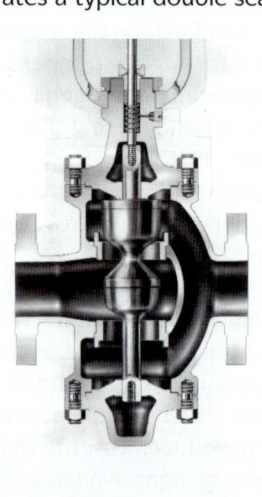

Figure PRCV12: Drawing of double seated Globe Control Valve with top and bottom guiding (Flowstream/Hunt & Mitton). Sectioned view of double seat arrangement (Koso Kent Introl)

For applications which require a control valve to provide a mixing mechanism of two individual streams in to a single fluid stream or alternatively to split a single fluid stream in to two individual streams a three way globe is used. These valves are normally manufactured in sizes up to 300mm nominal diameter and for ratings up to and including ANSI class 600. Figure PRCV13 illustrates a typical 3 way mixing control valve and a typical 3 way diverting control valve.

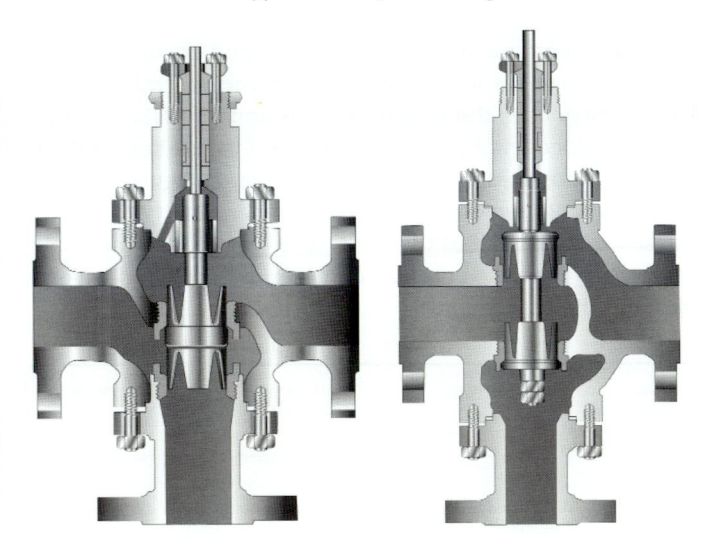

Figure PRCV13: Left – a 3 Way Mixing Control Valve and right, 3 Way Diverting Control Valve (Dresser Masonielan)

Rotary Control Valve Types

Rotary design control valves are manufactured and supplied normally for high capacity low duty applications. They tend to have less components, be more compact and lighter in weight compared to globe control valve designs however they are limited with respect to the fitting of low noise and anti-cavitation trim designs.

Butterfly Control Valves are sub-categorised as either concentric or offset designs, which relate to the mounting of the disc on to the shaft. The principles of each design type and the available body styles are described in the chapter on Butterfly valves. The discs are also supplied in swing through and seated types, the seated types having both metal to metal and soft seated options. The high performance (double offset) and triple offset designs operate and control over 90 degrees rotation whilst the concentric designs are usually restricted to controlling over 60 degrees. Butterfly valves are available with a configured trim design to improve the performance and anti cavitation / low noise characteristics. Figure PRCV14 illustrates a Butterfly Control Valve Assembly.

Ball Control Valves are sub categorised as either floating or trunnion mounted designs, which refers to the support method for the ball. The principles of each design and the available body styles are described in the chapter on Ball valves. The standard ball design provides a quick open characteristic and designs are available to provide an equal percentage characteristic using a notched ball, referred to as a V-notch ball. These valves also inherently have a high rangeability and the ball remains in contact with the seal during rotation to produce a shearing effect. The movement of these Ball valves are all 90 degrees for full rotation. A V-Notch Ball Control Valve is illustrated in Figure PRCV15. Similar to the butterfly control valve there are designs available that provide a degree of noise attenuation and cavitation resistance.

Figure PRCV15: V-Notch Ball Control Valve Assembly (Fisher / Emerson Process Management)

The eccentric rotating plug control valve has a partial spherical rotating plug which is non contacting apart from at the point of final closure, this is achieved by the plug being eccentric to the centre of the shaft. The arms connecting the plug face to the shaft are able to deform allowing for tight shut off. The design minimises plug and seat wear and friction between these two surfaces. With rotary motion it is possible to maintain a high integrity shaft seal. Figure PRCV16 illustrates an eccentric rotating plug control valve design.

Figure PRCV14: Butterfly Control Valve Assembly (Leeds Valve)

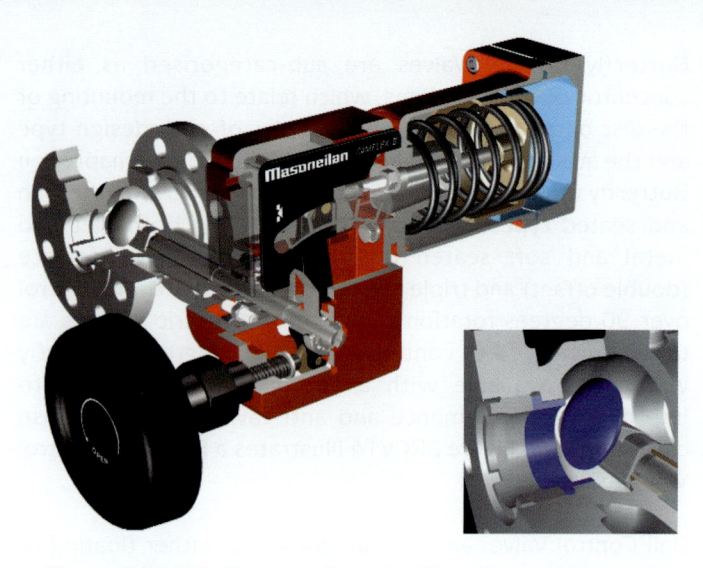

Figure PRCV16: Eccentric Rotating Plug Control Valve with [inset] close up view of the plug arrangement (Dresser)

Valve Characteristics

The characteristic of a control valve is the relationship between the capacity (C_v) of the particular valve / trim combination and the linear travel for a rising stem control valve or angular movement for a rotary control valve. There are standard characteristics available that meet the majority of applications encountered and custom engineered characteristics can be provided on some control valve designs. For certain designs the trim characteristic is inherent within the design. The standard characteristics most commonly used in the valve industry are Equal Percentage, Linear, Modified Parabolic and Quick Opening. These standard characteristics are illustrated in figure PRCV17 and defined below.

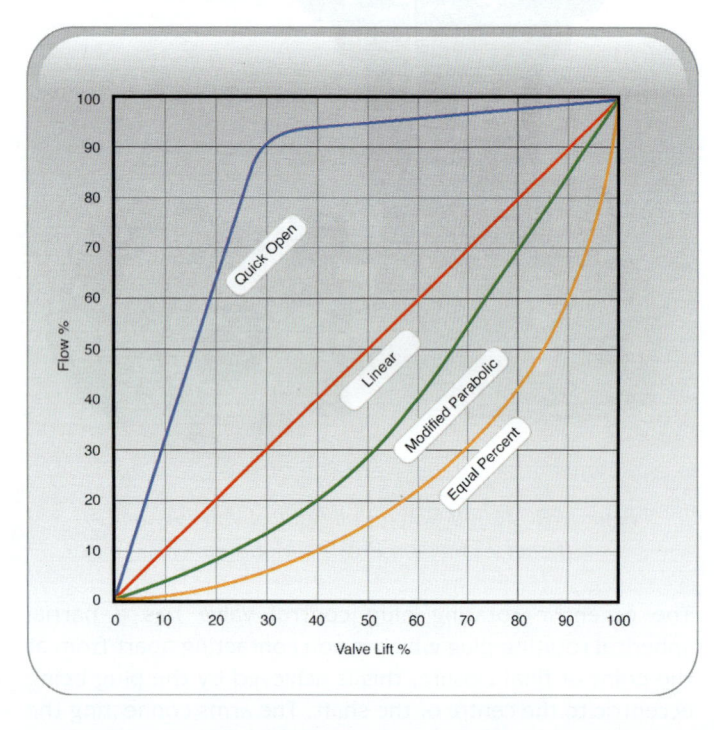

Figure PRCV17: Standard Valve Characteristics

- Linear – Equal increments of travel provide equal increments of the C_v.
- Equal Percentage – Equal increments of travel provide an equal percentage increase in the C_v.
- Modified Parabolic – This is a characteristic which falls between the above linear and equal percentage curves.
- Quick Opening – This is used for on/off applications and has no control characteristic. The maximum C_v is achieved with minimum travel of the control valve.

Seat Leakage

Control valves are not normally intended to be used for isolation purposes in conjunction with the control function. The industry uses the International Standard ANSI/FCI 70-2 2006 for specifying and testing control valves seat leakage to the various classes within the standard. Figure PRCV18 details a table which provides a summary of the different classes within ANSI/FCI 70-2, the valve type normally associated with each class, the maximum specified leakage rate and the test medium.

Leakage Class	Typical Valve Type	Maximum Leak Rate	Test Medium
I	Any	No leakage requirement	Not applicable
II	Double Seated Swing thro' Butterfly	0.5% of rated capacity	Water or air
III	Cage trim balanced metal seal	0.1% of rated capacity	Water or air
IV	Single seat unbalanced Cage trim balanced soft seal	0.01% of rated capacity	Water or air
V	Single seated and cage trim with balanced soft seal and special lapping	5 x 10^{-12} m^3/s of water per mm of seat diameter per bar differential or 11.1 x 10^{-6} standard m^3 per hour of air per mm of seat diameter	Water or air
VI	Single seat unbalanced with soft seat	Leakage as per para 5.4.4 of ANSI/FCI 70.2	Air

Figure PRCV18 : ANSI /FCI 70-2 Maximum Leakage Classifications

Control Valve Sizing and Selection

Control valves are required to handle a wide range of process conditions with a wide range of fluids being controlled. Control valve sizing and selection is undertaken using standard procedures and based on the fluid type. The two main fluid types are liquids (incompressible fluids) and gas/vapour (compressible fluids). Multi phase applications also need to be considered and these are a combination of the incompressible and compressible methods. For these multi phase applications it is recommended that you contact the respective control valve manufacturer.

Figure PRCV19 provides a flow chart for sizing and selecting a control valve on an incompressible (liquid) fluid application.

The following provides a review of the parameters used during the sizing and selection process for incompressible fluid applications.

Flashing is a single stage process that only occurs on liquid applications. The liquid pressure reduces as it flows through the control valve trim to the minimum static pressure within the control valve, which is known as the 'vena contracta' pressure. If the reduction in pressure is to below the liquid vapour pressure, vapour bubbles are formed (liquid boils) and the fluid becomes two phase. If the pressure then remains below the liquid vapour pressure as it exits the control valve this is known as 'Flashing'. As the fluid exits the control valve it is two phase, being a mixture of liquid and vapour. It is not possible to prevent flashing occurring for a specific application however it is possible to minimise damage by using a low pressure recovery trim design and using erosion resistant materials for the trim.

Cavitation is a two stage process that again only occurs on liquid applications. The first stage is the same as previously detailed for flashing whereby the static pressure within the control valve trim falls below the liquid vapour pressure and this results in the formation of the vapour bubbles. If the pressure within the control valve then rises back above the liquid vapour pressure the vapour bubbles originally formed implode and this is known as 'Caviation'. The implosions create shock waves which travel through the liquid and impact on the valve components causing extremely high local stresses and material removal. The latter is a characteristic of cavitation damage. In conjunction with the damage relatively high noise levels can be generated. Cavitation in control valves is predictable and can be eliminated by using use low pressure recovery and multi stage trim designs.

LIQUID SIZING AND SELECTION CHART

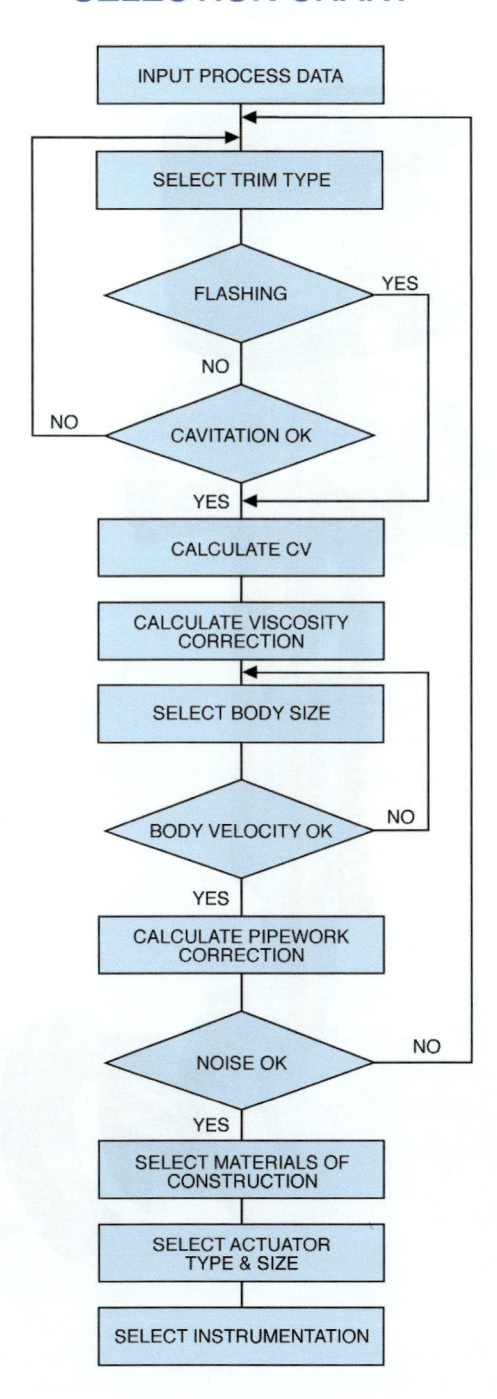

Figure PRCV19: Incompressible Fluid Sizing & Selection Flowchart

GAS / VAPOUR SIZING AND SELECTION CHART

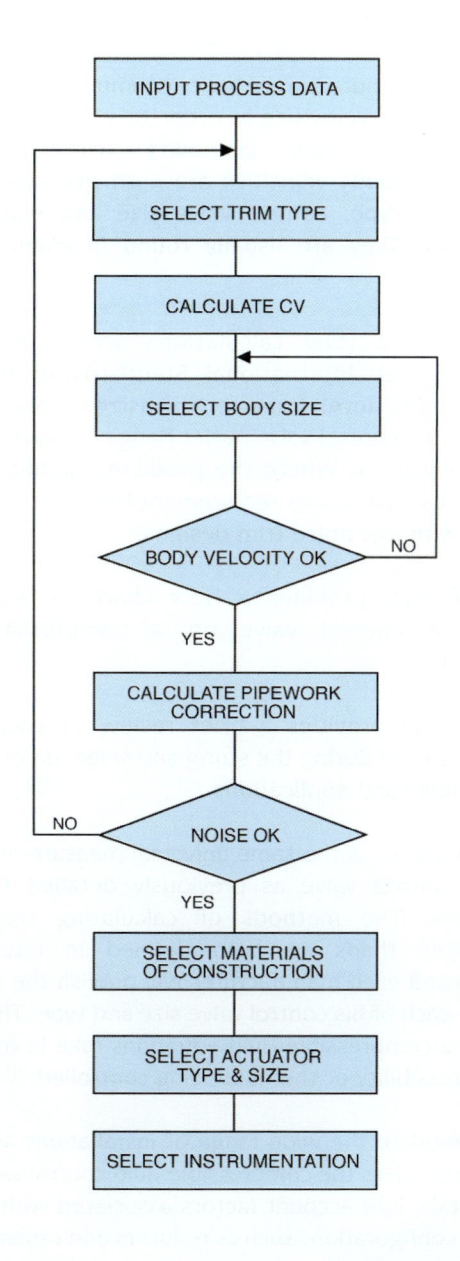

Figure PRCV20: Compressible Fluid Sizing & Selection Flowchart

The calculated C_v is the universal measure of capacity within a control valve and is based on a combination of the flow and pressure drop. The definition of C_v is the flow through the control valve assembly in US Gallons per minute with a pressure drop of $1 lbf/in^2$ using water with a specific gravity of 1. The methods of calculating the C_v are defined in International standards and each manufacturer will publish the design C_v values for each control valve size and type.

The liquid fluid sizing methods are based on the fluid being turbulent and where control valves are used for laminar flow applications they require a Viscous Flow factor to be applied.

To accommodate the wide range of installations associated with control valves the control valve sizing methods take into account factors associated with adjacent pipework configurations such as reducers and expanders. It is recognised good practise for control valves never to install a valve which is less than half the adjacent line size.

A number of manufacturers have recommended maximum liquid body inlet velocities in order to prevent premature erosion damage and instability. These maximum recommended body velocities are normally based on the control valve type, control valve size and materials of construction. They are usually found in manufacturers' literature.

Sound pressure level calculations are undertaken in accordance with International Standards to provide a predicted noise level 1 metre downstream and 1 metre away from the control valve outlet flange for each specified process condition. Where the predicted sound pressure level exceeds a particular requirement the control valve can be fitted with low noise trim designs.

Figure PRCV20 provides a flow chart for sizing and selecting a control valve on a compressible fluid application.

The following provides a brief review of each of the parameters used during the sizing and selection process for compressible fluid applications.

The calculated C_v is the same universal measure of capacity within a control valve as previously detailed for liquid applications. The methods of calculating the C_v for compressible fluids are also defined in International standards and each manufacturer will publish the design C_v values for each of his control valve size and type. The factors used in the compressible fluid equations take in to account the compressibility of the fluid being controlled.

To accommodate the wide range of installations associated with control valves the compressible fluid control valve sizing methods take into account factors associated with adjacent pipework configurations such as reducers and expanders. It is again recommended good practice for control valves to never install a valve which is less than half the adjacent line size.

A number of manufacturers have recommended maximum gas / vapour body inlet and outlet velocities in order to prevent instability and regenerated noise at the control valve outlet. These maximum recommended body velocities are based on the control valve type, control valve size and required maximum sound pressure level to be achieved. They are usually found in manufacturers' literature.

Sound pressure level calculations for compressible fluids are also undertaken in accordance with International Standards to provide a predicted noise level 1 metre downstream and 1 metre away from the downstream pipewall for each specified process condition. Where the predicted sound pressure level exceeds a particular requirement the control valve can be fitted with low noise trim designs.

Other Control Valve Designs

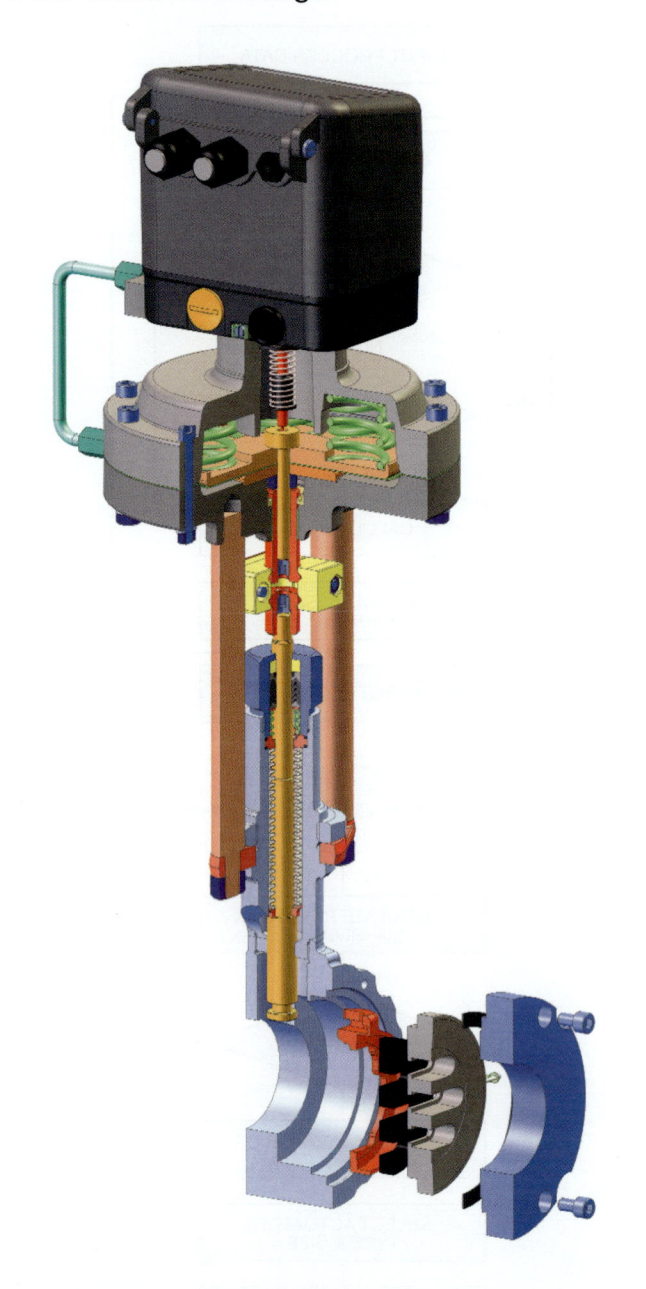

Figure PRCV21: A Sliding Gate Control Valve with pneumatic actuator and digital positioner. Slots in the movable disc pass over similar sized slots in the fixed disc, varying the orifice apertures to control flow (Schubert & Salzer)

Figure PRCV22: Innovative combination of a rotary motion plug and stem with a globe valve body (Metso Neles)

When faced with **hostile environments - high pressure, high temperature, deep water, high H$_2$S or CO$_2$ you need to know you are in safe hands.**

BEL Valves Ltd are leaders in the design and manufacture of critical subsea and surface valves, actuators and controls for oil and gas production. Our enabling technologies ensure the most complex fields can be developed reliably and safely.

We always operate:
- at extremes of **pressure**
- at extremes of **temperature**
- at extreme **depths**
- in hostile **environments**

BEL VALVES

Excellence in Valve Engineering

For more information on our **high integrity** range of **ball and gate valves** and **actuators for surface** or **subsea** applications visit **www.belvalves.com** or call **+44 (0) 191 265 9091**

Recent Developments

Figure RD1: Some companies can provide a 'one stop shop' service for HIPPS (BEL Valves)

High Integrity Pipeline Protection System (HIPPS)

There are regions across the globe containing hydrocarbons at great depth, high pressure and high temperature. As reserves of existing oil and gas deposits begin to deplete, the drive to develop these reserves has been growing and installers are having to deal with High Temperature High Pressure (HTHP) scenarios to access these reserves. HTHP scenarios can be costly in terms of upgrades required to equipment – if that route is taken. They can also have a catastrophic impact on infrastructure, human life and the environment, if the correct safety measures are either not put in place or not managed correctly. Because of the increase in arduous applications of this type, and the heightened focus on safety in the oil and gas industry, the spotlight has been thrown firmly on HIPPS. HIPPS are specified when pressure relief devices are impractical and there are environmental constraints.

A HIPPS is essentially an independently instrumented protective device to protect downstream applications, offering the ultimate in system protection. For example if you have a high pressure wellhead, and a lower pressure export link, you need to ensure there is a mechanism in place to protect equipment, personnel and the environment in the event of over pressurisation. There are other options, such as pressure relief valves, bursting disks, or a combination of the two, which work well in particular situations. However these options do have an environmental impact, as they have to relieve the overpressure somewhere. When there is an overpressure problem, the safest thing to do is to shut down the application altogether, and that is what the HIPPS does.

Ultimately the decision to install a HIPPS will depend on the assessment and documentation of the HIPPS' ability to mitigate risk versus other safety devices, which can be a complex and daunting task.

A HIPPS is a highly complex system and is illustrative of how far manufacturers have progressed in response to customer needs for this type of system. It is necessary to combine expertise in valve design, actuators and electronics plus that all important established track record.

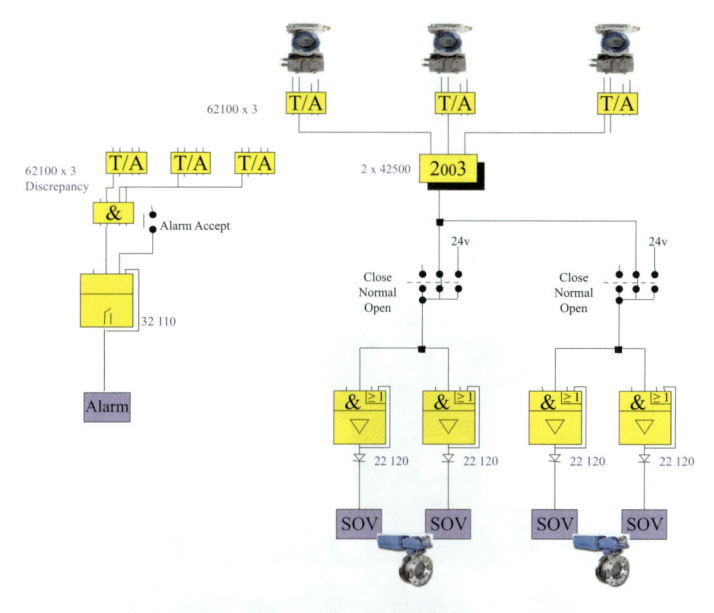

Figure RD2: Outline HIPPS configuration (BEL Valves)

Developing a HIPPS solution requires extensive communication between end user, engineering consultant and supplier so that the solution is designed to meet the stringent and specific requirements. It must also take into account the problems of the subsea installation.

A HIPPS system normally requires a Safety Integrity Level (SIL) of 3, and an experienced HIPPS supplier can provide a HIPPS which is warranted and certified for the specific application including all the unique calculations of risk and probability. The supplier should also be working closely with a third party safety expert in producing the SIL and other required certifications such as IEC 61508.

The valves within a HIPPS solution are small in number and highly specialised. When designing for high pressure, valves are necessarily large and heavy but trying to reduce weight and space is often a significant customer requirement. Material choice therefore is key when dealing with high pressures. Steels need to have high strength that can be heat treated to maintain the strength throughout the thickness of the metal and must also have excellent corrosion resistance. The best solution to this is to make the body and bonnet from a high strength steel and weld clad all of the internal surfaces with a corrosion resistant material.

Designing equipment for high temperature creates equally challenging but different problems, the most significant of these is sealing. Through conduit gate and ball valves typically have some form of non metallic seal on the seat ring and stem. These seals are made from thermoplastic compounds that can operate at temperatures in excess of 250°C but as the pressure increases the working temperature decreases creating the first hurdle. Secondly the seal materials used for high temperatures are generally rigid at room temperature making it difficult to achieve a gas tight seal at low temperature. Suppliers must therefore run intensive long term programmes of seal testing to find a sealing configuration that is suitable for both high and low temperatures.

The valves need to be fast closing, in order to respond to a sudden surge of overpressure. This ability to close rapidly is part of the SIL3 assessment. They may also need to be clearway design to allow for pipeline cleaning or inspection by pigging. Finally the HIPPS will need to be monitored once installed to ensure continued high performance. Remote monitoring enables this function to also be provided by the supplier of the system.

Safety Integrity Level (SIL)

Figure RD3: Vee-ball Valve with SIL3 capability (Fisher / Emerson Process Management)

The previous section has already introduced the term SIL. Safety Integrity Levels are targets applied to the reliability and performance of the safety systems used to protect hazardous activities such as hydrocarbon refining or production. There are 4 SIL levels. The higher the perceived associated risk, the higher the performance required of the safety system and therefore the higher the SIL rating number.

The International Electro-technical Commission (IEC) introduced the following industry standards to assist operators with quantifying the safety performance requirements for hazardous operations: -
- IEC 61508 Functional Safety of Electrical/Electronic/Programmable Electronic Safety-Related Systems
- IEC 61511 Functional Safety - Safety Instrumented Systems for the Process Industry Sector

IEC 61508 is a generic functional safety standard, providing the framework and core requirements for sector specific standards. IEC61508 does not specifically cover non electrical pressure relief valves, safety valves and check valves but these do have a Risk Reduction Factor and do contribute to reducing risk. IEC 61508 is the standard used by manufacturers wishing to have their equipment assessed for its capability to work in a Safety Instrumented Function (SIF) with a defined SIL level.

Figure RD4: An onboard Safety Function Control Module (SFCM) for electric actuators, approved for use in SIL certified systems (Rotork)

To date three sector specific standards have been released using the IEC 61508 framework, one of which is IEC 61511 for the process industry sector. Manufacturing processes, such as refineries, petrochemical, chemical, pharmaceutical, pulp and paper, and power are included in the process industry sector but nuclear power facilities or nuclear reactors are not. This standard has been widely adopted in the hydrocarbon and oil and gas industries. IEC61511 is concerned with the specification, design, installation, operation and maintenance of a Safety Instrumented System (SIS).

An SIS is a Safety Related System (SRS) performed by instrumentation. The SIS performs specified functions such that it can be confidently entrusted to achieve or maintain a safe state of the process when unacceptable or dangerous process conditions are detected. SISs are separate and independent from regular control systems but are composed of similar elements, including sensors, logic solvers, final elements (e.g. valve actuators) and support systems.

The SIF is implemented as part of an overall risk reduction strategy which is intended to reduce the likelihood of identified hazardous events involving a catastrophic release. The safe state is a state of the process operation where the hazardous event cannot occur. The safe state should be achieved within one-half of the process safety time. Most SIF are focused on preventing catastrophic incidents. An SIS may implement one or more SIFs.

Determination of SIL Rating
Once the scope of an activity is determined, the operator can identify the possible hazard(s) and then assess their potential severity. The risk associated with a hazard is identified by assessing the likely frequency of occurrence and the potential consequences if the hazard is realized. The operator must then assign a number for the severity of consequence and for the estimated frequency of occurrence. These numbers are then fed into a matrix to

allow the operator to assign the required SIL rating to protect against the hazard. An example of such a matrix is shown below in Figure RD5.

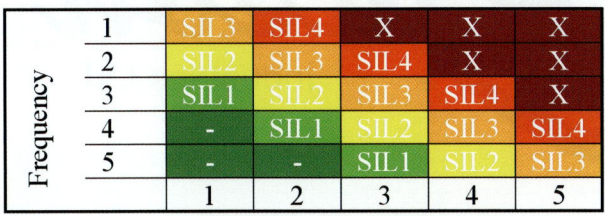

	1	2	3	4	5
1	SIL3	SIL4	X	X	X
2	SIL2	SIL3	SIL4	X	X
3	SIL1	SIL2	SIL3	SIL4	X
4	-	SIL1	SIL2	SIL3	SIL4
5	-	-	SIL1	SIL2	SIL3

Frequency (rows) vs Severity of Consequence (columns)

Figure RD5: Frequency Consequence Matrix

SILs are part of a larger scheme called Functional Safety that deals with techniques, technologies, standards and procedures that help operators protect against hazards. Functional Safety adopts a life cycle approach to industries that deal with hazardous processes that includes plans from concept through to final decommissioning of plants. This process is iterative and any phase is effected by the requirements of the previous phase so, subsequent phases must be revisited to assess the impact of a change to an earlier phase.

The Probability of Failure on Demand (PFD) is the measure used to define the level of protection offered by the SIF. EIC 61508 defines the maximum allowable PFDavg (the average probability between 0 to 1, that the safety function will fail to operate) for the SIF. The allowable level is dependant upon whether the SIF is deemed to be low demand or high demand. Low demand SIFs are defined as having an expected safety demand interval of greater than one year, and a proof test interval for the equipment that is at least twice that of the expected safety demand interval. The vast majority of actuated valves in SIF's fall into this low demand type. High Demand safety control systems are defined as those that are operated more frequently than once per year.

For the manufacturer of equipment there are two aspects to the process of attaining a SIL capability certificate. The first is assessing the design and failure rates of the equipment. This can be accomplished through either FMEDA (Failure Modes, Effect and Diagnostic Analysis) or "Proven in Use".

The second aspect is the auditing the vendor's manufacturing and quality processes. This audit proves that the vendor is capable of manufacturing the product to the designed performance standard. These assessments must be audited by an approved accreditation body such as exida or TüV.

FMEDA is a technique that assesses the performance of a device by evaluating the effects of the different failure modes of all components in the design. Every component is assessed for the type of failure (dangerous or safe) and the likelihood of failure (failure rate). All of this data is then collated to produce overall dangerous and safe failure rates that can be used in safety calculations. FMEDA studies can be conducted either by the vendor or a third-party body but, in both circumstances, must be audited by an accredited body to prove that best practices have been used.

It may not be possible, practical or cost effective to conduct an FMEDA on a product, particularly if it is of an old or complex design. In these cases, products may be certified by using "Proven In Use". "Proven In Use" as defined in the IEC 61508 standard is a documented assessment that has shown that there is appropriate evidence, based on previous use history of the component, that it is suitable for use in a safety system. This documented evidence must include the following:

- The manufacturer's quality and management systems.
- The volume of the operating experience with statistical evidence to show that the claimed failure rate is sufficiently low.

Once the studies have been completed, the operator is presented with the failure rate data. This data falls into two fundamental categories: dangerous failure rate (λD) and safe failure rate (λS). The dangerous failure rate (λD) data relates to failures that will result in the SIF being unable to perform the required safety function upon demand. The safe failure rate (λS) data relates to those failure modes that will put the safety function in its safe state (e.g., shutdown). SIL is only concerned with the dangerous failure data but the safe failure data is important as this provides the operator with a measure of how likely the SIS is to spuriously trip.

Figure RD6: A standalone Smart Valve Monitoring Unit for pneumatic actuator partial stroke testing and valve performance monitoring, which assists with SIL compliance (Rotork)

Associated with each SIL is the maximum level that the PFDavg is permitted to reach. It is important to maintain the PFDavg below this maximum level using Proof Tests and Diagnostic Tests.

A proof test is a manual test performed during shutdown that tests the entire functionality of the SIF from sensing to actuation. It must be suitably configured to test all aspects of the safety function to prove that the SIF is "as good as new" The proof test necessitates a process shutdown which is costly event. A way of extending the proof test interval is to carry out a diagnostic test. This is an automatic test performed online that does not necessitate process shutdown. This type of test must be performed at least ten times more frequently than the expected SIF demand rate. A diagnostic test will test only a percentage of the total possible failure modes of the SIF; this percentage is called the Diagnostic Coverage (DC). The higher the DC, the greater the benefit gained from the test. For valve actuators this type of test is called a partial stroke test.

Colson
Industries Ltd

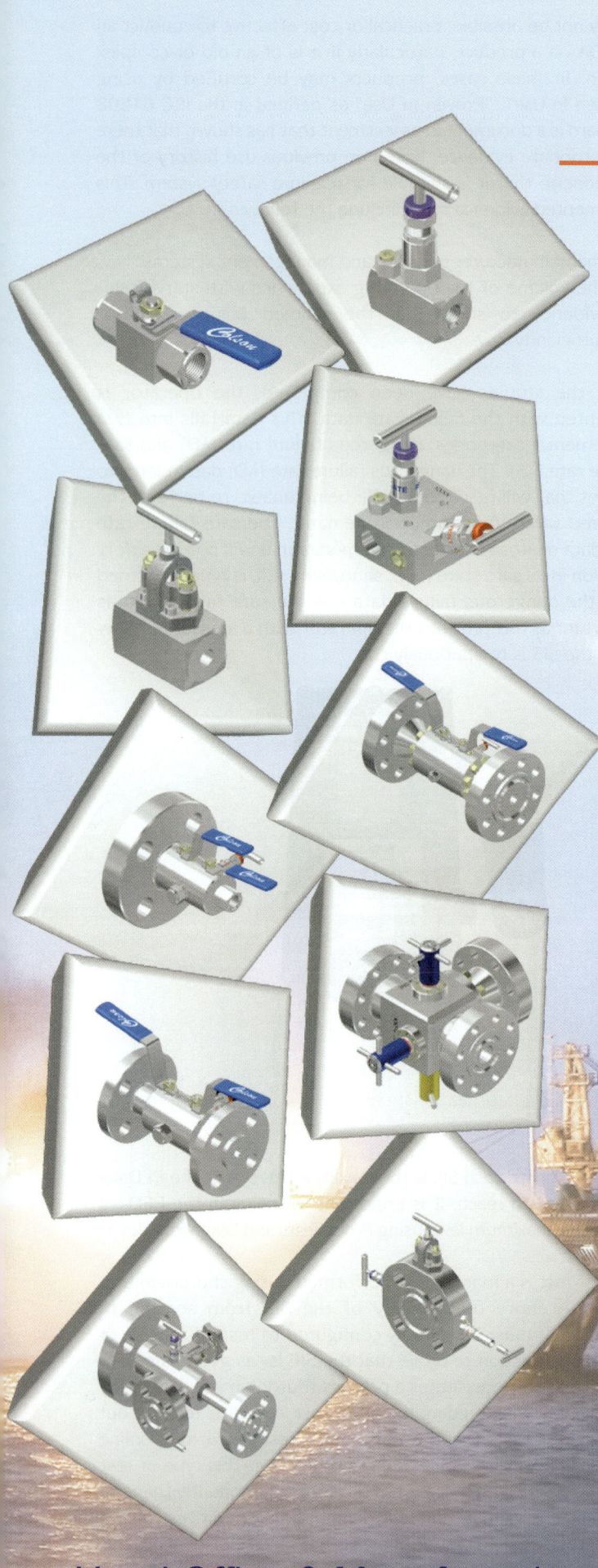

Your specialist for Instrumentation, Process and Bespoke Valves. Engineered to your specifications in 316ss and Super Alloys.

Valve Product Range
Ball
Check
Needle
Plug
Manifold
Double Block & Bleed
Mono Flange
Modular
Integral
Sample Probe
Chemical Injection
Wafer

Head Office & Manufacturing
Park Works, Park Road, Elland, West Yorkshire, HX5 9HN, England
Tel +44 (0) 845 094 3780   Fax +44 (0) 845 094 379
sales@colson.co.uk   www.colson.co.uk

Double Block and Bleed Valves

A double block and bleed valve (DBB) is a manifold that combines one or more isolation valves, usually ball valves, and one or more bleed/vent valves, usually a needle style globe valve, into one assembly for interface with other components e.g. pressure measurement transmitters, pressure gauges, switches.

The purpose of the DBB valve is to provide double isolation so that the process fluid from upstream of the DBB valve does not reach components of the system that are downstream. The bleed or vent ensures that fluid is drained from the cavity between the two isolating valves or at least the pressure in the cavity is reduced to the pressure of the vent system. The vent is also intended to drain the fluid from the downstream component's side of the secondary isolating valve before that valve is closed allowing maintenance, calibration, repair or replacement of the component.

DBB ball valves are also used on chemical injection or sampling duty. In this application a quill or sampling probe is mounted on the process side of the valve. This extends into the process flow stream and facilitates the injection of chemicals into or the taking of a sample from the flow stream. The DBB valve allows the downstream activity to be isolated from the process fluid and process pressure and the cavity between the two isolating valves drained of either the chemical or used to flush through the process fluid to ensure a good sample.

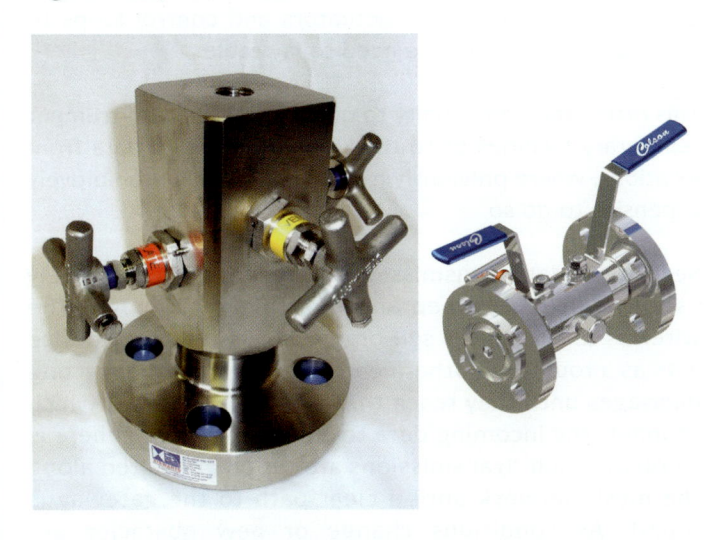

Figure RD7: Double Block and Bleed Valve designs can vary (left, Stewarts, and right, Colson)

These manifolds/valves can be manufactured with a combination of Isolation and Vent valves thus making them suitable for numerous applications and clients' requirements, either for Process or Instrumentation use. Valve configurations can use a combination of ball, plug, butterfly and needle valve styles as either primary isolation, secondary isolation or Vent Valves.

DBB valves replace the previous traditional technique employed by pipeline engineers to create a double block and bleed configuration in the pipeline, usually by fabricating three valves using flanges, 'Tee' pieces and associated bolting. This was time consuming and also produced an assembly that was not only bulky, but also heavy. The manifold style of DBB valves reduces weight, the space required to accommodate the valve(s) and has the added advantage of reducing the potential leak paths for environmental leakage of the fluids.

Materials of Construction
Typically DBB valve bodies are integrally forged or made from bar stock. This ensures excellent pressure envelope integrity although there can be issues with through thickness material properties in the centre of the section of the large bar stock diameters unless the heat treatment is carefully controlled.

Applications
DBB Valves are used by both piping (process) and instrumentation disciplines: -

- Instrumentation - These valves have a smaller bore as flow volume is not the primary consideration and they are normally supplied with a screwed bonnet inside screw design for the vent valve.

- Process - These valves have a full or reduced bore as flow is considered the primary consideration. A full bore and a clearway bore are identical in the DBB valve type so full bore valves would be specified when pipelines are to be pigged. Vent valves are normally bolted bonnet outside screw and yoke design.

Styles of Construction
DBB valves can be manufactured in a variety of styles and configurations, each providing a unique solution for the system requirements.

- **Barstock Style DBB Valves** - These are used within small bore pipe systems normally ≤DN25 and are popular for control circuits for hydraulics. Standard pressure ratings are 6,000 psi, (414 bar), 10,000 psi (689 bar) and 15,000 psi (1034 bar) at ambient temperature.

- **Mono Flange Style DBB Valves** - These are normally used for instrumentation requirements, have a small bore and are the smallest and lowest weight design. The design incorporates three needle style globe valves mounted at 90° intervals round the flange OD. The flange is drilled with a series of small bore holes to allow the process fluid to flow from the inlet to the outlet or the vent via the isolating valves. For critical services the primary isolation valve is bolted bonnet outside screw design. The small bore holes of the mono flange are ideal for clean service and where there is little or no flow. The mono flange mates with process pipe flange at the inlet and has a threaded outlet connection.

- **One Piece Block Style DBB** - These are also known as Cartridge style. This type of DBB Valves is where the valve internals are fitted from either or both ends to provide no body joint leak paths, except for the flange termination connection. Overall lengths can be either custom or to International Standards (as specified in ASME B16.10 or API 6D for valves ≥DN50).

Figure RD8: Mono flange style DBB (Colson)

Figure RD9: One piece block style DBB (Colson)

- **Two/Three Piece Style DBB Valves** - These feature the traditional style of two or three piece ball valve body and are available with a variety of body inlet and outlet end connections including ANSI B16.5 flanges, hub connections, welded ends and screwed ends to suit the pipeline systems into which they are to be installed. They feature all the benefits of the Integral DBB valve, with the added benefit of a bespoke face-to-face dimension, if required, or to an International Standard length (ASME B16.10 or API 6D for valves ≥DN50) to provide a means of replacing older single block and bleed valve configurations with the later DBB styles. They also offer the provision of mixing alternative connections within a system without the need for additional adaptors (more leak paths) i.e. Flange/Hub type terminations.

Locking Devices

It is common to have locking devices fitted to prevent the unauthorised operation of any of the valves. Needle valves can be fitted with a lockshield arrangement where the operating key is removed from the valve. Various designs for locking the position of ball valves exist. In the open position there are no issues about the small amount of movement that is permitted by the manufacturing clearances. In the closed position the design needs to ensure that there is sufficient seat ball overlap such that the slight rotational movement that must be possible due to manufacturing tolerances and clearances does not result in seat leakage.

Wireless Actuators and Instrumentation

Figure RD10: Wireless pressure transmitters provided a cost-effective and reliable method of monitoring this temporary gas fuel supply system (Emerson Process Management)

Data output from instrumentation and control of actuators has until recent times always involved hard wiring. Recent developments in wireless technology are now making it possible for instruments, actuators and control loops to communicate without the need to lay cable.

This offers the opportunity to quickly and easily instrument temporary facilities or to think about obtaining data from locations, where previously it would have been prohibitively expensive to do so.

Security of data transmission and reception is absolutely essential to the acceptance of this technology. The wireless technology is self organizing so that each device acts as a router for other nearby devices, re-transmitting messages until they reach the network's gateway, which channels the incoming data to a control point. If there is an obstruction, transmissions are simply re-routed along the mesh network until a clear path to the gateway is found. As conditions change or new obstacles are encountered in a plant, such as temporary scaffolding, new equipment, or a parked construction trailer, these wireless networks simply reorganize and find a way to deliver their messages.

All of this happens automatically, without any involvement by the user, providing redundant communication paths and better reliability than direct, line-of-sight communications between individual devices and a receiver. This self-organising technology optimises data reliability while minimising power consumption. It also reduces the effort and infrastructure necessary to set up a successful wireless network, because many devices can be served by one gateway. New instruments can normally be added to a network in just minutes.

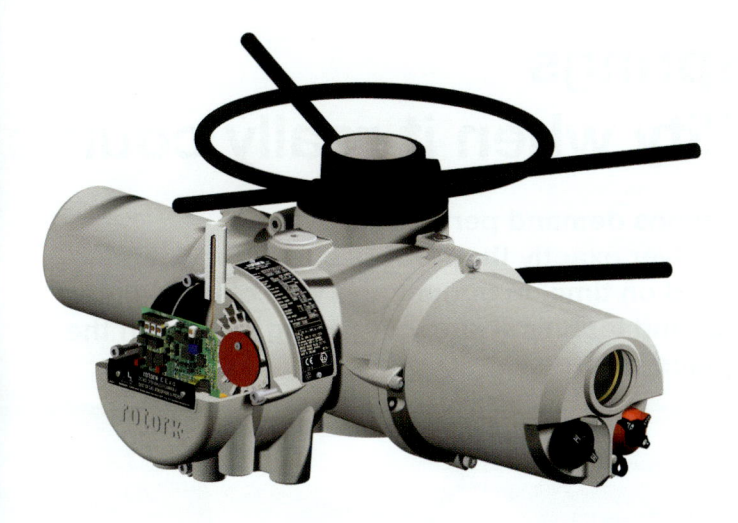

Figure RD 11: Cutaway showing an Electric Valve Actuator fitted with a wireless digital control module (Rotork Controls)

Electric actuators are being developed with this capability. The user has the choice of a fully wired loop for control and monitoring, a fully wireless control and monitoring system or wired control with wireless monitoring. A major advantage of the wireless system is the ability to access actuator data logger and configuration files, which up until now have only been downloadable locally, using hand held tools. Built in web pages make it possible to easily extract actuator data logger and configuration files from the control room.

The wireless system can also increase the number of actuators that can be controlled from a master station. Typical line-of-site operating ranges of approximately 70 metres indoors and 1000 metres outdoors are quoted by manufacturers. The use of meshing and repeaters further increases the range to individual units.

Most process plants have hundreds, or even thousands, of valves that are not connected to the control system because of high wiring costs. These valves currently provide no feedback on their actual positions, even though incorrectly positioned valves represent a significant cause of safety-related incidents. Unfortunately, users have typically not been able to access the position data that is valuable to the performance and safety of their plant.

Figure RD12: Wireless position monitors deliver previously unavailable equipment position data (Fisher / Emerson Process Management)

Wireless position monitors are now available for valves in standard and hazardous area locations. Non-contacting designs are available for both rotary and linear motion valve monitoring.

Position monitoring is battery powered and therefore battery life is a concern. Manufacturers claim a 5 to 10 year battery life and wireless solutions generally are claimed to be highly reliable. Early indications from plant trials are that the technology really works, and the operating range exceeds expectations.

BVAA 70 years of service to industry

DOIG SPRINGS

- Formula 1 & Automotive
- Aerospace & Defence
- Medical
- Oil & Gas

Valve springs
Reliability when it really counts

Critical applications demand perfection. Doig Springs deliver exactly that. Our bespoke products are manufactured on time, in budget and with clear, regular communication and technical knowledge engaged from the moment you place an order.

Diversity and Expertise

Products that require precision in their performance and significant research, design, prototyping and development are our strengths. This has made us a global leader in the manufacture of springs and related products.

From precise and critical medical applications, through to the technical performance of Formula One Racing and the robust and reliable requirements of the oil and gas industries, all our products carry with them a wealth of experience in their development.

We supply a wide range of custom designed springs and pressings to industries worldwide from our UK manufacturing bases in Buckinghamshire and Aberdeenshire. Our comprehensive service includes; design, tooling, development and manufacture drawing from our in-house R&D tooling centre.

We manufacture in accordance to **BS EN ISO 9001:2008** standards.

Expertise: Aerospace, Defence, Oil & Gas, Medical, Valves, Pumps, Actuators, Wind Turbines, Propulsion Systems, Rotary Seals, Electronics, Automotive, Solar Power, Furniture, Gas & Pipe Fittings, Telecommunications, and many more.

Springs: Seal Energising, Garter, Helical Seal, Canted Coil, Flat Leafed, Formed Lead, Flat Waved, Compression, Conical, Wireforms/Clips, Waved, Cantilever, Drawbar, Extension, Spiral, Tension, Torsion, Bespoke Pressings.

Materials: We manufacture using both exotic and standard materials including; Martensitic & Austenitic Stainless Steel, Nickel based allows, Inconel X750, Alloy 90, Allow C276, Copper Alloys, Beryllium & Phosphor Bronze.

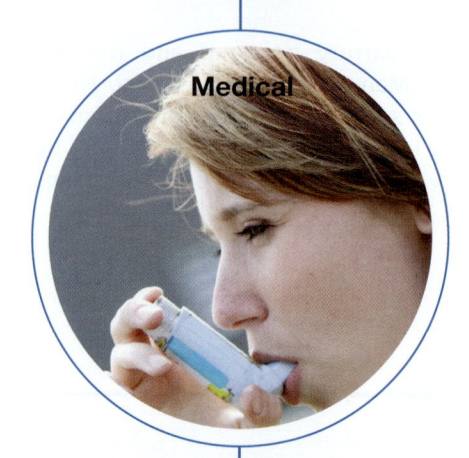

Contact the spring experts today

High Wycombe, Buckinghamshire:
Tel: +44 (0)1494 556700 Fax: +44 (0) 1494 511002

Fraserburgh, Aberdeenshire:
Tel: +44 (0)1346 518061 Fax: +44 (0) 1346 516817

Email: enquiries@springs.co.uk
Web: www.springs.co.uk

twm

Doig Springs is a trading name of Turner Workflow Manufacturing Ltd

Valve Operating Torques

Figure OT1: Stainless Steel Worm Type Gearbox (Opperman Mastergear)

The means of safe operation is an essential part of the valve and this must be part of the consideration when selecting a valve or a valve and actuator package.

The options available are :-

- Direct manual operation via a handwheel or lever
- Assisted manual operation using a gearbox with handwheel or lever
- Actuator with or without gearbox
- Actuator with auxiliary manual override in the event of failure of the power source.

Direct Manual Operation

Rotary Motion Valves - Manual operation of rotary motion valves is the most efficient. The rotary motion can be directly provided by the application of a force tangentially to the rim of a handwheel or more usually at 90 degrees to a lever. A consideration for butterfly valves is that fluid forces either static or dynamic or both can create a torque on the disc which may exceed the gland and shaft bearing friction torque and cause the disc to rotate either closed or open depending on flow direction. To prevent this, a means of locking the lever in the desired position needs to be provided.

Linear Motion Valves - Linear motion valves require a thrust applied axially to the stem to open and close them. This thrust can be applied directly by an actuator using air or hydraulic fluid but for manual operation an operating mechanism is required which translates rotary motion to linear motion. Theoretically it is possible for manual linear thrust to be provided directly but only for the very smallest valves and lowest pressures. There is also the problem that for the valve to remain closed the force must be continually applied by some form of locking mechanism restraining the stem. Manual linear motion valves are therefore almost always supplied with an operating mechanism.

It is important to understand the fundamentals of this mechanism. The stem thread converts the torque to thrust. The pitch of the thread determines the number of turns and torque required to operate the valve for a given thrust. As the pitch increases the number of turns reduces and torque increases. As pitch decreases the number of turns increases and torque reduces. For the rotating stem globe valve the thrust is resisted by the yoke bush or the yoke, if no yoke bush is provided. Outside screw non-rotating stem gate and globe valves have a yoke sleeve, which resists the axial thrust through contact with the yoke. Inside screw non-rotating stem gate valves have a stem nut, which resists the axial thrust through contact with the bonnet.

There is an inherent inefficiency in every stem thread in that one component of the force generated by the stem thread acts radially and therefore makes no contribution to the axial thrust required for opening or closing. There are additional losses, which are due to friction as the various surfaces slide over each other. The choice of materials for the stem nut, yoke bush and yoke sleeve are crucial to the reliability of the operating mechanism. The material choice for the stem is generally determined by the process conditions and the operating forces that have to be transmitted to the obturator. The stem nut, yoke bush or yoke sleeve are required to have good strength but also good wear properties and low coefficients of friction in sliding contact with the stem material. Copper alloy and austenitic nickel SG irons are popular choices. Sliding contact surfaces must be kept well lubricated to ensure that operating torques remain as they were when the valve was new and also that wear is kept to a minimum.

A significant reduction in friction between the yoke sleeve and yoke can be achieved by the fitting of needle roller thrust bearings. These are typically fitted as standard in valves with a DN x PN ≥ 4000.

An important consideration for the valve designer in his choice of stem thread for linear motion valves is that it is self locking. These means that when the force is removed from the operating element, the thrust is still locked into the stem or the weight of obturator is not able to drive the valve in the opposite direction. The use of a self locking stem thread places an upper limit on the efficiency of the stem thread.

EN12570

A crucial assumption that the valve designer must make in his design of a manual operating mechanism for both the rotary or linear motion valve type is the force that a person is capable of applying to a handwheel or lever. To ensure a consistency of approach within European standards EN12570 was developed. This standard gives guidelines to the designer to ensure that the correct diameter of handwheel or lever is selected for the valve. The standard has to compromise between a range of capabilities. If the standard assumes that a person has a low capability then a person of above average capability can apply forces to the operating mechanism, which might do structural damage to the valve. Equally by assuming too high a capability, the below average person is not able to operate the valve.

It is also important to understand that the values included in EN12570 are based on assumptions namely that the handwheel or lever is at a height such that a person can apply their best effort and that the environmental conditions i.e. temperature, low wind, good footing, no space restrictions are all optimal.

EN12570 considers two criteria namely the operating torque that is required at closure and the operating torque that is required to move the valve between the closed and open position sometimes known as the running torque. Typically the running torque is lower than the seating torque because it does not include the force required to force the seating faces together. However the running torque is applied over a much longer time span and therefore a person can not be expected to maintain a high force input over this timescale. The standard therefore specifies a running torque capability of 0.4 to 0.5 times the seating torque capability. The designer considers both running torque and seating torque and makes his selection of handwheel diameter or lever length based on the largest he calculates for the two cases.

In the case that the calculated lever length or handwheel diameter is impractically large the designer must provide additional mechanical advantage in the form of a gearbox. The criteria of EN12570 are also applicable to the handwheel or the lever fitted to the input shaft of a gearbox when the gearbox is part of the operating device of a valve.

Assisted Manual Operation Using a Gearbox

Figure OT2: Worm gearbox with spur gear input operating a large ball valve (Rotork)

Gearboxes consist of a set of gears, shafts and bearings that are factory mounted in an enclosed lubricated housing. Gearbox materials include stainless steel, cast aluminum and cast iron. They can be used in a diverse range of environments, such as chemical, power, waterworks, gas pipelines, HVAC and most general industrial applications. Gearbox types are available for part-turn and linear valves of all types and sizes.

Gearboxes are highly versatile and can be adapted to suit any environment, such as buried or offshore duty, high or low temperatures etc. In addition, there is available a wide range of gearbox mounting accessories, including mounting brackets, switch boxes, valve position monitors and other adaption accessories.

Figure OT3: Position Limit Switchbox for manual gearbox (Rotork)

Worm Gearboxes for Rotary Motion Valves

Suitable for part turn valve types such as ball, plug, butterfly and dampers, "worm" gearboxes are most commonly used as they have the advantage of being inherently self locking (valve cannot back drive the gearbox). A worm gear is a threaded input shaft meshed with a worm gear that is mounted to the output sleeve. These gearboxes can cater for small or large torque ranges, and are suitable for manual and motorized applications (see Electric Actuators Chapter).

For manual operation, available ratios range from 40:1 up to 80:1, however with multi-turn gearboxes added to the worm gear input, ratios of up to 1000:1 are possible (see multi-turn gearboxes below). Efficiency of worm gearboxes generally ranges from 30% to 40%, which means that real gain in mechanical advantage between the input and output shaft of 40:1 worm gearbox is between 12:1 and 16:1.

Part turn gearboxes usually include mechanical stops to limit the available travel to 90° (+/-10°), important as the valve may not include stops such as is the case with rubber lined butterfly valves. The output drive may be removable to allow easy machining or integral to the quadrant in which case the valve shaft detail must be supplied with the order to allow the gearbox manufacturer to machine during gearbox assembly.

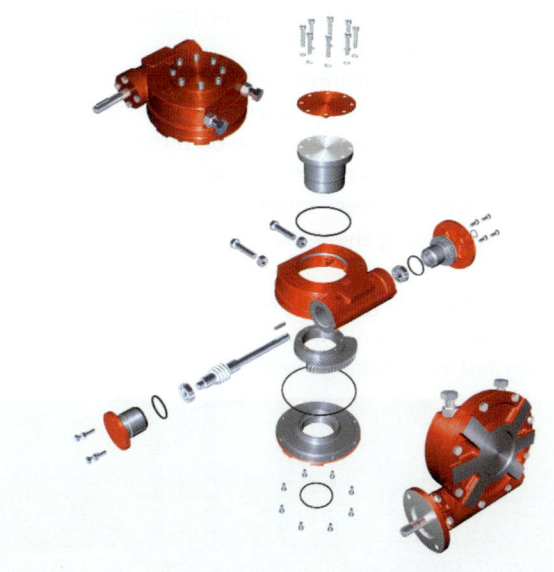

Figure OT4: Exploded diagram of the component parts of a typical worm gearbox (Rotork)

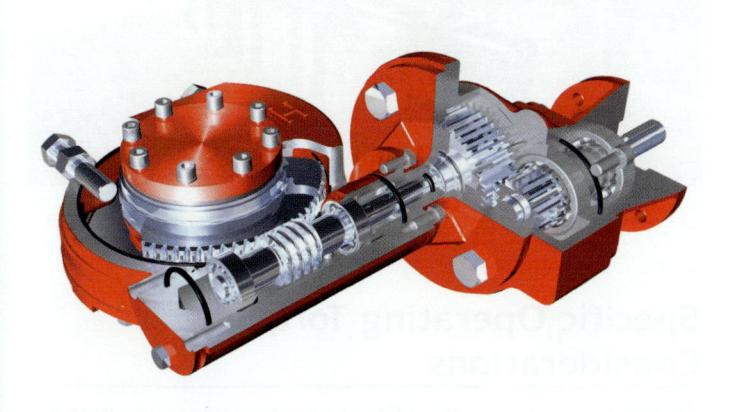

Figure OT5: Cutaway diagram of a worm gearbox with spur gear input (Rotork)

Figure OT6: Assembly of Worm Gear Unit (Opperman Mastergear)

Figure OT7: Worm Gear gearboxes come in a wide range of sizes (Opperman Mastergear)

Bevel and Spur Gearboxes for Linear Motion Valves

Suitable for linear valves such as wedge gates, parallel slide, knife, and sluice/penstocks, multi-turn spur or bevel gears are used (see Figures OT8 and OT9). The gearbox output is a drive nut which rotates and is threaded to the valve stem thread – turning the nut causes linear movement of the valve stem. Gearboxes must be capable of withstanding thrust generated in applying torque to the stem thread nut and therefore will have a thrust rating as well as a torque rating. Bevel and spur gear types are not self locking, this aspect being catered for by the stem thread/nut inefficiency.

145

BVAA VALVE & ACTUATOR USERS' MANUAL - 6th Edition

Bevel gears orientate the input drive 90° from the output drive which may be a distinct advantage in providing good operator accessibility. Spur gears input and output drives are in the same plane but offset. For bevel gears, ratios are available between 1:1 and 8:1 while spurs can go up to 24:1. For both types, ratios can be extended with the addition of a second spur or bevel applied to the input of the main gearbox. Typically, the efficiency of bevel and spur gearboxes is between 80% - 90%. Bevel and spur gearboxes generally have a removable drive nut for the valve manufacturer to machine.

Figure OT8: Exploded diagram of the component parts of a typical spur gearbox (Rotork)

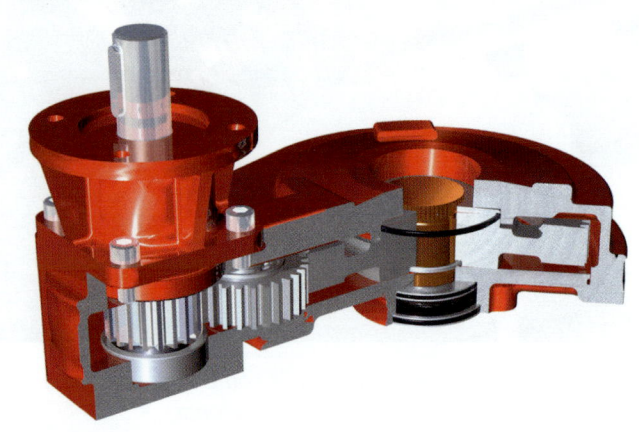

Figure OT9: Cutaway diagram of a spur gearbox (Rotork)

Sizing Gearboxes

The decision to fit a gearbox is normally made by the valve manufacturer in order to provide sufficient mechanical advantage to safely operate the valve. However more and more attention is being paid by users to health and safety and to changes in the physical capability of operators such that users are increasingly specifying the requirements for gearboxes. For example EN12570 specifies an operator rim pull capability of 400N for a running torque. If this is reduced to 250N by user specifications the manufacturer has the option to fit a larger operating element or a gearbox.

The combination of operating mechanism, gear ratio, gearbox efficiency and operating element size will determine the operating force that has to be applied to the operating element. The number of turns to open or close the valve is a combination of the operating mechanism and gear ratio. In general, reductions in operating force have to be paid for by increases in the number of turns to open or close the valve.

Gearboxes can be obtained with a specific ratio and handwheel size that optimizes the operating force and number of turns to open or close the valve but these bespoke products are expensive. Therefore a valve manufacturer will normally select from the gearbox manufacturer's standard catalogue and this generally means that the choice is the best one available. Selection must also consider the valve gearbox mounting interface and the gearbox maximum stem acceptance diameter.

What must not be forgotten is that satisfying the requirement to make the valve easier to operate provides additional mechanical advantage and an increased opportunity to over load the valve operating mechanism or damage the internal components such as seats and discs. If the valve does not close, or in some cases does not open, operators have been known to apply excessive force or indeed regrettably make use of a large lever to apply additional force to try to overcome the problem.

The accuracy of manufacturers quoted gearbox efficiency is important because if it is over stated the operating element will be undersized and the valve may be difficult to operate. If it is under stated then the operating element will be oversized, which further increases the potential for structurally overloading the valve.

Valves and gearboxes in general are robust and will withstand a high level of abuse but as with everything mechanical there is a limit beyond which things will fail.

Figure OT10: River sluice gates are common applications for manually operated gearboxes (Rotork)

Specific Operating Torque Considerations

Whether a valve requires a linear or rotary operating motion, in selecting and specifying the correct operating device the valve manufacturer must include detailed consideration of the valve's torque or thrust requirement during the valve stroke from the fully open to the closed position, and vice versa under the pressure and/or flow conditions of the particular application. The material of the seating surfaces, the function of the valve i.e. isolating, regulating, control or mixing and the required seat leakage rate for isolating valves are essential factors to be taken into account. Zero visible leakage on metal to metal seat isolating valves requires a higher seating torque than for soft seats. Running torque requirements are more important for regulating and control valves. The requirements of actuators for control valves are considered in detail in a later chapter.

Fluid temperature and whether the fluid is a gas or clean liquid or liquid with solids can have an important effect on operating torques. With the exception of diaphragm and pinch valves all types need to overcome sliding friction in the gland. With exception of the diaphragm, pinch, globe valve and triple off-set butterfly valves all types need to overcome sliding friction at the seating surfaces. The coefficients of friction at these sliding surfaces vary with temperature and, in the case of the seating faces, whether the fluid is dry or lubricating. Seat wear or seat damage can cause increases in the coefficient of friction between seating surfaces.

In all types of valves, the operating device has to provide a torque to overcome the friction caused by the gland packing acting on the valve stem or shaft. The friction is a function of the coefficient of friction between stem or shaft and packing, the compressive load applied by the gland bolting and the stem or shaft diameter. Again the valve designer is presented with two conflicting requirements, namely sealing around the stem or shaft as it moves up and down or rotates and ease of operation of the valve. In service the user can create a significant increase in operating torque by tightening the gland bolting in an attempt to stop a fluid leak from the gland. Normally the manufacture will specify a maximum torque that should be applied to gland bolting such that operating torque remains within design limits. For low pressure applications, the majority of running torque will be that required to overcome this friction.

The seat bore diameter is a major influence on the operating torque. Reduced bore valves particularly rotary motion types can have operating torques of 50% of the full bore valve. If valves are to be actuated then significant savings can be made all though these savings need to be offset against higher energy costs associated with increased pressure drops across the valves.

Linear Motion valves

The line pressure acting on the cross-sectional area of the valve stem, where it passes through the gland, creates an upward force on the stem. In the majority of applications this force tends to balance the weight of the obturator. However where fluid pressures are in excess of 60 bar the additional running torque to close the valve should be taken into account.

- **Solid wedge and flexible wedge gate valves** - With these types of valves, shut off of the flow is achieved by the gate travelling across the bore of the valve. As the gate travels in the closing direction, the throttling effect on the flow increases, causing an increase in the differential pressure across the gate. This differential pressure forces the gate onto the seat and the actuator has to provide the necessary thrust to overcome this friction. Additional thrust is required in order to finally wedge the gate into the seats for tight shut off. Higher friction factors are applied when gate valves are used on steam and other gases because of the reduced lubricating effect compared to liquids, and are increased when the fluid temperature is in excess of 400°C. An important consideration for wedge gate valves is also the unseating torque. Valves that have been closed with the pipework hot can require significantly greater forces to unseat them if the pipework has now cooled. The size of gates valves means that access to the operating element can present problems and extended stems, chain wheels or cable drives may be necessary to bring the operating element to a safe location.

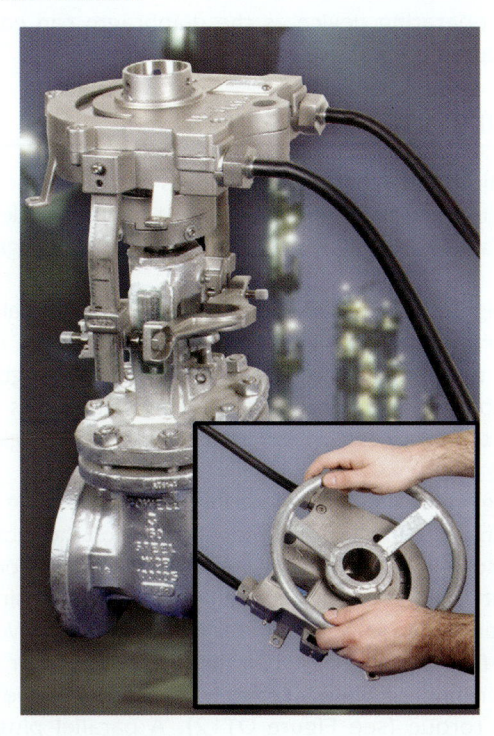

Figure OT11: Flexible Cable Drive System for remote manual operation (Smith Flow Control)

- **Parallel slide gate valves** - The forces required for operating parallel slide valves are similar to that required for the wedge gate valves, except that running torque must overcome the sliding of the gate across the seat over the full travel and no allowance has to be made for forcing the gate onto the seat. This latter is an advantage when fitting actuators with the parallel slide design allowing smaller sized actuators to be used than with a similar sized wedge gate valve, resulting in cost savings.

- **Through conduit gate valves** - The forces required for through conduit gate valves are similar to that required for the wedge gate valves, except that the running torque of the non-expanding type must overcome the sliding of the gate across the seat over the full travel.

- **Globe valves** - The globe valve is normally installed in the pipe so that flow is from under the disc. In this case the stem thrust has to overcome the pressure acting on the underside of the disc plus providing the additional force necessary to clamp the seating faces together. As this type of globe valve usually has metal to metal seating faces a substantial additional thrust allowance is made in order to effect tight shut-off. If the flow direction is reversed then a lower operating torque is required to seat the valve as the pressure load now contributes to the seat face clamping load. Unseating the valve now requires more effort.

- **Diaphragm and pinch valves** - Diaphragm and pinch type valves are by definition soft seal valve types. The stiffness of the materials used for the diaphragm or the tube in the pinch valve varies depending on materials chosen to suit the service conditions. Therefore the thrust or torque requirements for these types of valves have to be established empirically. As with soft seat valves in general care needs to be taken not to oversize the operating device as the diaphragm can easily be ruptured if excessive force is used.

Rotary Motion valves

The majority of rotary motion valves are of the nominal 90 degree travel type, though there are a number of designs for special applications which travel more than 90 degrees.

The weight of the obturator is supported by internal bearings all of which have coefficients of friction that create operating torque to rotate the obturator. Part of the weight of the obturator is off set by line pressure acting on the cross-sectional area of the valve stem, where it passes through the gland.

- **Plug valves** - The operating torque of plug valves mainly depends on whether they are of the lubricated or non-lubricated type. The torque requirements are relatively consistent between opening and closing. In a tapered plug valve the operating torque must overcome the friction force between the plug and seat or sleeve. The running torque is typically 90 percent of the break-out torque (see Figure OT12). A parallel plug valve design requires less torque overall but running torque is still 90 percent of the breakout torque. Eccentric plug valves with one seat and expandable plug valves have lower running torques as the plug is no longer in contact with the seat or body for most of the opening cycle. The operating torque of the balanced plug valve is lower than for the unbalanced but it also increases with line pressure due to the difference between the diameters of the two ends of the taper plug.

- **Ball valves** - In the floating ball valve, line pressure acts on the outside diameter (OD) of the seat. Therefore the unseating torque from the closed position is the maximum operating torque. Valves fitted with reinforced PTFE, PEEK, or metal seats have a much higher un-seating torque than virgin PTFE due to the different coefficients of friction. There is also a tendency for a film of PTFE to transfer to the surface of the ball. This leads to a further increase in un-seating torque if the valve is not operated for sometime as the seat and the PTFE film on the ball adhere to one another and this connection has to be broken.

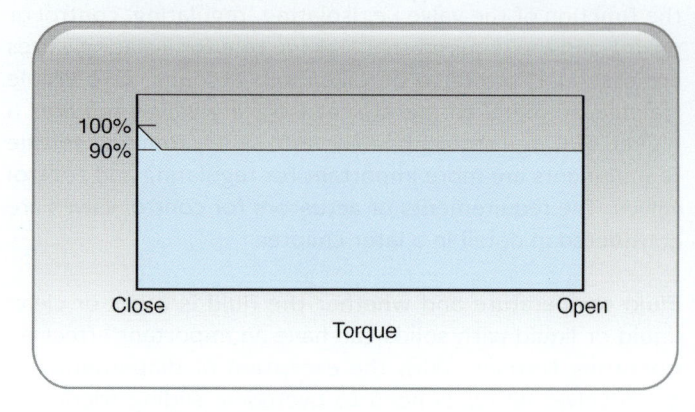

Figure OT12: Operating Torques for Plug Valves

Figure OT13 (blue line) shows the typical variation in operating torque for virgin PTFE floating ball valve moving from the closed to open to closed position. In the open position the ball is not subjected to the line pressure so the un-seating torque from the open position is approximately 60 percent of that from the closed position. The running torque and seating torque remain at 30 and 60 percent respectively.

For PEEK and metal seats the operating torque is reasonably constant regardless of whether the valve is open, closed, or running. This is shown by the horizontal red line in Figure OT13.

Trunnion mounted ball valves have lower operating torques than the equivalent floating ball design because the line pressure is only acting on the annulus of the seat ring rather then the OD of the seat. If the valve is to be actuated it may be a cheaper overall package for ≥ DN150 to use a trunnion mounted design rather than a floating ball.

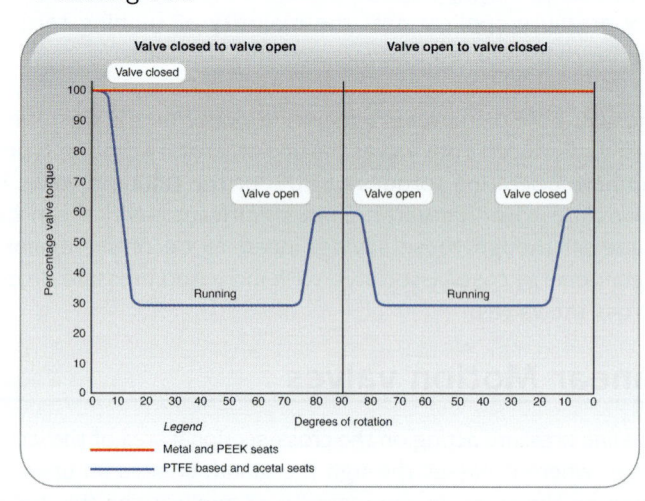

Figure OT13: Operating Torque curve for Ball Valves

- **Butterfly valves** - The butterfly valve is the most difficult of all the valve types to determine the operating torque because the torque is made of a number of factors. The fact that the disc of the butterfly valve is always in the flow path means that aerodynamic, momentum and pressure differential fluid forces continue to act on the disc throughout its travel. The effect of these varies due to the shape of the disc, the flow velocity and the design of the valve offset or not. Computer aided fluid dynamic software packages have helped to calculate these forces and to design discs that are least affected by the flowing fluid. Manufacturers should always be consulted for recommendations on operating torques.

The triple off-set design metal-to-metal seated valve requires a clamping force between the seating faces to achieve the specified leakage rate. Again manufacturers' recommendations on the required operating torque should be sought.

Figure OT14: Subsea Manual Gearbox for operation by Remotely Operated Vehicle (ROV) with hi-visibility paint scheme (BEL Valves, Rotork Gears)

Figure OT15: Intelligent Compensation Subsea Valve Actuating Gearbox (Opperman Mastergear)

BVAA 70 years of service to industry

BVAA VALVE & ACTUATOR USERS' MANUAL - 6th Edition

rotork®

50 YEARS AT THE FOREFRONT OF VALVE ACTUATION

rotork® Controls

rotork® Fluid Systems

rotork® Process Controls

rotork® Gears

rotork® Site Services

Rotork worldwide. 87 countries. 15 manufacturing plants. 279 offices and agents

visit www.rotork.com for your local agent

Actuators

Figure A1: Electric Multi-turn Actuator (Rotork)

The alternative to manual operation of a valve is to use an actuator as the operating device.

The decision to actuate a valve is usually based on one or more of the following considerations: -

- Safety
- Reliable operation
- Control and process system performance
- Inaccessible or remote valve location
- Excessive valve operating forces
- Emergency shutdown/fail-safe requirements.

An important consideration when deciding to fit an actuator is the differential pressure against which the valve has to be opened or closed. In general the valve pressure envelope is designed for the PS of the flange or end connection. In practice the design or working pressure of the valve may be well below that. For example a valve operating at 250°C will have a lower operating pressure than the PS at 20°C simply as a result of the Pressure/Temperature curve for the material of the shell. Sizing the actuator to open and close the valve at the design conditions of 250°C will result in a lower cost valve and actuator package than sizing it for the PS at 20°C. Detailed consideration of the potential working conditions of the valve and actuator need to be made and the necessary information passed to the valve and actuator manufacturer such that an appropriate solution can be provided.

Figure A2: Pneumatic Rack & Pinion Actuator, Quarter Turn, with Position Indicator (Emerson Elo-Matic)

Figure A3: Gas Hydraulic Quarter Turn Valve Actuator (Emerson Bettis)

BVAA VALVE & ACTUATOR USERS' MANUAL - 6th Edition

Very often in applications involving actuators, safety factors are added because of concerns regarding the potential for the actuator not to open or close the valve. These safety factors may be applied to the energy source i.e. electric voltage, air or oil pressure, the valve operating torque, the friction coefficients of the sliding components in the operating mechanism and operating device, generally based on user experience or custom and practice. Care needs to be given to the total level of safety factors applied and users and manufacturers should ideally co-operate on the appropriate total level of safety factor. It is not good practice in critical applications to apply large safety factors based on either ignorance or perceived valve or actuator reliability issues.

Excessive over sizing of actuators adds to the cost of the valve and actuator package, it creates installation and maintenance issues due to the size and weight of the package and there may be structural problems for the valve and actuator mounting kit in the event that some or all of the perceived negative effects on operating torque do not in practice exist.

Figure A5: Typical valve/actuator mounting kit (Tyco)

Figure A6: Mounting kit designs can be analysed using computer packages and Finite Element Analysis to calculate likely stress, strain and deformation (Heap & Partners)

Figure A4: Control Valve with Pneumatic Diaphragm Actuator, fitted with Digital Positioner (Emerson Fisher)

Figure A7: Quarter Turn Pneumatic Actuator complete with Mounting Kit and Geared Manual Override (Flowserve)

Actuator power comes from compressed air (pneumatic, e.g. Figures A2, A4, etc.), compressed liquid (hydraulic – Figure A3) or electricity (Figure A1). The actuator can deliver the power either as a thrust (linear – Figure A10) or a torque (part-turn – Figure A2, A9 etc. or multi-turn – see Figure A1).

A key feature of the valve and actuator package is the interface between valve and actuator. It must be capable of transmitting the thrusts or torques provided by the actuator and ideally be standardized to allow users and valve manufacturers to select from a range of competitive products.

Two international standards EN ISO 5210 'Multi-turn actuator attachments,' EN ISO 5211 'Part-turn actuator attachments' and one European standard EN 15081 'Mounting kits for part-turn actuators' have been developed to meet this need.

Figure A8: Pneumatic Single Acting, Spring Return Actuator (Ebro Armaturen)

When reviewing an actuator design, the following should be considered:-

- Capability to provide thrust or torque to seat the valve under the minimum power supply condition
- Capability to hold the valve gate, disc, plug, or ball in the required position under the worst flow conditions
- The power required to drive the valve through its full travel at the required speed
- The fail-safe mode in the event of either power or control signal failure
- The forces that could be applied to the valve under unfavourable power supply conditions
- Power source availability
- Whether the valve is an isolating or control valve
- Frequency of operation
- Type of valve
- Valve operating torque characteristics
- Environmental conditions such as ambient temperature range and hazardous area classification.

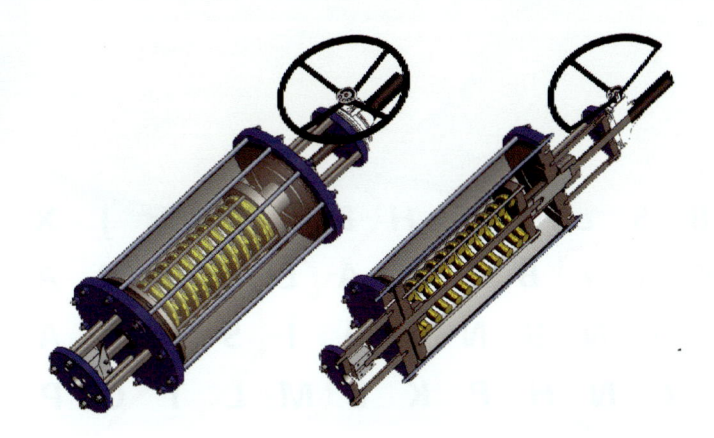

Figure A10: A 'Fail Closed' Pneumatic Piston Actuator (Valve Solutions)

Figure A9: Scotch Yoke type Quarter Turn Pneumatic Actuator with Spring Return, with Switch Box and Controls (Emerson Bettis)

A major feature of many pneumatic and hydraulic actuators is the provision of a spring or springs, which act to either close or open the valve in the event of the fluid pressure being removed. The springs are sized such that the thrust they generate is sufficient to move the valve to its closed or open position. In the case of valve types that require forces to clamp the seating faces together e.g. wedge gate, globe, triple offset butterfly, this thrust must also be supplied by the spring if the valve is fail closed (Figure A10). Valve types that rely on line pressure to provide the seating face clamping load e.g. ball or parallel slide gate do not require this additional force. With the operating fluid depressurised the spring is least compressed. To move to the opposite end of the valve travel the air pressure must further compress the spring plus providing the operating torque or thrust required by the valve. The spring compression is a maximum at the end of travel therefore for a given air pressure the actuator net output is then at a minimum. Whatever the valve type, the actuator output less the force or torque required to compress the spring must be sized to exceed the required valve operating torque at all positions between closed and fully open.

IMPORTANT

If an actuator is fitted with a spring to provide return energy it must be remembered that this spring is in its most compressed state when the valve is being powered by the air supply. This spring load is reacted by the bolting that holds the spring cover to the actuator body. It is therefore vital that this bolting remains in good condition and does not deteriorate due to environmental conditions such as corrosion (see Figure A11). If the bolting were to fail the spring compression would be released, which would be a significant hazard to personnel or equipment nearby and the valve would lose its fail closed or open capability.

Maintenance of actuators that have compressed springs in their designs must be done strictly in accordance with manufacturers' instructions to prevent the sudden release of stored energy and injury to personnel.

Figure A11: A Quarter Turn, Spring Return type pipeline Emergency Shut Down (ESD) Valve Actuator, where the spring end plate tie rods failed due to corrosion (HSE Safety Bulletin 3-2010)

BVAA VALVE & ACTUATOR USERS' MANUAL - 6th Edition

Looking for a valve automation supplier with all the right answers?

It's time you had a word with us.

Emerson offers the most comprehensive and proven valve automation portfolio available anywhere, encompassing pneumatic, hydraulic, electric and gas-hydraulic products and services. By bringing together best-in-class brands such as **Bettis**™, **Dantorque**™, **El-O-Matic**™, **FieldQ**™, **Hytork**™, **Shafer**™ and **EIM**™, all of your valve automation requirements – on any make of valve – can be solved. See for yourself at **www.EmersonProcess.com/ValveAutomation**

For enquiries, kindly contact us at
Emerson Process Management, Valve Automation
6 Bracken Hill
South West Industrial Estate
Peterlee SR8 2LS, United Kingdom
T (0) 191 518 0020, F (0) 191 518 0032
enquiries-peterlee@emersonprocess.com

The Emerson logo is a trademark and a service mark of Emerson Electric Co. ©2010 Emerson Electric Company.

EMERSON. CONSIDER IT SOLVED.™

EMERSON
Process Management

Pneumatic Actuators

Figure PA1: Linear Spring Opposed Diaphragm Actuator and a Linear Double Acting Piston Actuator (Koso Kent Introl)

Pneumatic power has been used for the actuation of valves for many years. It is particularly suited to the operation of rotary motion valves, with their limited stroke requirements, but it can also be applied to the operation of linear motion valves. Pneumatic power is easy to store and pneumatic actuators are, therefore, commonly the most suitable choice for installations where electric power sources are of limited capacity or unreliable. Pneumatic actuators are available to suit valves of almost all sizes except the very largest and highest differential pressures. In particular for those actuators requiring spring-to-close the spring force needed to achieve the maximum acceptable level of seat leakage can become impractically large.

Pneumatic power supplies

Pneumatic actuators are normally designed for use on pneumatic supplies of 4 barg to 6 barg nominal pressure. The supply is usually compressed clean dry air, but designs are also available for use with natural gas. Air supply systems require a compressor working in conjunction with an accumulator plus appropriate filters to ensure moisture, oil and dirt are removed. The number of actuators, duty cycles and types of actuator taking air from the system need to be assessed in order to achieve the most economical overall system.

Actuator output torques are proportional to the gas supply pressure. The required valve operating time for large valves with large air requirements could be a major factor in determining the capacity of the air supply system and the size of the piping needed to deliver the required volume of air.

Pneumatic Actuator types

There are four basic types of actuator which convert pneumatic power into mechanical output: -

- **Piston actuator** - The piston actuator consists of a cylinder and piston which is sealed with an elastomer 'O' ring (see Figure PA2). Air is supplied either to both sides of the piston (called double acting) and movement is achieved by one side being at a higher pressure than the other, or air is supplied to one side (single acting) and the return energy comes from a spring. Piston actuators generate linear force by the air acting on the piston. They can be used to operate linear motion valves by directly attaching them to the valve stem. However the stroke length - particularly for gate valves - and the area of the cylinder to give the necessary operating thrust require huge volumes of air to operate the valve. This generally means that rotary motion valves are given preference over linear motion valves when pneumatic fail closed or open operation is required.

BVAA **VALVE & ACTUATOR USERS' MANUAL** - 6th Edition

The piston actuator requires some form of mechanism for use on rotary motion valves to convert linear force to torque. This is made easier because the majority of part-turn valves require 90 degrees of movement only and therefore only a short piston movement. There are four types of mechanism available to convert linear to rotary motion.

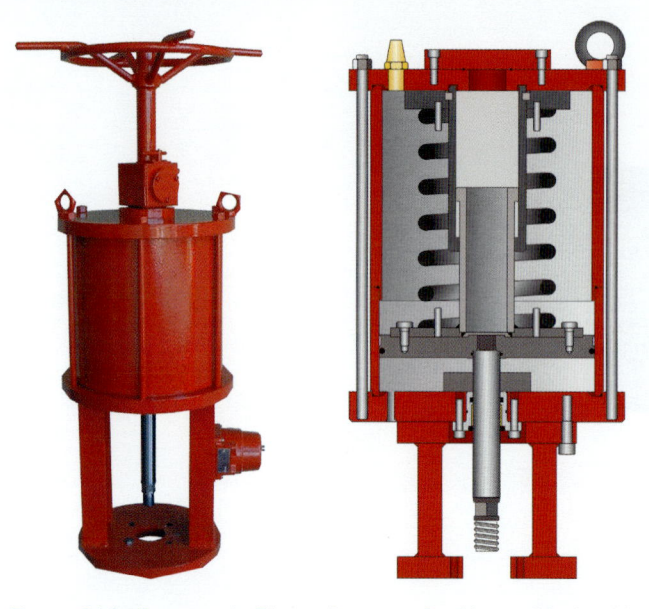

Figure PA2: Pneumatic Piston Actuator for Linear Motion Valves

(a) A scotch yoke actuator consists of a piston, connecting shaft, yoke, and rotary pin (see Figure PA3). The yoke is off set 45° from the axis of the piston at the two ends of travel. At the mid travel position the yoke is at 90° to the piston shaft and the centre of rotation is closest to the shaft. The fact that the moment arm is the shortest means that the torque output at mid travel i.e. 45° open is a minimum. Figure PA4 shows the output torque characteristics of the scotch yoke actuator both as double acting and as single acting with spring return to the closed position. The torque curve for the single acting is the net curve of the double acting minus the spring return curve. The minimum torque on the spring return curve is no longer at the mid travel because of the fact that the spring force is reducing as the valve closes thereby moving the minimum towards the 30° open position.

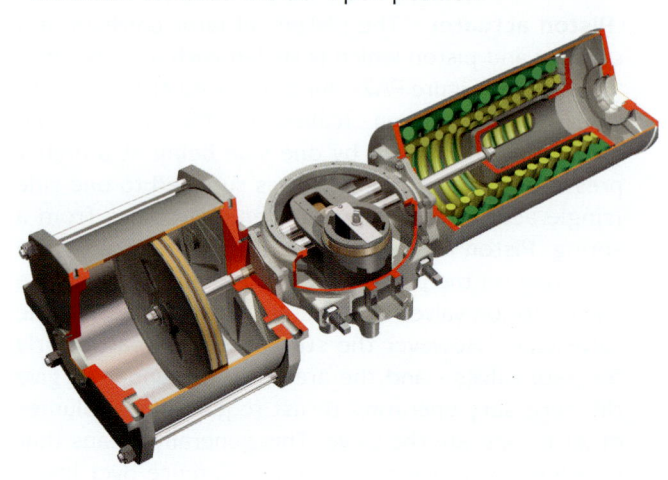

Figure PA3: Section view of a typical Scotch Yoke Pneumatic Actuator (Flowserve)

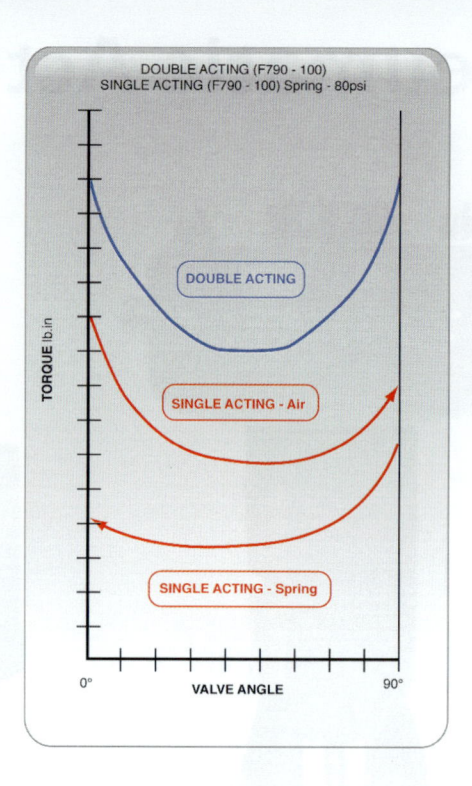

Figure PA4: Torque Characteristics for Scotch Yoke Actuator

A variation of the scotch yoke is the canted scotch yoke (see Figure PA5). By offsetting the axis of the slot in the yoke such that it no longer passes through the shaft axis, the output torque is slewed towards the closed position. This torque curve is particularly suited to the offset butterfly valve.

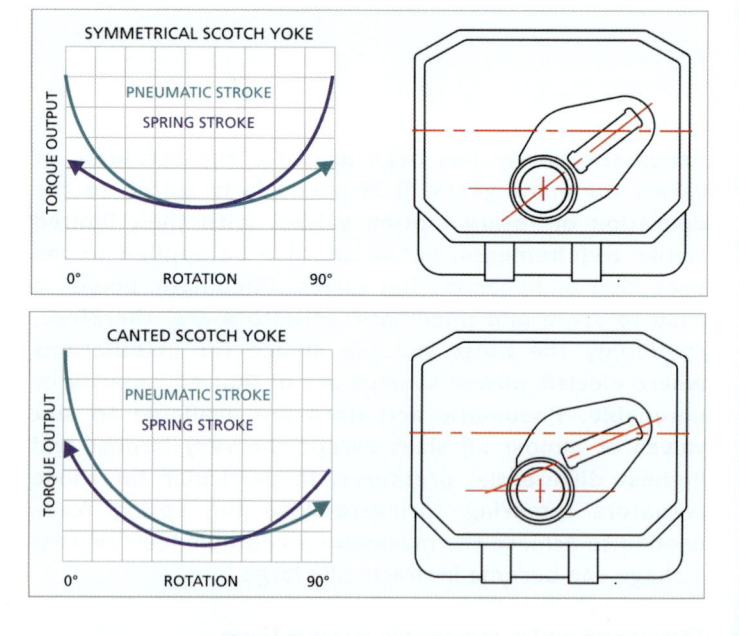

Figure PA5: Symmetrical and Canted Scotch Yoke (Rotork)

(b) The rack and pinion actuator consists of a single or double piston coupled with an integral rack which drives the pinion. The majority of actuators are designed for producing a 90 degree turn on the pinion (see Figure PA6). However, rack and pinion actuators are available which produce 180 degrees output. The rack contacts the pinion at a constant dimension from the centre of rotation and, therefore, the torque output of a double acting actuator is constant throughout the stroke. This makes this type particularly suitable for plug valves.

Figure PA6: Detail view of a dual opposed, single acting spring return rack and pinion pneumatic actuator (Emerson Hytork)

A rack and pinion actuator may be of the single or double piston design. The double piston is referred to as 'dual opposed' because, as the actuator is stroked, the pistons move inwards towards one another (see Figure PA7). The use of two pistons in this way enables more torque to be provided, as the air pressure is applied to two piston areas simultaneously.

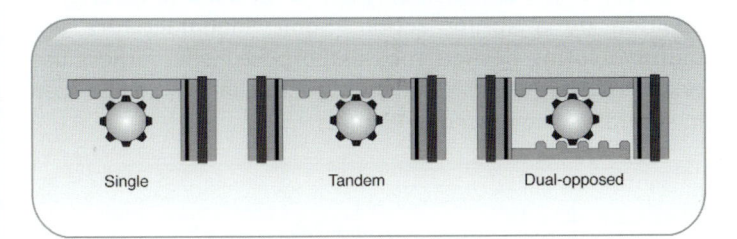

Figure PA7: Rack and Pinion Actuators with Single, Tandem and Dual Opposed Pistons

(c) Trunion/lever arm actuators normally consist of a simple trunion-mounted cylinder with a piston which acts directly upon a lever attached to the valve shaft. Generally speaking, this type of actuator is not popular in view of the fact that there are exposed moving parts which may cause injury to personnel. However, it is possible to use them when they can be enclosed within a suitable guard. The torque characteristics are similar to that of the scotch yoke mechanism in that the length of the moment arm varies with angular position.

(d) A cam mechanism actuator consists of double pistons connected by spacer bars with a cam and shaft between them (see Figure PA8). The rotational stroke is determined by the cam profile, which is in contact with the centre of the pressurized piston face throughout the stroke. The cam profile can be designed to give an exact 90 degrees rotation, alleviating the need for mechanical end stops.

The output torque of double acting cam actuators remains constant throughout the stroke.

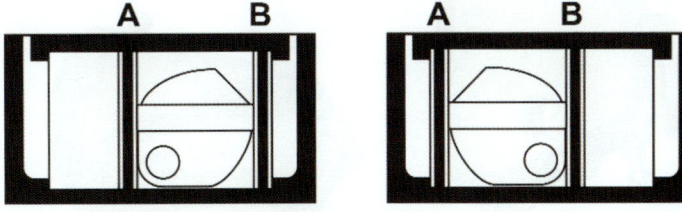

Figure PA8: Double Acting Cam Actuator (Matic Actuators, Div. of Imtex Controls Ltd)

- **Diaphragm actuators** - Another form of linear motion pneumatic actuator is the diaphragm actuator (see Figure PA9). It comprises a rubber diaphragm and stem contained in a circular pressed steel housing. They are usually single acting with air being supplied to one side of the diaphragm and springs providing the return energy. The nature of the flexible diaphragm means that travel is relatively restricted compared to that of the cylinder actuator. They are ideally suited to the shorter maximum travels of diaphragm and globe valves and they can also be used on small gate valves.

For globe valves spring thrust and travel limitations start to create difficulties if flow is under the disc and spring to close action is required. By reversing the flow direction and thereby using line pressure to assist closure of the globe valve, the range of valve sizes and differential pressure for which this type is suitable can be extended.

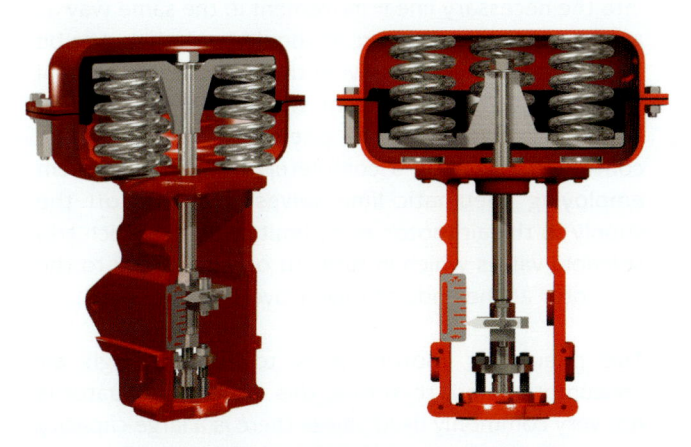

Figure PA9: Pneumatic Diaphragm Actuators, Air Fail Open to LH side, Air Fail Closed to RH side (Dresser)

- **Vane actuators** - Vane actuators are simple quarter-turn devices using a vane with an integral rotary output shaft to produce the torque (see Figure PA10). When air is applied to the actuator, it acts upon the area of the vane to produce force. As the distance from the centre of the vane to the axis of the output shaft is fixed, the torque output of a double acting vane actuator is constant (See Figure PA11). A coil (clock) spring can be mounted on top of the actuator to provide spring return.

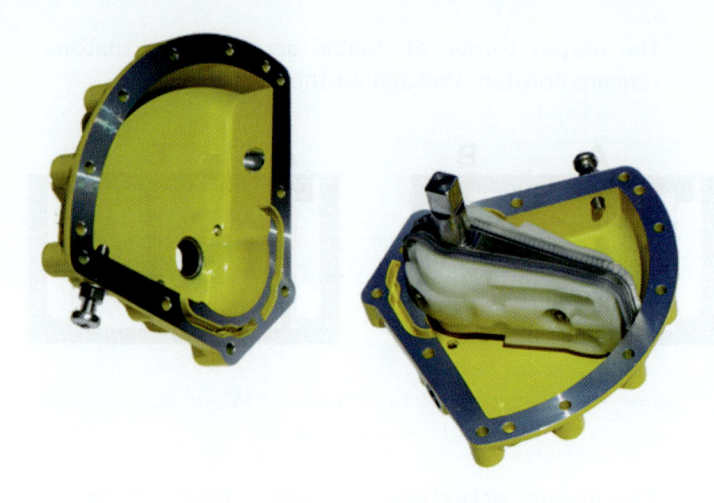

Figure PA10: Vane Actuator (Kinetrol)

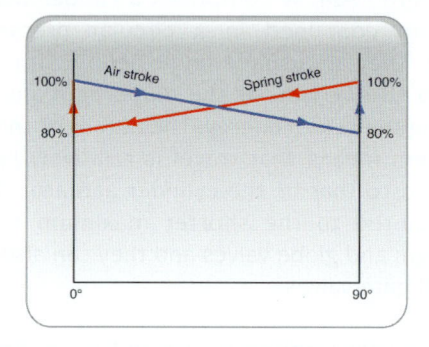

Figure PA11: Torque Characteristic for Spring Return Vane Actuator

Pneumatic motor actuators - The pneumatic motor multi-turn actuator requires the provision of a threaded stem on the valve in order to convert the rotary output into the necessary linear movement in the same way as an electric motor. The thread should be designed to be 'self-locking' so as to provide the necessary positional stability that is not inherent with the other types of pneumatic actuators. The pneumatic motor can be controlled either by a completely pneumatic system employing pneumatic limit valves which cut off the supply to the air motor, or by limit switches which trip solenoid valves which in turn cut off the supply to the air motor at the ends of valve travel.

The pneumatic motor tends to have a high air consumption and, therefore, this design of actuator is not very commonly used unless there is a large capacity gas supply available, such as would be the case in gas transmission pipelines. Pneumatic motor actuators are also susceptible to seizure due to ingress of dirt or due to corrosion of the internal parts if there is any moisture in the air or gas supply system.

Sizing double-acting actuators

All actuator manufacturers publish the torque output for their actuators and show a wide variation between double acting, spring return, scotch yoke, rack and pinion, and vane type designs. Actuator thrust or torque output is proportional to air supply pressure. When sizing actuators for linear motion valves, maximum travel must also be taken into account. Most actuators are sized on a compressor standard of 5.5 barg air. However, if the valve and actuator are situated at a remote part of the plant, or in an area where there are many items of equipment feeding off the air supply, the supply pressure could fall to 4 barg, which will reduce the thrust or torque output of the actuator. Due consideration should be given to this and other environmental and operating factors when considering safety factors to add to the valve manufacturer's specified torque. Actuators are selected from a range of available sizes. The chosen actuator will usually have a torque or thrust output at least 15% higher than the valve operating torque at the minimum air supply pressure.

Care must be taken with the application of double acting piston actuators to gate valves. In the closing direction the air acts fully on the piston diameter. In the opening direction the stem connecting the actuator to the valve reduces the area on which the air can act when applied to the underside of the piston. Hence the thrust generated to unseat the valve will be less than that used to seat it at constant air pressure. It may therefore be necessary to restrict the air pressure on the seating stroke below the maximum available and use the maximum for the unseating stroke.

Sizing spring return actuators

A spring return actuator has air pressure producing torque or thrust in one direction only and is reversed to its original position by a spring action. Some manufacturers use multiple-spring designs for this purpose, while others use one or two large ones. Such spring combinations can be seen in several of the preceding Figures.

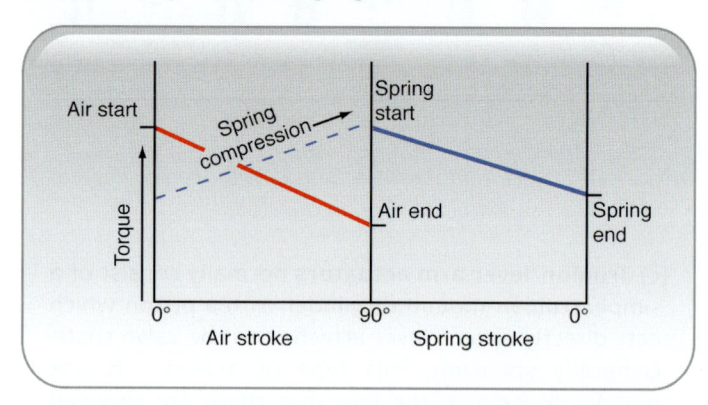

Figure PA12: Torque Characteristic for Rack and Pinion Actuator

There are four key values of output torque to consider when sizing spring return actuators (See Figure PA12). These are:-

Spring in least compressed state air applied to actuator (air start)

Spring in most compressed state air applied to actuator (air end)

Spring in most compressed state air supply pressure zero (spring start)

Spring in least compressed state air supply pressure zero (spring end)

The spring return design of the cam type piston actuator produces a constant torque on the air applied stroke as the design incorporates an additional piston to cancel out the spring force completely on the air stroke. At the start of the spring return stroke there is a 15% increase in torque output, which is potentially beneficial when using this type of actuator on emergency shutdown valves (see Figure PA13).

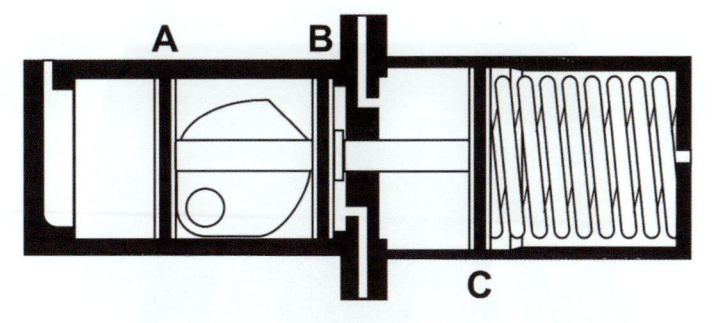

Figure PA13: Spring Return Cam Actuator (Matic Actuators, Div. of Imtex Controls Ltd)

Fire-safe pneumatic actuators for rotary motion valves

Most actuators for rotary motion valves are manufactured from extruded or cast aluminium, and, therefore, are unacceptable in situations where fire may occur. Fire proof coatings can be applied, which limit the temperature rise inside the actuator for at least 15 minutes, or the whole actuator can be enclosed in a fire-proof box. However, this may prove undesirable if space is at a premium, or uneconomical. The alternatives are actuators manufactured from SG iron castings, or fabricated in carbon steel or stainless steels.

These actuators will operate satisfactorily from ambient up to 150°C and, when exposed to a fire of 1110°C for a period of 15 minutes, should remain functional both during and after exposure.

These actuators are used extensively on oil rigs and refineries on hydro-carbon service in highly hazardous areas. The importance of their fail-safe capability is paramount and they may be sized up to four times the normal valve torque requirements to ensure the spring return stroke operates satisfactorily in an emergency. The actuator may be supplied with an air accumulator vessel to guarantee enough air pressure locally to operate the actuator and valve at least once to achieve emergency shut down.

Figure PA14: Methods of fire-proofing valves and actuators

Control accessories for pneumatic actuators used for isolation valves

The following accessories are commonly available for all types of pneumatic actuators: -

- **Solenoid valves** - Solenoid valves are used as a pilot valve to control the flow of air to and from the actuator. The valves are categorized by the number of port openings which govern the directions in which air may flow, e.g., a three-way port valve has a pressure port, an output port, and an exhaust port, making it an ideal choice for a spring return actuator. A four-way port valve has a pressure port, two output ports, and an exhaust port, and is used in conjunction with double acting actuators (see Figure PA15).

Figure PA15: Intrinsically Safe Solenoid Pilot Valve used to control the flow of air to and from an Actuator (Asco Numatics)

- **Limit switches** - Limit switches are used for remote position indication circuits, usually for the beginning and end of the stroke. These switches are commonly contained in a switch box (see Figure PA16) which houses the switch elements, cams, terminal strip and cable connector but can sometimes, on larger actuators, be individual, metal clad switches (mounted off the main body of the actuator). With these switch box designs, a rotary input shaft is driven directly from the actuator output, and the box usually includes a mechanical position indicator (see Figure PA18). Individual switches or switch box designs are available to suit environments where there is a risk of explosion; the switch enclosures being either flameproof (Exd) or intrinsically safe (Exi) certified, depending upon the loads that are to be switched.

Figure PA16: Switch box internals showing four inductive proximity sensors (K Controls)

BVAA VALVE & ACTUATOR USERS' MANUAL - 6th Edition

Figure PA17: Switchboxes come in many different combinations (K Controls)

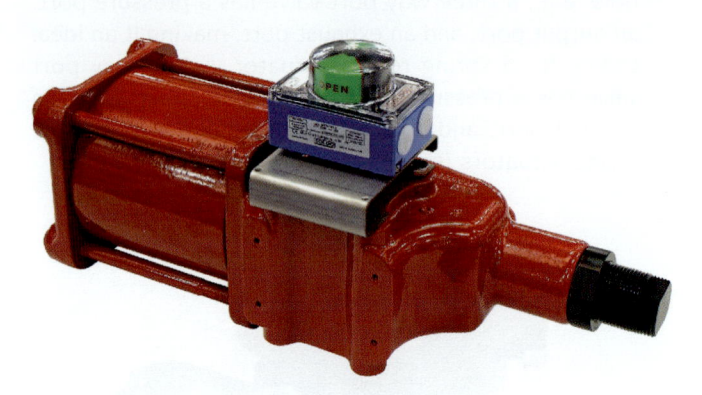

Figure PA18: Mechanical Position Indicator (Rotork)

- **Manual override** - This is a facility to operate the valve manually with the use of a handwheel and gear in the event of failure of the actuator, because of loss of air pressure or of electrical power to the solenoid valve (see Figure PA19). The gearbox should be mounted between the actuator and valve so that the valve can be operated if the actuator has to be removed for repair. However, the auxiliary manual override should have a de-clutching arrangement so that the hand-wheel remains stationary under power operation of the actuator, and this must be sized to cope with the valve torque/forces plus the actuator torque/forces where necessary.

Figure PA19: A Quarter Turn Pneumatic Actuator complete with Geared Manual Override (Rotork)

- **Adjustable opening and closure stops** - As it is extremely important for a ball valve not to over-travel or under-travel about its 90 degrees of operation, some actuator manufacturers have provided adjustable travel stops to ensure the ball stops in the exact position on the valve seats to provide 100 percent opening and 100 percent closure of the valve (see Figures PA20 and PA21).

Figure PA20: Adjustable Limit Stops on a Rack & Pinion Actuator mounted at the end of piston stroke (Flowserve)

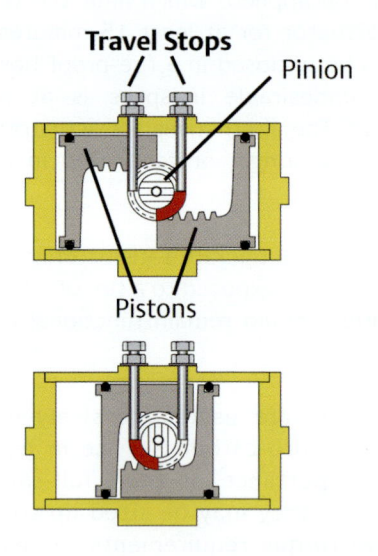

Figure PA21: Schematic of Adjustable Limit Stops directly applied to the Actuator's Pinion (Emerson Hytork)

- **Position indication** - All actuators should provide a clear visual indication of the valve position via a position indicator, which should show the 'Open' and 'Closed' positions as a minimum, unless such an indicator is already provided on the valve.

- **Accumulators** - Accumulators or reservoirs consisting of small storage cylinders can be mounted on the actuator to provide either single- or multi-shot fail-safe operation on applications where a spring return solution is either undesirable or unacceptable.

Advantages of pneumatic actuators

- Compressed air is an economical and convenient source of power
- The different failure options of closed, open and last position can easily be provided
- Air, when subjected to heat will still act satisfactorily as an energy transfer medium and any increase in pressure due to thermal expansion will tend to assist valve operation
- Adjustment of the air supply pressure is possible thereby more closely matching actuator operating torque output to valve operating torque requirements with minimum potential for exceeding the valve strength torque
- Can be installed in hazardous areas providing any auxiliary equipment is also rated for the same area classification
- Can operate satisfactorily from ambient up to 150°C when housings are manufactured from an appropriate material
- When exposed to a fire of 1110°C for a period of 15 minutes, should remain functional both during and after if manufactured from SG iron castings, or fabricated in carbon steel or stainless steels
- Exhaust air can be vented to atmosphere.

Disadvantages of pneumatic actuators

- In most cases no kinetic energy to provide a 'hammer blow' effect to unseat wedging valves is available, although some manufacturers are introducing hammer-blow designs
- Compressibility of air limits its ability to hold a valve position and may result in drift of position with gate and butterfly valves. It also limits distance for pneumatic control circuits to a few hundred yards, so that electric controls are often required for remote circuits
- The relatively low operating pressure of pneumatic supplies limits the size of valve that can be operated to within the range of practical cylinder diameters and, where spring return designs are required, the available output torques will be further reduced
- The materials normally used for piston rings and seals in the pneumatic elements of the actuators are vulnerable to heat and the equipment must therefore be provided with adequate thermal insulation if operation during a fire is to be assured
- The provision of manual operation on very large valves can present problems due to the high locked-in forces that need to be overcome when de-clutching to permit the piston to move during power operation.

BVAA 70 years of service to industry

rotork®

IQ EVOLVING RELIABILITY

IQ Pro

Rotork - over 50 years at the leading edge of actuation technology

- Multi-turn and quarter-turn isolating/regulating duty
- Rugged construction - double-sealed to prevent water and dust ingress even during site wiring
- Simple, non-intrusive, infra-red setup and adjustment
- Multilingual display
- Digital, analogue or bus system remote control and status reporting.

• Three-phase, direct current and single-phase actuators • Direct drive quarter-turn (IQT models) • On-board datalogger included as standard • IrDA™ compatible for local and remote actuator analysis via InSight PC software • Clear, user friendly controls and indication • Multilingual text display for status and setup • Simplified torque and position control for increased reliability • Comprehensive control and flexibility • Approved for use in SIL applications (with additional SFCM control module).

Rotork worldwide. 87 countries. 15 manufacturing plants. 279 offices and agents
visit www.rotork.com for your local agent

Electric Actuators

Figure EA1: Cutaway diagram of Multi-Turn Electric Actuator (Rotork Controls)

The conventional modern electric-motor-powered valve actuator provides a versatile method of valve operation.

For linear motion valve types (gate, globe, diaphragm), actuators with a "multi-turn" output drive raise and lower the valve obturator via a drive nut - threaded stem arrangement.

Rotary motion valve types (plug, ball and butterfly) use actuators with a direct drive 90° output or, for higher torque applications, a multi-turn gearbox driving into second stage part–turn output gearbox.

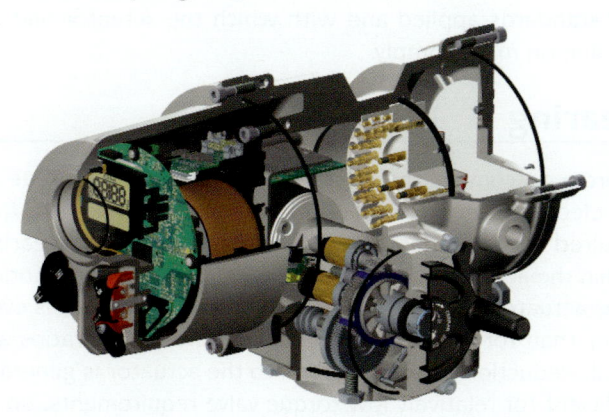

Figure EA2: Cutaway diagram of Rotary Motion Electric Actuator (Rotork Controls)

Facilities for local and remote electrical operation plus handwheel operation are normal features and sizes are available to suit the complete spectrum of both linear and rotary motion valve types.

Electrical supplies

Small actuators often use motors suitable for either single phase AC or DC power supplies. The three-phase squirrel cage induction motor, however, provides the most efficient, robust and simply controlled form of electrically powered prime mover for larger valve applications. It is important that the power supply type and voltage is clearly specified, including tolerances as this will determine the actuator type and size selected by the manufacturer.

Environmental Protection

Actuators for valves on isolating duty often remain idle for long periods and may be installed in harsh environmental conditions. It is essential therefore that the design of electric actuators is such that the electrical components are housed in enclosures which provide watertight protection to at least IP67. It is also desirable that the actuator has a segregated, "double sealed" compartment for cable termination as this will protect the actuator should terminal covers be left off during installation or if cable glands are poorly sealed (see Figure EA3).

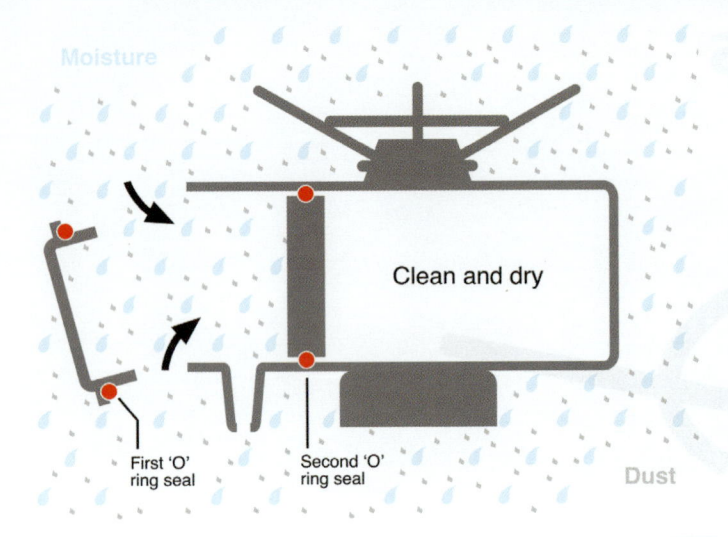

Figure EA3: Diagram showing the benefits of double-sealing (Rotork Controls)

In addition, modern "non-intrusive" electrical actuators (see Figures EA4 and EA5) eliminate the need for access to the electrical compartments during installation, commissioning and subsequent adjustment when the ingress of dust and moisture might cause deterioration of the electrical and mechanical components.

Figure EA4: Non-intrusive communication using a wireless link (Rotork Controls)

Figure EA5: Non-intrusive Multi-turn Electric Actuator (Emerson EIM)

Non-intrusive set up may be by infra-red or wireless system tools supplied with the actuator or by using the local control selectors in "set-up" mode. Another advantage of non-intrusive set up is that it removes the need for specialist knowledge of different manufacturers' electro-mechanical limit and torque systems, making the process of commissioning and adjustment, safer, simple and faster.

Hazardous Areas

Statutory requirements governing the use of electrical equipment in hazardous areas, where there is a likelihood or possibility of a flammable atmosphere being present, is onerous and requires special actuator design, construction and quality arrangements.

The power requirements of electric actuators are such that actuator designs are based on a "flameproof" - "Exd" construction for the enclosures housing all the electrical components. The concept of this protection is that an explosion within the actuator housing cannot propagate to the surrounding atmosphere (it is assumed that a flammable atmosphere can be present within the actuator enclosure regardless of sealing arrangements) and therefore preventing a major site explosion or fire. Sometimes an "increased safety" – "Exe" termination method is specified, a system of terminating power and control wiring at the actuator that minimises the risk of sparks or hot-spots occurring and therefore allows termination in a watertight only enclosure. In this case actuators will be certified "Exde", sparking equipment being located within a "d" enclosure and non sparking cable termination in an "e" enclosure. Additionally, the maximum surface temperature of the actuator must be limited to a temperature below the flashpoint of the surrounding gas or dust atmosphere, an important consideration due to motor heating during operation.

Some manufacturers are able to offer a few components such as limit switches to suit 'intrinsically safe' – "Exi" external circuits but these tend to be very limited due to the difficulties of assuring the segregation needed from higher power circuits within the actuator.

Depending on the geographical location of the plant into which the actuator will be incorporated, there are many national and international "hazardous area" directives, codes and standards applied and with which the actuator and its installation must comply.

Gearing

In order to convert the relatively high speed/low torque of the electric motor to the relatively low speed/high torque required for the operation of the valve, reduction gearing within the actuator is necessary. In the case of rotary motion valve actuators, the speed required at the valve stem is even lower than for linear valves and therefore higher ratios are usual. Reduction gearing integral to the actuator is generally sufficient for relatively low torque valve requirements, up to around 3000Nm, allowing the actuator to be fitted "directly" on to the valve, Figure EA6.

Figure EA6: Direct Drive Quarter-Turn Electric Actuator operating a Plug Valve (Rotork Controls)

For higher torque/thrust valve applications, multi-turn output actuators are combined with a "secondary" part- or multi-turn gearboxes incorporating further gear reduction to achieve higher torque outputs (Figure EA7). This provides actuator and gearbox combinations with torque outputs up to 1 million Nm for rotary motion valves and 50,000 Nm (thrusts up to 3000kN) for linear motion valves.

Figure EA7: A large Plug Valve operated by a Multi-Turn Electric Actuator with intermediate Quarter-Turn Gearbox (Rotork Controls)

For rotary motion valves it is essential that the gearing provided by the actuator or actuator-gearbox combination is self locking to ensure the actuator cannot be back-driven and therefore the valve position is maintained irrespective of valve loadings. Linear motion valve actuators achieve self locking at the valve stem-drive nut interface, the efficiency of the thread being low enough to ensure irreversibility.

Manual Operation

Hand wheels are normally provided for manual operation during actuator set-up and in the event of power failure. For linear motion valve actuators, hand wheels are normally of the direct drive type on smaller actuator sizes, with drive through reduction gearing being introduced for higher torque applications. On rotary motion actuators hand wheels normally operate through reduction gearing. The handwheel is normally engaged by a lever operated clutch which automatically returns to motor drive when the motor is powered. This is a vital feature as it avoids the risk of an actuator being left electrically inoperative by mistake after manual operation.

It is important that the gearing used for the motor drive is not used for handwheel operation, as failure of the gearing would then leave the valve inoperable both electrically and manually.

Travel & Torque limitation

All valves operate between defined positions corresponding to the valve being fully open or closed and the maximum travel is usually defined by mechanical stops. Additionally, the valve seating faces may require either a defined force to be applied as in the wedge gate, globe or triple off-set butterfly designs or to attain a precise position as in the parallel slide gate valve, plug valve or ball valve. All but the smallest, low cost actuators therefore must incorporate position sensing to monitor movement and stop at the set position limits and torque sensing to measure applied force and stop when the set torque limit is reached (see Figure EA8). These systems may be electro mechanical or with more modern designs, electronically based while employing a mechanical pick up derived from the actuator's output movement and applied torque.

Figure EA8: Electronic Torque-sensing device in a modern Electric Actuator (Rotork Controls)

The torque sensing system is factory calibrated to the actuator catalogue torque rating and is usually adjustable down to 40% of rated. Independent torque settings can be made for the closed and open valve strokes.

Irrespective of the sensing system types, the actuator has position limits that must be set in synchronization with the valve's open and closed positions and torque levels that are set to meet the operating forces required by the valve. In addition, to suit the particular design of the valve seating faces, the actuator can be adjusted to stop as a result of reaching a set position, known as "limit", or as a result of delivering the set torque, known as "torque". A combination of "torque" to close and "limit" to open would normally be used on a globe valve. It is usual that the torque sensing also provides valve and actuator overload protection at end of travel and during the valve stroke, stopping should the valve be stuck or be become obstructed. For example a plug valve may become jammed in either direction due to lack of lubrication.

Position limit contacts are usually made available to provide end of travel indication to give the user feedback of valve position (open, intermediate or closed).

Control

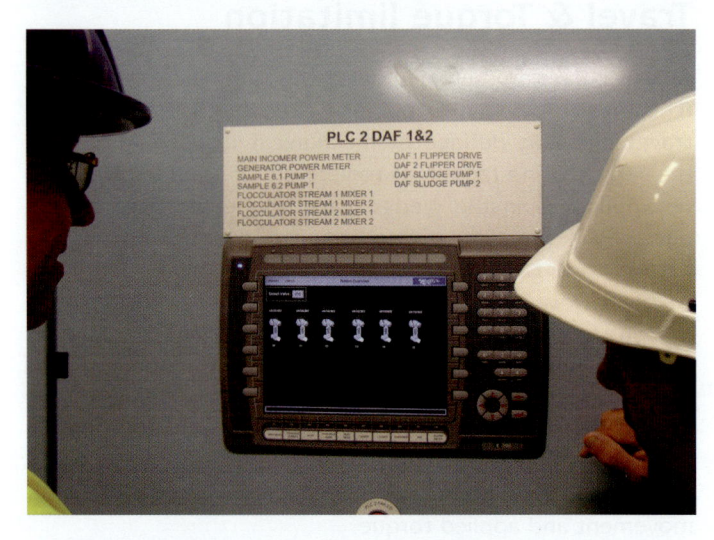

Figure EA9: Actuator control and monitoring HMI screen in a modern water treatment works (Rotork Controls)

Actuators provide a means of achieving centralised process control and electric actuators specifically can interface directly with modern computer based systems. There are two basic types of actuator control systems; integral and separate.

Integral control actuators contain all the necessary motor switching gear and ancillary circuits within the actuator enclosure and are therefore self-contained. Actuators utilising separate control comprise of the motor, limit switches and the mechanical equipment and therefore require the motor switch gear to be located remote to the actuator, usually located in a motor control centre provided by the user.

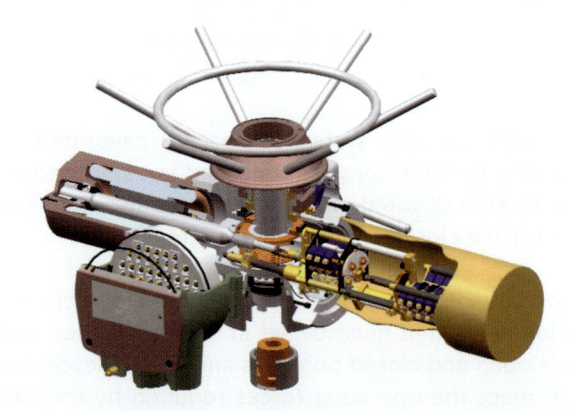

Figure EA10: Cutaway diagram of an Electric Actuator for the Nuclear Power generation industry, one of the few areas where actuators without integral controls are preferred due to specific seismic and operational requirements (Rotork Controls)

The advantage of the integral system is that the actuator manufacturer designs and builds the control package appropriate to the actuator and is therefore totally responsible for it. All interconnections are made by the manufacturer and can be tested as a "unit". Those in the supply chain such as valve makers and contractors benefit as they can carry out package set-up, acceptance testing, site commissioning etc. with no more that the availability of an electrical power supply.

Separate controls require third and even fourth parties to source and supply the necessary motor control gear and make all the necessary interconnections, blurring the areas of responsibility. Separated control packages may have an advantage where limiting the equipment in the actuator may be beneficial such as when the valve is located in a high temperature environment or where vibration is significant.

The ability to integrate with process control systems is key in valve actuation. Commonly, these systems utilise digital data transfer for control and indication. In these cases, actuators include field control units which convert actuator status to digital outputs for transfer to the control system, and digital command inputs from the system for control of the actuator.

Figure EA11: A two-wire digital control system designed to control up to 240 actuators on a single fault-tolerant bus loop of up to 20 kilometres in length (Rotork Controls)

Duty cycle

As the valve load imposed on the actuator together with the time the actuator has to drive the valve causes heating of its motor, the performance required of electric actuators is related to the duty required by the valve in achieving the required process control. Four categories of duty are recognized as per EN15714-2 'Industrial valves – Electric actuators.'

- **Isolating** – valves for full open/closed duty where operation will be few times per day, week or even year.
- **Regulating** – as isolating but also required to move the valve a number of times in mid stroke by inching or coarse positioning followed by fairly long periods of inactivity, where the actuator is expected to operate between 30 and 60 starts per hour.

- **Modulating** – generally associated with true valve process control, where the actuator is expected to operate between 60 and 1200 starts per hour in accurately positioning the valve.
- **Continuous Modulating** – as modulating but where the actuator is expected to operate above 1200 starts per hour.

The majority of electric actuators designed for isolating duty can also accommodate the requirements of regulating duty, but this ability should be checked with the individual manufacturers.

Electric actuators for modulating duties are normally designed with special motors and solid state motor starting, allowing rapid switching and frequent operation and as a result, are usually more costly than those designed for isolating duty.

Torque Characteristic

Generally, the torque output characteristic of electric actuators is a "flat line" across the valve stroke (see Figure EA12), the level of which is governed by the setting of the torque switches – usually in the range of 40% to 100% of the actuator rated (catalogue) torque. It is important to distinguish the actuator torque characteristic from that of the actuator motor torque characteristic. A particular motor will be selected by the actuator manufacturer with a torque-speed characteristic that, when combined with the actuator gearing, provides that catalogue rated output torque at the required speed and meets the required duty. The motor characteristic is largely irrelevant to the user who need only focus on the actuator output performance. Aside from some DC motor driven actuators, the majority of actuators can be considered single speed machines and therefore typically a manufacturer will specify the actuator performance in terms of a rated torque at a speed with a duty cycle. For example, an actuator specified as having 150Nm rated torque at an output speed of 24rpm, is capable of operating for 15 minutes at an average of 1/3 rated torque.

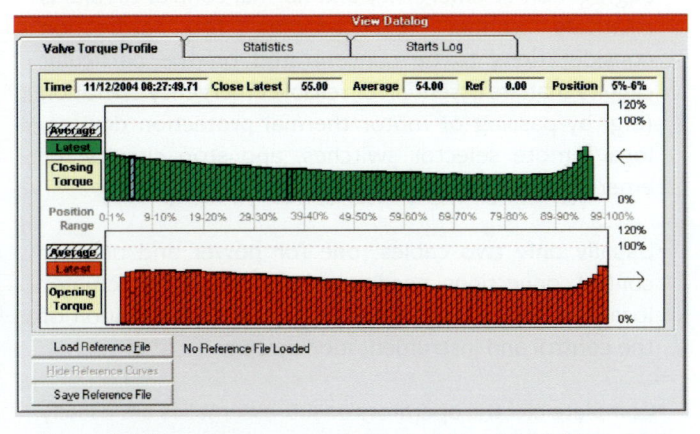

Figure EA12: Actual torque characteristic of 36" Class 150 ball valve taken from the actuator datalogger. The bottom red graph is the opening torque profile, top green graph is closing. Dash lines define the actuator open/close rated torque (100% torque switch setting)
(Rotork Controls)

This flat line torque characteristic simplifies the actuator selection. When operating times full open to close are typically less than 5 minutes only the valve operating torque relative to the actuator rated torque has to be considered in sizing the actuator. For longer operating times the requirement of the actuator to operate at an average of 1/3 rated torque must also be considered and the valve running torque needs to be assessed against this lower actuator running torque.

Provision may be made in the control of the actuator to bypass the torque switches during unseating of the valve from its open or closed positions when additional torque to that provided for a "clean" valve may be required. In bypassing the torque switch protection, the motor is capable of delivering torque up to the point it stalls. For AC actuators, stall torque will be in the region of 1.3 to 3 times the actuator rated torque. This option should only be considered when the valve manufacturer has confirmed that the valve and actuator mounting kit are capable of accepting the higher stall torque without damage.

General rules for actuator sizing

In selecting the correct size of actuator for the specified supply voltage and enclosure requirement, the following factors must be considered as a minimum.

The actuator rated torque or in the case of long operating times the actuator run torque must meet or exceed the valve operating torque.

- The actuator must mechanically accept the valve stem/shaft and have a compatible mounting flange to that provided on the valve or by the adaption kit.
- Actuator speed or operating time should meet the stroke time for the valve as required by the process.
- The actuator stall torque/thrust shall not cause damage to the valve or any adaption kit.

Special issues for sizing linear motion valves

Multi-turn electric actuators are used for operation of linear motion valves and their output performance is defined in terms of output torque and speed at a specified supply voltage and frequency. As output speed or supply frequency increases the rated torque for a particular actuator size reduces.

The main difference between an actuator for rotary motion and the multi-turn actuator for linear motion valves is that the multi-turn actuator must also have a thrust rating capability.

The actuator speed required to achieve a particular valve operating time will be governed by the thread lead of the valve stem. It is important to note that an increase in the thread lead will achieve a shorter valve travel time far more efficiently than increasing the actuator speed by the appropriate amount, thus avoiding an increase in the inertia of the motor.

Thrust is the force transmitted from the valve stem thread to the valve obturator by the action of applying torque to the actuator stem nut and is proportional to the applied torque for a given stem factor. The thrust generated in moving the valve must also be reacted by the actuator and the actuator mounting kit. It is important to understand that while the torque required by the actuator to operate the valve can be relatively modest; the thrust generated can be very high, often measured in many tons. Any failure of the actuator, mounting kit or valve in withstanding the thrust could lead to a catastrophic and potentially dangerous event. All multi-turn actuators have a maximum thrust rating which needs to be greater than the maximum thrust required by the valve and greater than the thrust generated by the actuator maximum torque applied to the valve stem. This is especially important to consider if safety factors are applied to the operating torque and/or operating device.

Figure EA13: Exploded diagram of separate drive bush and thrust base in an Electric Actuator (Rotork Controls)

For example, a gate valve requires a maximum operating thrust of 137kN which, with a defined stem thread and stem factor, gives a required actuator torque of 500Nm. If the actuator and operating torque are subject to a combined safety factor of 1.5, a torque of 750Nm is required. Due to actuator speed requirements and sizing breaks, an actuator with a rated torque of 850Nm may then selected. As the stem factor is the same, the selected actuator will develop a resultant thrust of 233kN at its rated torque or 1.7 times the thrust required by the valve. If this resultant thrust is not considered by the actuator manufacturer, or is unknown to the valve maker there is a real danger of damage occurring – a bent stem at best or an actuator-valve mounting failure and an ejected actuator at worst.

It is important to note that if the actuator motor stalls (energised but unable to move) as can happen in a jammed valve situation, the torque developed can be up to 3 times the rated torque. The thrust developed at stall torque must therefore be considered in the design and construction of the valve and any interface components.

Advantages of electric actuators

- Electric actuators have a "flat line" torque output making sizing easier. Providing the actuator rated torque meets or exceeds the maximum valve required torque, no further consideration of the valve stroke torque characteristic is required

- Energy efficient operation

- Motors have high rotational speed providing high kinetic energy to help the unseating of valves

- Valve position stability is inherent due to the reduction gearing and/or nut/screw arrangement normally employed for transmitting the drive to the valve

- Ability to interface directly with modern computerized and telemetric remote control systems

- Fewer system components and low maintenance

- Electric circuits do not transmit dirt or moisture, and operate instantly over great distances

- Handwheel operation is simple to provide, together with a hand/power clutch mechanism having power preference

- Interface between motor and remote control is a contact, which has better reliability than solenoid pilot valves

- The complete actuator and control system, apart from the final drive to the valve, can be tested without moving the valve position, by holding the handwheel engagement clutch lever in the 'hand' position

- Where safe plant shut down demands a particular sequence of valve operation, the necessary interlocks can be provided simply by interconnection of limit switches on the appropriate apparatus

- Segregation of emergency and normal control circuits is relatively simple and enables fire protection considerations to be concentrated on the particular circuits that are essential for reliable emergency operation (e.g. by-passing of motor thermal protection devices, local/remote selector switches, and stop circuits for emergency operation)

- Usually only two cables, one for power and one for control connections, need to be brought into the actuator location, irrespective of the degree of sophistication of the control and instrumentation system

- Complete electric operators or sub-assemblies are usually readily interchangeable, though they may have initially been supplied for completely different applications. This leads to the possibility of quick turn-round on repair or replacement of a faulty actuator on emergency duty, if electrical actuators are chosen for both emergency and normal applications on the same installation.

Disadvantages of electric actuators

- Stay-put failure mode in the event of power failure

- Vulnerability of electrical equipment and cabling to high temperatures

- For a given actuator the range of torque output between maximum actuator torque and valve operating torque can be significant. If safety factors in terms of voltage supply are applied to valve operating torque and actuator stall torque is considered as the maximum torque, ratios of 5:1 are possible. For small valves ≤ DN50 this can present considerable difficulties to the valve manufacturer to ensure that the strength torque of the valve is greater than the maximum actuator torque

- Electrical energy is difficult to store and the need for gearing between the motor and the valve usually precludes mechanical means of energy storage in the form of a spring for emergency one-shot operation

- Vulnerability of electrical equipment to moisture requires high integrity sealed enclosures.

Figure EA14: Electric valve actuators controlling the injection plant at an underground storage depot for natural gas (Rotork)

ADVANCED ACTUATORS

ACTUATORS – HYDRAULIC – PNEUMATIC – ELECTRIC

Unit 4, Ryefield Way, Silsden, Bradford BD20 0EF
Tel : 08707 576664 Fax : 08707 576665
email : sales@adcomtec.co.uk   www.adcomtec.co.uk

Hydraulic Actuators

Figure HA1: Double Acting Linear Hydraulic Actuator (Rotork Fluid Systems)

Pneumatic and hydraulic actuator drives are very similar in terms of the basic design concepts. The differences between pneumatic and hydraulic actuated systems is the power medium used. Pneumatic actuators operate from a compressed air supply (typically 5.5 barg) and hydraulic actuators operate from pressurised liquid at generally higher pressures (100 to 200 barg). Hydraulic actuators can be offered to operate at lower pressures when used in self contained actuators packages.

Hydraulic actuators are mostly used when very high torque, thrust, fast stroke speeds or fail safe action is required to operate the valve. The self-contained electro-hydraulic actuator is particularly suitable for critical fail-safe applications, extremes of ambient temperatures and remote locations when electrical operation is preferred and pneumatic or hydraulic power is not available. Some models are also suitable to be powered from Solar panels on remote installations.

With pneumatic actuators the fluid is typically air which does not present a hazard to the environment when discharged during valve movement or should the actuator or air supply system leak. Hydraulic actuators are normally installed in a closed loop systems that would not discharge the fluid unless there is a seal failure. Any leakage of hydraulic fluid could present a hazard but over recent years various biodegradable hydraulic fluids have become available for more sensitive applications. The pressurising unit (hydraulic power unit - HPU) draws the hydraulic fluid from a reservoir and the fluid is pumped around the closed loop system with the fluid being circulated back to the reservoir eliminating potential contamination of internal components that may occur with pneumatic systems.

The incompressible nature of liquid and higher operating pressures have the advantage of generating small volume changes of the fluid inside a fixed space. Thermal expansion of the pressurised fluid needs to be taken into consideration due to ambient temperature changes, both increasing and decreasing, when designing the actuator system. Due to the smaller volumes and higher pressures within hydraulic systems, potential leakages are magnified but providing the reservoir is correctly sized and pressure monitoring is designed into the system leakage can be detected and minimised.

The higher pressure within hydraulic systems requires the various seals and valves in the hydraulic system to be maintained. Good maintenance is therefore more critical than in pneumatic systems. The hydraulic fluid must be filtered to maintain cleanliness to minimise deterioration of the components in the system. Providing the system is designed as a closed loop the hydraulic fluid acts as barrier to protect the internal components from the environment.

Hydraulic actuators are more suited to installations with locally mounted hydraulic power units. The selection of the type and viscosity of the hydraulic fluid is critical to suit the application and performance of the actuator. Pneumatic actuators are more suited to installation with long pipe runs from the air compressor providing they can provide sufficient torque or thrust to operate the valve.

Also as pressures are higher, hydraulic cylinder diameters are smaller than pneumatic cylinder diameters for a given force. This brings substantial savings in size for double acting actuators but for spring return actuators the spring will remain the same whether the system is hydraulic or pneumatically operated.

The speed of stem movement or shaft rotation is a function of how quickly the delivery system can supply the hydraulic fluid, the viscosity of the fluid in the cylinder, piping size and the size of orifices in the return valve to discharge the pressure in the opposite direction. By placing restrictors in the hydraulic piping piston speeds can be varied.

Hydraulic power supplies

The source of hydraulic power falls into one of three categories namely manual hand pump, central power unit with hydraulic ring main, or self contained. Both the central and self contained systems require electric power to run the pumps that move and pressurise the fluid.

Manual
The manual hand pump can be the sole means of operation. It is also common to have a manual override on a spring return hydraulic actuator. This is an effective means of overcoming the high spring forces that are typically found in this type of actuator (see Figure HA2).

Figure HA2: Hydraulic linear, spring-return actuator with optional Hand Pump (Emerson Bettis)

Central Electro–Hydraulic System with hydraulic ring circuit
Traditional electro-hydraulic systems operate with a hydraulic ring circuit for multiple hydraulic actuators, powered from a central electro-hydraulic power unit. The HPU provides an interface between electrical controls and is powered from a suitable electric supply to provide the hydraulic fluid pressure to drive the actuators. These systems are generally used on large valves with high torque or thrust and when fast stroke speeds are required.

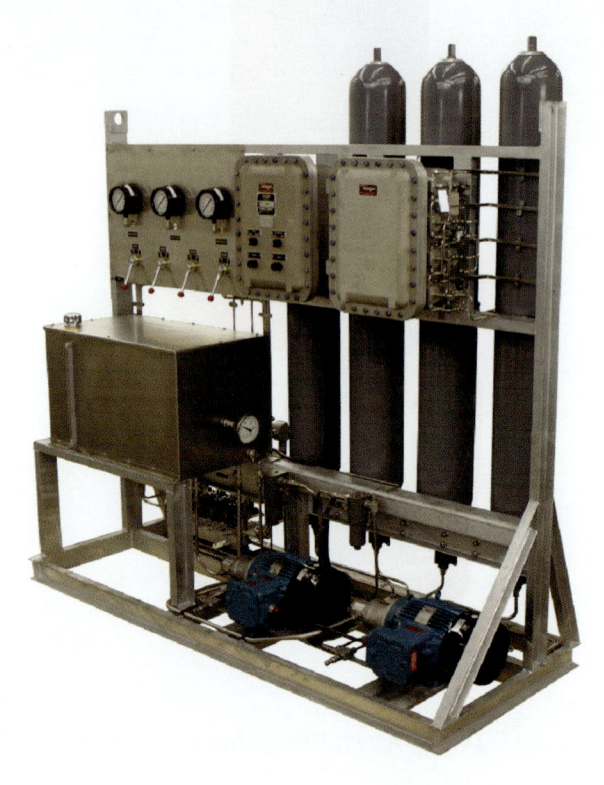

Fig HA3: Central Hydraulic Power Unit - HPU (Rotork Fluid Systems)

Self Contained Electro-Hydraulic Actuators
Over recent years the advancement of electronic, motor and hydraulic control and together with the requirements for remote communication has seen a growth in compact self contained electro-hydraulic (EH) actuators which provide a more cost effective and reliable solution.

Self contained EH actuators provide an electrically operated actuator solution with the reliability of fail-safe emergency shutdown and modulating control, not normally associated with conventional electric actuators.

Typically a self contained EH actuator consists of an EH control module, motor, hydraulic pump and quarter turn or linear actuator drive that is designed to operate one valve. They are particularly suited to safety critical applications providing stored energy in the form of a spring return hydraulic drive, or double acting drive with hydraulic accumulator back-up, to drive the valve to the safe position to suit the process conditions. These actuators can be supplied for fail-safe to close, open or lock-in-position on loss of power or control signal. Self contained EH actuators are also suitable for modulating control applications. See the chapter on Actuators for Control Valves.

They can also be supplied with the capability to work within a defined Safety Integrity Level (i.e. SIL 2 and 3), with hardwired emergency shutdown (ESD) signals to operate dual fail-safe solenoids to provide redundancy in a critical application.

EH actuators can also be provided to monitor the valve position and hydraulic pressure applied to the actuator drive which in turn will monitor and limit the torque applied to the valve. Remote alarm and diagnostics are available for the actuated valve assembly through the on board processer and Fieldbus communications.

Self-contained EH actuators are available covering torque and thrust requirements for both small, low pressure and large high pressure valves.

Low pressure self contained EH actuators (LP-EH)
For the lower torques and thrusts the LP EH actuator: -

Consists of an EH power module with integral hydraulic reservoir, motor pump and electronic controls. The module can be either mounted on quarter turn (see Figure HA4) or linear hydraulic actuator (Figure HA5). Optional extras could include position transmitter, electro mechanical limit switches and manual override.

Figure HA4: LP EH power module with Quarter Turn Scotch Yoke Piston Actuator (Rotork Fluid Systems)

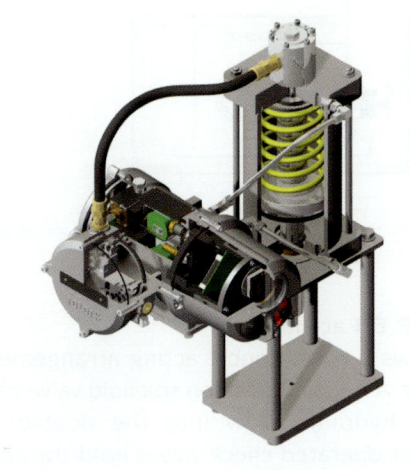

Figure HA5: LP EH power module with Linear Piston Actuator (Rotork Fluid Systems)

The actuators are hydraulically filled, sealed to the environment and operate on a pump and bleed principle with a low internal hydraulic pressure (10 bar). Utilising an internal motorised pump to provide hydraulic pressure in one direction and spring return in the opposite (bleed) direction; the position is controlled via internal manifold solenoids.

High Pressure Line
Low Pressure Line

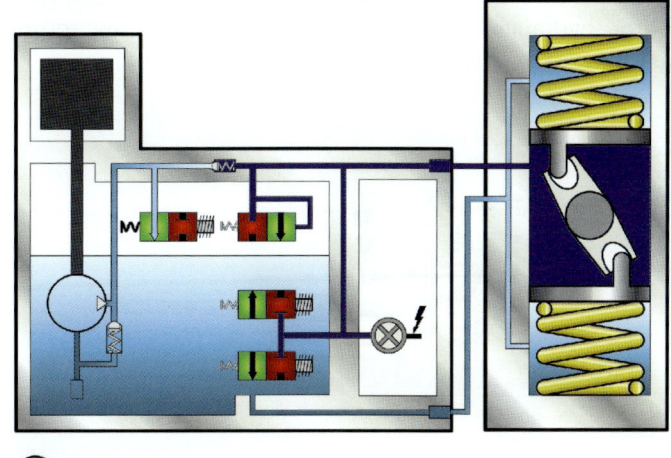

Figure HA6: LP-EH Schematic diagram of a Quarter Turn Actuator (Rotork Fluid Systems)

When an electrical supply is applied, the motorised pump is switched on and the solenoid valves are energised closed; pressurised hydraulic fluid is supplied into the cylinder resulting in linear or rotary motion. As the actuator travels through the stroke, the hydraulic fluid on the low pressure side of the actuator is transferred back into the reservoir to maintain the level (Figure HA6).

When in position or at the end of travel, the motorised pump is de-energised and the hydraulic fluid is held under pressure in the spring opposed side of the actuator.

To close, the bleed solenoid valve(s), and motor are de-energised. The pressure is released and the hydraulic fluid returns to the reservoir resulting in the springs forcing the drive shaft and valve to the safe position. This will also apply to loss of power or emergency shutdown signal; the actuators can be supplied to fail close, open, or Lock-in-position.

The actuator could be supplied with a pressure transmitter (to monitor) and pressure relief valves (to limit) the hydraulic pressure generated in the actuator and thereby restrict the torque or thrust applied to the valve.

An option for critical fail safe applications can be provided to partially stroke the valve, which will allow the actuator and valve to move to a pre-determined position without affecting the process, thereby verifying the actuated valve assembly is available to operate in an emergency.

High pressure self contained EH actuators (HP-EH)
For the higher torques and thrusts the HP EH actuator offers a flexible and customised actuator package to suit the specific application and eliminate the installation and maintenance cost

associated with conventional hydraulic systems with central hydraulic power units. Accumulators and various hydraulic pumps sizes are available to suit the required operating speed and application.

The HP EH (Figure HA7) consists of an integrated EH control module with hydraulic manifold (Figure HA8), power unit with motor, hydraulic pump and reservoir and a spring return quarter turn or linear hydraulic actuator.

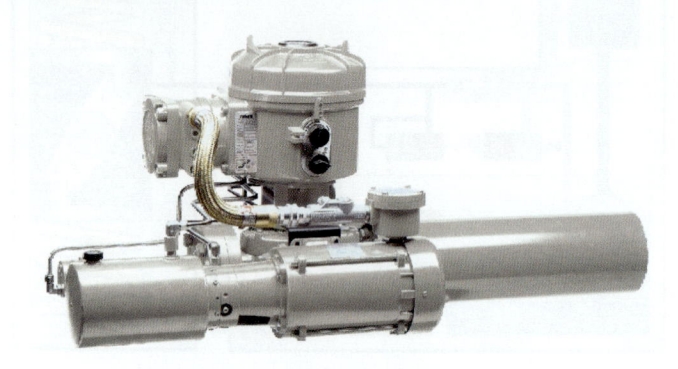

Figure HA7: HP EH unit shown mounted on a Quarter Turn Actuator (Rotork Fluid Systems)

Figure HA8: EH Control Module (Rotork Fluid Systems)

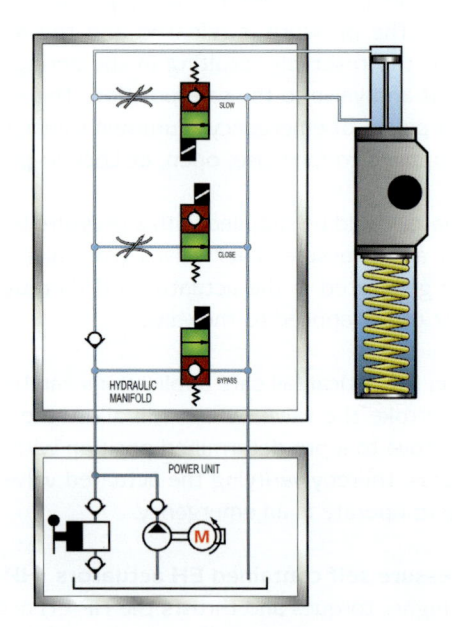

Figure HA9: Schematic - Spring Return HP EH (Rotork Fluid Systems)

Spring Return HP EH actuator

Figure HA9 shows a basic but typical spring return arrangement for an HP-EH actuator. The actuator operates on a similar principle to the LP-EH, but as a result of operating with a higher internal hydraulic pressure will provide a higher torque or thrust output. The actuator would be supplied with a pressure transmitter and pressure relief valves to monitor and limit the hydraulic pressure generated in the system (Note: These are not shown in the schematic Figure HA9).

The actuator is hydraulically pressurised by the motorised pump. Solenoids and pilot valves control the flow of pressurised hydraulic fluid into the cylinder side of the actuator. The actuator is returned to the safe or closed position by the actuator return spring under control of the bleed solenoids and flow restrictors.

Open Stroke:
Pump starts. Bypass valve opens to circulate hydraulic fluid back to tank while motor starts

Bypass and bleed valves close (de-energised) forcing hydraulic fluid into cylinder

Bypass valve opens to stop actuator, pump stops and the hydraulic fluid is held in the system by non-return and bleed valves.

Close Stroke:
Bleed solenoid de-energized, allows oil to flow to tank and the actuator spring returns the valve to a failsafe position.

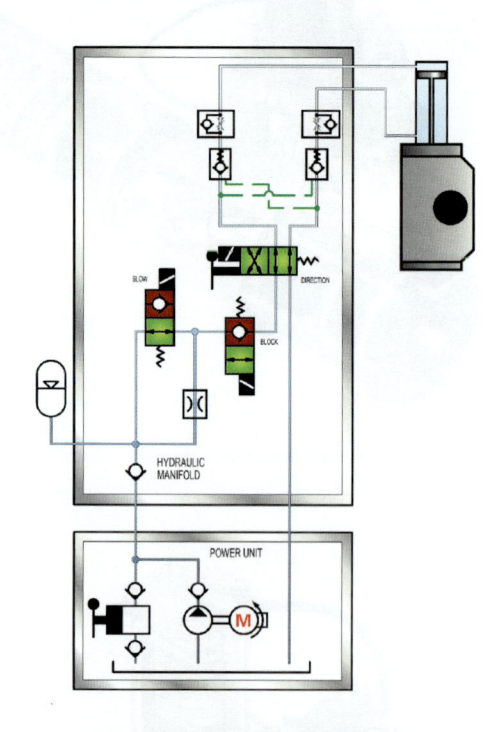

Figure HA10: Schematic - Double Acting HP EH with Accumulator (Rotork Fluid Systems)

Double Acting HP EH actuator

Figure HA10 shows a basic double acting arrangement for an HP-EH actuator. A 4-way 2 position solenoid valve controls the direction of hydraulic flow into the double acting actuator. Dual pilot operated check valves hold the actuator in position upon loss of hydraulic pressure and flow control valves provide adjustment of the stroke speed. The actuator would be supplied with pressure transmitter, and pressure relief valves to monitor and limit the hydraulic pressure generated in the actuator (Note: These are not shown in the schematic Figure HA10).

The actuator includes one or more accumulators to provide stored energy in the form of pressurised hydraulic fluid to operate the double acting actuator and to drive the actuator to a predetermined position on loss of power supply or control signal. The accumulators are sized to provide enough pressurised hydraulic fluid to return the actuator to the pre-determined position or to operate the actuator for multiple strokes when power has been lost. They can also provide reduced stroke speed by pre charging the accumulators from the motorised pump.

Hydraulic Actuator types

There are four basic types of actuator which convert hydraulic power into mechanical output.

Piston actuator - The piston actuator consists of a cylinder and piston which is sealed with an elastomer or PTFE seal. Hydraulic fluid is supplied either to both sides of the piston (called double acting) and movement is achieved by one side being at a higher pressure than the other or fluid is supplied to one side (called single acting) and the return energy comes from a spring. Piston actuators generate linear force and can be used to operate linear motion valves by directly attaching them to the valve stem. The combination of hydraulic fluid and piston cylinder provides accurate and repeatable modulating control of globe type control valves. The small area of the cylinder is particularly useful for the long stroke length of a gate valve as the quantity of fluid needed to move the valve is reduced significantly. The piston actuator requires some form of mechanism for use on rotary motion valves to convert linear force to torque. This is made easier because the majority of part-turn valves require 90 degrees of movement only and therefore only a short piston movement. There are four types of mechanism available to convert linear to rotary motion.

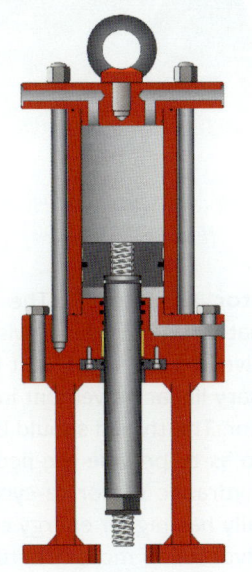

Figure HA11: Double Acting Linear Hydraulic Actuator (Rotork Fluid Systems)

(a) **A scotch yoke actuator** operates in exactly the same way as the pneumatic version. The canted scotch yoke is also available in hydraulic form (see Figure HA12).

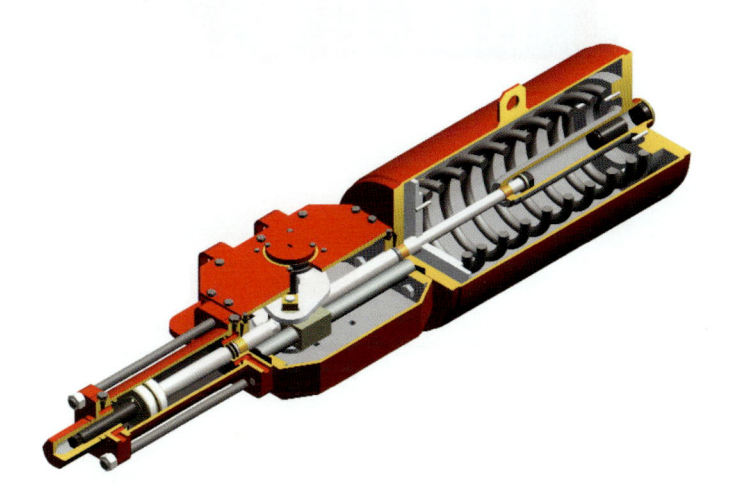

Figure HA12: Single Acting, Scotch Yoke, Quarter Turn Hydraulic Actuator (Rotork Fluid Systems)

Figure HA13: LP EH power module with Quarter Turn Spring Return Dual Opposed Scotch Yoke Actuator (Rotork Fluid Systems)

(b) **The rack and pinion actuator** consists of a single or double piston coupled with an integral rack which drives the pinion. Standard hydraulic rack and pinion actuators operate with a single piston for double acting or spring return action as previously explained. Double piston designs are normally used on low pressure self contained electro hydraulic actuators and operate as described in the pneumatic actuators chapter of this manual. The majority of actuators are designed to producing a 90 degree turn on the pinion (see Figure HA14).

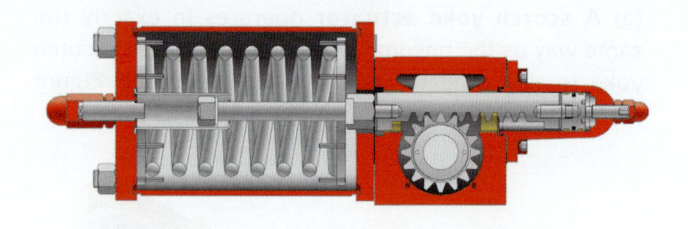

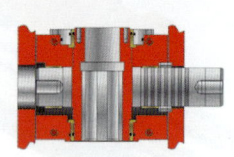

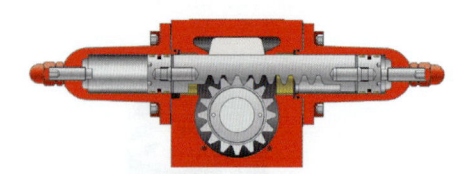

Figure HA14 – Section view of a Rack and Pinion spring return (top) and double acting (bottom) Hydraulic Actuator (Rotork Fluid Systems)

(c) **Trunnion/lever arm actuators** normally consist of a simple trunnion-mounted cylinder with a piston which acts directly upon a lever attached to the valve shaft. Generally speaking, this type of actuator is not popular in view of the fact that there are exposed moving parts which may cause injury to personnel.

(d) **A cam mechanism actuator** consists of double pistons connected by spacer bars with a cam and shaft between them (see Figure PA8, Pneumatic Actuators chapter). The rotational stroke is determined by the cam profile, which is in contact with the centre of the pressurized piston face throughout the stroke. The cam profile can be designed to give an exact 90 degrees rotation, alleviating the need for mechanical end stops. The output torque of double acting cam actuators remains constant throughout the stroke.

Helical Spline Rotary actuators - The double acting rotary piston type (Figure HA15) is a variation of the cylinder actuator that is only available as a hydraulic actuator due to the high axial thrusts required. The large pitch multi start thread machined on the inside diameter of the housing forces the piston to rotate as it moves. This rotation is transferred to the output shaft by the axial splines machined on the inside diameter of the piston and on the outside of the output shaft. The large pitch thread generates a large torque from the force applied to the piston.

Figure HA15: Helical Spline Rotary Hydraulic Actuator (Emerson Bettis)

Vane actuators - Vane actuators are simple quarter-turn devices using a vane with an integral rotary output shaft to produce the torque (Figure HA16). They are multi vane and of much smaller diameter than the pneumatic equivalent.

Figure HA16: Rotary Vane type Hydraulic Actuator (Emerson Schafer)

Hydraulic motor actuators - The hydraulic motor multi-turn actuator requires the provision of a threaded stem on the valve in order to convert the rotary output into the necessary linear movement in the same way as an electric motor. The thread should be designed to be 'self-locking' so as to provide the necessary positional stability. The hydraulic motor re-cycles the hydraulic fluid but typically has higher energy consumption than a cylinder. Hydraulic motor actuators are also susceptible to seizure due to ingress of dirt so the hydraulic fluid must be kept clean.

BVAA VALVE & ACTUATOR USERS' MANUAL - 6th Edition

Sizing Hydraulic actuators

The method for sizing hydraulic actuators is similar to that used for sizing pneumatic actuators except that the fluid pressures are much higher. An important feature of hydraulic actuators is the ability to control valve position by the use of incompressible fluid compared with pneumatic or electric actuators.

Fire-safe hydraulic actuators for rotary motion valves

The smaller size of the hydraulic actuator compared with pneumatic actuator makes fire protection easier and more cost effective. However the incompressible nature of the fluid means that thermal expansion will create a large rise in pressure if the system is unprotected and exposed to fire. Mineral type hydraulic fluids are also inflammable. Many fire-resistant hydraulic fluids now exist, and there has been a recent resurgence in water hydraulics, but these applications often require specialist materials and seals.

Control accessories for hydraulic actuators used for isolation valves

Hydraulic actuators have similar accessories to pneumatic actuators.
- Solenoid valves
- Limit switches
- Manual override - A common form of manual override for hydraulic actuators is a hand pump connected to the hydraulic reservoir
- Adjustable opening and closure stops
- Position indication
- Accumulators.

Advantages of hydraulic actuators

- High operating forces can be achieved enabling the largest size valves to be operated
- Hydraulic pressure can be stored so fail-safe operation can be arranged
- The incompressibility of the fluid enables the necessary position stability to be achieved for the operation of control and gate valves, where required, a hydraulic lock can be applied
- The hydraulic operator is extremely compact, which enables fire protection insulation to be conveniently and economically provided
- The fluid from the high pressure side of the cylinder can be exhausted rapidly to give a high spring return speed.

Disadvantages of hydraulic actuators

- In most cases no kinetic energy to provide a 'hammer blow' effect to unseat wedging valves is available, although some manufacturers are introducing hammer-blow designs
- Costs of providing a hydraulic power supply (HPU) system could be high
- Piping pressure drops limit the distance for hydraulic control and therefore an electrical interface is needed for remote systems
- Hydraulic fluid has a relatively high thermal expansion rate and, therefore, any hydraulic control circuit piping must be heavily insulated against fire. It may also have to be provided with suitable pressure relief devices to avoid the risk of pipe fracture or pilot valve malfunction caused by excessive pressures resulting from thermal expansion
- Some hydraulic fluids are also inflammable
- The close machining tolerances associated with high pressure hydraulic components makes a hydraulic system susceptible to malfunction due to damage to external piping or heat from a fire if no thermal insulation is provided.

Advantages of self contained electro hydraulic actuators

- Reduced installation costs
- Most reliable means of positioning a valve in a safe position on loss of power or ESD signal
- Designed for safety critical applications
- Totally sealed system suitable for harsh environments
- Suitable for Zone 1 / Division 1 hazardous areas.

BVAA 70 years of service to industry

rotork®

FLOW CONTROL INNOVATION

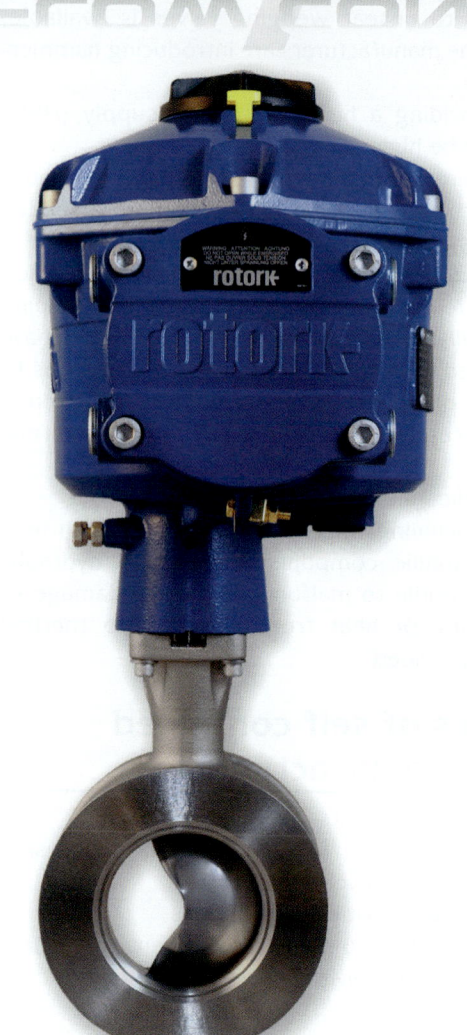

Control Valve Actuator

Rotork - over 50 years at the leading edge of actuation technology

- Linear and quarter-turn actuators to automate control valves
- High performance, continuous unrestricted modulation duty - S9
- High resolution and repeatability • Non-intrusive Bluetooth™ setup and adjustment
- Rugged construction - double-sealed to prevent water and dust ingress even during site wiring
- Watertight IP68 and explosionproof enclosures • Comprehensive datalogging
- Bus system remote control and status reporting.

• Single-phase or direct current power supplies • Linear drive action (CVL models) • Quarter-turn drive action (CVQ models)
• On-board datalogger included as standard • Bluetooth™ compatible for local setup / control and diagnostics
• Accurate and repeatable positional control using 4-20 mA signal • Digital communication options include HART®, Profibus and Foundation Fieldbus • Direct torque / thrust measurement for protection and monitoring • Scalable control input characterisation
• Configurable fail-to-position option using supercapacitor technology.

Rotork worldwide. 87 countries. 15 manufacturing plants. 279 offices and agents
visit **www.rotork.com** for your local agent

Control Valve Actuators

Figure CVA1: An Electric Control Valve Actuator (Rotork) fitted to a wafer pattern, V-port Segment Control Valve (Metso Neles)

General

An actuator is the operating device, which transforms a signal into a corresponding movement of the internal regulating mechanism (closure member) of the control valve using an external power source.

The basic types of control valve actuators available are:

- Pneumatic spring opposed diaphragm
- Pneumatic double acting cylinder
- Pneumatic spring opposed cylinder
- Pneumatic rotary spring opposed cylinder
- Electric
- Hydraulic
- Electro Hydraulic

Control Valve Actuator Terminology
The actuator is an essential element of the control valve and so IEC 60534-1 defines terminology associated with actuators with particular regard to their application to control valves.

Actuator – Device or mechanism which transforms a signal into a corresponding movement controlling the position of the internal regulating mechanism (closure member) of the control valve. The signal or energising force may be pneumatic, electric, hydraulic, or any combination thereof.

Actuator Power Unit – That part of the actuator which converts fluid, electrical, thermal or mechanical energy into actuator stem motion to develop thrust or torque.

Positioner – A position controller servo mechanism that is connected to a moving part of a final control element or its actuator that automatically adjusts its output to the actuator to maintain a desired position in proportion to the input signal.

Direct Acting – A diaphragm actuator in which the actuator stem extends with increasing diaphragm pressure.

Reverse Acting – A diaphragm actuator in which the actuator stem retracts with increasing diaphragm pressure.

Regulating Valves (Manual Operation)
Manually operated valves that are intended to be used in any position between closed and fully open are called regulating valves. There are many applications where such valves are fitted with suitable trim and are used for fluid control either when pneumatic power is not available or in a bypass situation.

A typical operating mechanism for a linear motion regulating valve is a handwheel. For high output linear motion applications or rotary motion regulating valves a gear unit is incorporated into the mechanism. A graduated travel indication plate should always be provided.

Figure CVA2: A Handwheel Operator (Emerson Fisher)

Pneumatic Diaphragm Actuator Types
Pneumatic spring opposed diaphragm actuators are the most common type used for actuation of linear motion control valves. This is mainly due to the simplicity of the design together with its reliability and operational performance. The design comprises of a spring and opposing diaphragm

connected to the control valve by means of the actuator stem. They are available as both direct acting as typically illustrated in figure CVA3 and reverse acting as typically illustrated in figure CVA4. For the direct acting design air is applied above the diaphragm and compresses the spring, which results in a downward linear movement. Similarly for the reverse acting air is applied under the diaphragm, also compresses the spring and this results in an upward linear movement. These type of actuators are only suitable for relatively low air supply pressures (typically 4 barg maximum), provide a failsafe mechanism and are used for relatively short linear strokes.

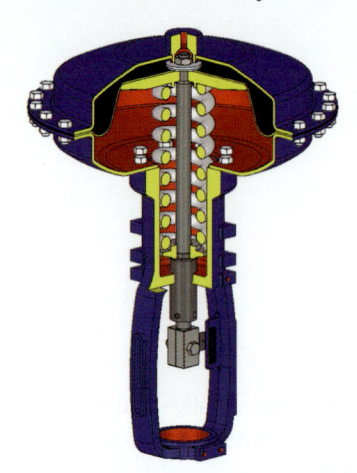

Figure CVA3 Direct Acting Actuator (Weir Blakeborough)

Traditionally the diaphragm actuator uses a single spring with a wide variation of spring forces, 0.2 to 1.0 barg and 0.4 to 2.0 barg being the most popular. These springs can be adjusted by using a spring tensioning mechanism within the construction. More recently alternative designs are also available with multiple springs sets which are normally either set at 0.2 to 1.0 barg or 0.4 to 2.0 barg. These multi spring designs are normally restricted to small actuator sizes and shorter travels than the single spring designs.

The force exerted by these actuators is the difference between the air force applied to the diaphragm and the opposing spring force. The total force is a function of the effective area of the diaphragm which will be specified by the respective manufacturer to provide the necessary net thrust.

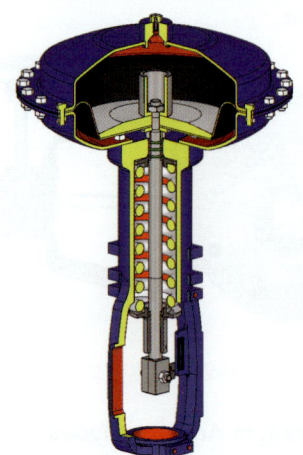

Figure CVA4: Reverse Acting Actuator (Weir Blakeborough)

Some manufacturers produce a spring opposed diaphragm actuator design that is 'field reversible' which means that the actuator can be changed from direct acting to reverse acting and vice versa with no additional components. The manufacturer would normally specify this in the respective technical literature.

The spring opposed pneumatic diaphragm actuator can be fitted with a number of accessories including side or top mounted manual override, maximum limit stop, minimum limit stop and a wide variety of instrumentation.

Linear Pneumatic Piston Actuator Types

For longer travels and higher thrust requirements the pneumatic piston actuator was developed. This is available in both a double acting design as illustrated in figure CVA5 and also a spring opposed design as illustrated in figure CVA6. These actuator types are designed for up to 10 barg maximum air supply pressure and linear travels up to 350mm. For the double acting unit air is applied on either side of a piston plate in order to move the actuator in a linear direction. The piston actuator can also be supplied with a spring, which opposes motion in one direction and drives motion in the reverse (no air) direction. Unlike diaphragm actuator designs, which have standard spring ranges, the spring for the piston type would be designed for the specific thrust required by the valve manufacturer.

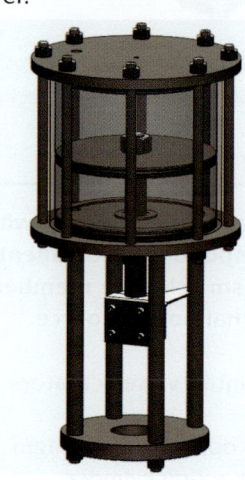

Figure CVA5: Piston Actuator Double Acting (Valve Solutions)

The pneumatic piston actuator can also be fitted with a number of accessories including side or top mounted manual override, maximum limit stop, minimum limit stop and a wide variety of instrumentation.

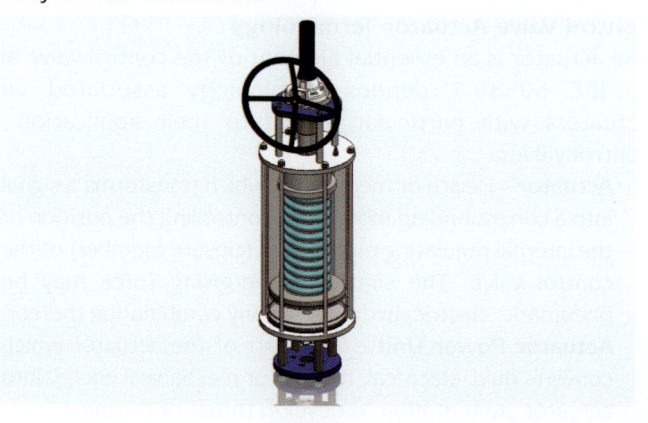

Figure CVA6: Single Acting Piston Actuator with bevel gearbox manual override (Valve Solutions)

Rotary Pneumatic Actuator Types

Both the pneumatic spring opposed diaphragm and pneumatic piston actuator designs can be adapted to incorporate a mechanism, which converts axial thrust to torque enabling them to be used on rotary butterfly, ball and other segmented control valve designs. Please refer to the chapter on Pneumatic Actuators for a full description of the different types of mechanisms. Typical rotary pneumatic actuator designs are illustrated in Figure CVA7.

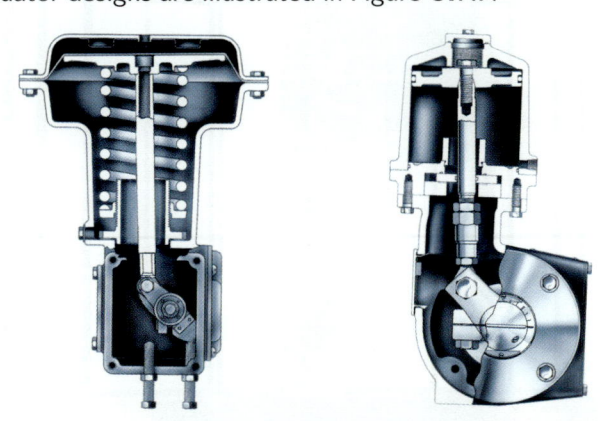

Figure CVA7: Typical Rotary Diaphragm and Rotary Piston Actuators for Rotary Shaft Valves (Emerson Fisher)

Electric Actuators

Electric actuators are frequently used to power control valves in remote locations, where only an electrical supply is available. One of the advantages of using an electric actuator is the ability to transmit signals and power over long distances with minimum transmission loss. Electric actuators can be provided with the relevant certification to meet most requirements including units for explosion proof environments and other hazardous areas. The units can be provided with positional control units and position feedback devices. They can be provided with rotary to linear devices for use on linear rising stem control valves as well as being suitable for rotary control valve applications.

Until recently, the electric actuators that were designed for control valve duties provided a slow and limited performance that lacked the precision of pneumatic equipment and made failsafe operation difficult. Recent technological advances in electric motor design, motor controls and electric energy storage design have transformed the situation.

Although the speed of operation of an electric actuator is still longer than can be typically achieved using pneumatic actuators, for standard applications the speed of modern electric actuators is sufficient for the majority of control valve requirements.

High efficiency, continuously rated brushless DC motor technology facilitates continuous and unrestricted modulating performance. Combined with this, the use of two, independent position sensors can eliminate any backlash and inertia effects from the gearing and achieve a performance resolution of 0.1%.

The introduction of super capacitors now enables electric control valve actuators to provide the same failsafe functions as their pneumatic equivalents, in either fail open, fail close of fail to position configurations.

The development of an electric actuator that effectively meets the needs of control valves also enables many of the advantages of contemporary, 'smart' electric on-off actuation technology to be applied to the control valve environment. These advantages range from simplified control circuit interface and integration to 'non-intrusive' set-up and configuration, using Bluetooth for example, and diagnostic capabilities for enhanced valve care and reduced maintenance.

Figure CVA1 illustrates a purpose-designed electric control valve actuator. Figure CVA8 shows conventional electrically actuated control valves.

Figure CVA8: Conventional electrically actuated control valves (Valve Solutions)

Hydraulic and Electro Hydraulic Actuators

Hydraulic actuators are used for actuating control valves normally where either no pneumatic supply is available or the pneumatic actuators cannot provide sufficient thrust or travel to control the valve. These units are generally compact and use either a spring return single acting hydraulic cylinder or double acting hydraulic cylinder. They provide a stiff and smooth operating actuator mechanism and they can be designed to operate with extremely fast response times. Since very few installations have a hydraulic ring main a self contained electro hydraulic actuator is normally supplied to provide the motive power. Further details can be found in the chapter on Hydraulic actuators. Figure CVA9 shows a typical electro hydraulic actuator fitted onto a control valve.

Figure CVA9: Self-contained Electro-Hydraulic Actuated Control Valve (Paladon)

BVAA VALVE & ACTUATOR USERS' MANUAL - 6th Edition

How do you keep up with valve industry developments?

We all have difficulty getting out sometimes.

Time is a precious resource after all. But if you're a significant user or buyer of valves and actuators, you really do need to keep up with new technology and product developments, and keep an eye out for new suppliers.

BVAA has the Answer!
We bring the exhibition to you!

For many years the BVAA has been organising 'desktop exhibitions' for major users, inside their own premises.

These zero cost, hassle-free events are managed by BVAA and are customised to suit your needs at your convenience.

Solve your supply chain issues over lunch

Designed for rapid set-up and breakdown, 'desktops' typically fit around your lunch period, to minimise downtime. We demonstrate the latest products, provide unrivalled industry advice, and have over 100 leading UK companies to choose from.

"We have had very positive feedback from exhibition attendees. We are already looking forward to doing it all over again"
- Dave Anderson, Score.

Previous hosts include:-
Ministry of Defence, Foster Wheeler, AMEC, MW Kellogg, Stone & Webster, Snamprogetti, British Energy, Score, Aker Kvaerner, KBR, Parsons...

BVAA

British Valve & Actuator Association
9 Manor Park, Banbury, Oxfordshire OX16 3TB
Tel: 01295 221270 Fax: 01295 268965
Email enquiry@bvaa.org.uk
www.bvaa.org.uk/exhibitions.asp

Installation and Operation of Valves and Actuators

Figure OP1: The delivery of a 36", Class 900 'U' Top Entry Ball Valve to Sarawak required the first ever flight of a wide body cargo plane to Bintulu. Inset: A tight squeeze, even for a Boeing 747-400F aircraft (Schuck Valves).

Installation

The installation procedure for both valves and actuators is critical to ensuring both long life and satisfactory performance. Valves stored on site awaiting installation should be kept in their original packing, in dry conditions, where damage will not occur. Before carrying out the installation, it is important to follow the basic procedures described below.

- Carefully unpack the valves and actuators and ensure that no obvious damage has occurred during transit or handling
- Check tags, identification plates, direction of rotation of handwheels, accessories, and so on, against bills of materials, specifications, and schematics. In particular ensure that pressure temperature rating marked on the valve is compatible with the intended service conditions
- Read all the literature and note any special warning tags or plates attached to or accompanying the valve or actuator and take appropriate action
- Check the valve for any marks indicating flow direction; appropriate care must be taken to install the valve for proper flow orientation.
- Inspect the valve interior through the end ports to determine whether it is reasonably clean and free from foreign matter and harmful corrosion. Remove any special materials which may be used as packing, both inside the valves, such as wooden blocks in gate valves to prevent disc movement during transport and handling, and outside, such as tape on stem threads
- It is not recommended to use operating levers or handwheels as lifting points
- Ensure that the valve is installed so that it can be safely operated and maintained
- If possible, cycle the valve by hand and inspect any functionally significant features, such as guides or seat faces, made accessible by the cycling. Following such inspection, it is usually desirable to leave the valve closure member in the position in which it was transported.
- With actuated valves, the following additional checks should be made on the actuator
  - Check the data given on the actuator name plate regarding electric, pneumatic or hydraulic supply requirements to match the site supply
  Note whether any mounting orientation restrictions are indicated on the actuator or its accessories

- Check that the actuator enclosure is suitable for the intended valve location, i.e., that actuators having electrical components are suitable for indoor, outdoor, or flammable vapour or hazardous environments, and that specified accessories are fitted correctly
- Check that the proposed mounting location/orientation will permit satisfactory routing of cabling or pipework, and will allow sufficient room for removal of covers.

Figure OP2: A risk of flooding was highlighted when these Electric Actuators were specified. They were indeed subsequently submerged in 3m of water during a flood, but double sealing prevented water from reaching any electrical components (Rotork)

Installing threaded-end valves

For tight sealing, threaded pipe joints depend on the fit between the male and female threads with the use of a soft or viscous material between the assembled threads. Where parallel threaded fittings are used in parallel threaded valve ports, some form of seal on the end face of the valve should be used. To ensure a leak-free system, the following points should be observed: -

- Check the threads for dirt, grit, or any visible damage to the threads
- Note the internal length of threads in the valve ends and compare with the length of thread on the pipe. Observe any need for care regarding how far the pipe is threaded into the valve
- Make sure the threads are aligned at the point of assembly. Tapered pipe threads are inherently loose at entry and a substantial wrenching force should not be applied until it is apparent that the threads are properly engaged
- Apply an appropriate tape or thread compound to the external pipe threads, except when dry seal threading is specified
- Assemble the joint wrench-tight by applying the wrench to the end of the valve where the joint is being made
- Repeat the process at the second valve end and again apply the wrench to the end of the valve to which the pipe is being assembled.

Installing flanged-end valves

Flanged pipe joints depend on the compressive deformation of the gasket material between the facing flange surfaces for tight sealing. The mechanical force required to maintain the compressive stresses on the gasket, as well as resisting the normal pressure forces tending to separate the joint, should be provided by the bolting. In order to obtain satisfactory flange joints, the following points should be observed: -

- Check the mating flange faces of the valve and pipework for correct gasket contact surface finish and condition
- Check the bolting for proper size, length, and material
- For Cast iron and copper alloy flanges the use of lower-strength steel bolting and full face gaskets on flat face flanges is recommended, to help prevent flange breakage or bending by over tightening of the flange bolts
- Check the use of the correct gasket material for each type of flange and that gaskets are free from damage
- Ensure good alignment of flanges being assembled and use suitable lubricants on bolt threads. Sequence tighten the bolting to make the initial contact of flanges and gaskets as flat and parallel as possible, and ensure bolting is tightened to correct torque
- In the assembly of wafer type valves between pipes, the installation should be checked for possible interference of moving parts between valve and adjacent pipe.

Figure OP3: 2" Single Isolate, ROV operated, Sub Sea Ball Valve with 'pup weld ends' (Alco Valves)

Installing welded-end valves

All welding should comply with the appropriate pipe system or application code. Welded joints should provide a structural and metallurgical continuity between the pipe and valve body and should not constitute a notch or weak link in the pipe valve pipe assembly. Thus, for socket weld joints, the weld fillet should always have more cross-sectional area than the pipe. Butt welds require full penetration welds and thickness at least equal to that of the pipe. Considerable distortion of the valve body may occur if valves are not welded into the line with care and body extensions known as 'pup weld ends' are commonly used to avoid heat from the weld being transferred to the valve and damaging seating faces particularly soft seats. Weld splatter inside the pipe or valve body should also be avoided, as its presence can be detrimental to the performance of the valve.

Installation of actuators

When mounting actuators on valves, particular care should be taken to check that the torque output gives the correct direction of rotation for the valve, e.g., most valves are required to operate clockwise from the fully open position to the fully closed position. It is possible with some designs of part-turn valves for the actuator to be mounted in an orientation which will achieve the opposite mode of operation for the valve. It is also particularly important to note that, where electric actuators have been removed from a valve and then re-fitted, the limit switches may no longer be correctly aligned with the respective ends of valve travel. The settings should, therefore, be checked and adjusted where necessary.

Figure OP4: Installation and commissioning of valves and actuators should only be undertaken by fully trained and competent personnel (Blackhall / Rotork)

Actuator mounting brackets

An automated valve package is constructed of three major components — the valve, the actuator, and the mounting adaptor and drive coupling. Improperly designed mountings are dangerous, can reduce the performance life of the package by causing premature failure of the valve stem seals and actuator bearings and seals, and can cause valve leakage.

The mounting bracket, by design and material selection, should be sufficiently strong enough to produce no visually discernible movement as a result of maximum actuator torque or thrust, or the maximum torque or thrust rating for the specific ISO pattern — whichever is the greater. The strength of the bracket must be adequate to support the weight of the actuator plus its accessories when mounted in any mode, and its design must ensure suitable stem/shaft coupling alignment and squareness without special skill or tooling during assembly.

The drive coupling must also be capable of accepting full actuator torque/thrust or the maximum torque/thrust rating for the specific ISO pattern — whichever is the greater without distortion. The total backlash on the drive coupling between valve input and actuator output shall be within the recommended tolerance for the related valve/actuator system.

Figure OP5: Mounting adaptors and drive couplings are major parts of an automated valve package (Pro-kits)

Mounting kits should have suitable corrosion-resistant coatings and must allow access for routine valve servicing, gland adjustment, plug lubrication, leak testing, and so on. The kits should be supplied with all fastenings and a complete set of instructions for fitting. The components should not become part of the valve pressure envelope, and any stem leakage must be free to vent to atmosphere.

It is important that the machining and drilling of any holes in the two mating surfaces of the mounting interface are carefully manufactured to ensure that the faces are parallel and the holes give concentricity of the driving and driven components of the coupling. Misalignment causes eccentric loads on the actuator and valve drive components and gives rise to additional torque requirements for operation.

International Standards ISO 5210 and ISO 5211 specify dimensions for the mounting interface on multi-turn and part-turn actuators, respectively. It is recommended that compliance with these standards is specified as this will then maximize the interchangeability of actuators on valves for maintenance and/ or replacement purposes.

Valve and actuator testing and adjustment

Following installation, all manual operated isolation valves should be operated to check they are functional by cycle testing the valve from closed to full open to closed. Actuated valves should not be operated over their full stroke without first checking that in the case of electric actuators the motor rotation is correct and in all cases that end of travel limit and/or torque switches have been correctly set according to manufacturers' instructions.

After the installation of pipework systems, it is common practice to clean the systems by blowing through with nitrogen or steam, or flushing through with a liquid to remove debris and dirt. It should be recognized that valve cavities may form a natural trap in a pipework system and any material not dissolved or carried out by the flushing fluid may settle in valve cavities and damage seat surfaces, adversely affecting the valve operation.

As actuated valves are often commissioned initially without any line pressure, it may be found, once the plant is on stream, that torque and limit settings should be checked and increased where appropriate. If after carrying out these adjustments, satisfactory valve operation is still not achieved, the valve manufacturer should be contacted as soon as possible for advice. The problem could be caused by foreign matter trapped in the valve, or by a damaged seat, and prolonged operation under these conditions could cause serious further damage to the valve and/or the actuator.

Once satisfactory commissioning of an actuated valve has been achieved it is important to check that all actuator covers are securely fastened with healthy gaskets or 'O' ring seals, as appropriate, that all cable entries are properly sealed, and that any pneumatic or hydraulic piping is free of leaks.

Figure OP6: Pneumatic Actuator and Ball Valve package being tested. The ball can been seen in the near-closed position (Apollo Valves)

Valve layout and position

Valves should be provided with adequate supports when mounted in the pipe, and be accessible for operation, adjustment, maintenance, and repair. It is preferable for linear motion valves to be mounted with the stem in the vertical position so that gravity helps the obturator centrally into the seat. Should critical valves need to be removed for overhaul, provision should be made for doing so without having to shut down the whole system, by fitting isolating valves at appropriate locations.

Specific Considerations

Check Valves - Care should be taken when fitting check valves as some models are suitable for horizontal pipelines while others are only suitable for vertical pipelines. It is also important to ensure that the flow entering and leaving the check valve is stable and uniformly distributed across the flow area. Manufacturers will make recommendations on the number of pipe diameters upstream and downstream between the check valve and either other valves or pipe bends. If these are not followed then the flow may not be uniformly distributed and the disc or discs may not reach full open, which will increase the pressure loss and operating cost of the plant. Disc chatter may also occur.

Control Valves - Control valves need occasional adjustment of actuator and accessory settings, as well as periodic maintenance. It is particularly important to ensure that the valve is easily accessible for these purposes, and that there is sufficient space around the valve for access. The orientation of the valve is also very important as this determines how difficult maintenance on the valve is going to be. The preferred orientation is with the actuator vertically up. Equally important is the need to be able to install lifting gear on larger valves where heavy components have to be handled during maintenance.

> **Piping** - The upstream and downstream piping can have a considerable impact on the valve's performance. Bends located too close to the inlet of the valve may create disturbances in the fluid stream that can impact the valve trim, giving rise to stability problems. A bend too close to the valve outlet can have an impact on the discharge flow and result in shock waves being set up that may affect the discharge characteristic of the valve. Multiple bends in the vicinity of the valve should also be avoided as the turbulence created can influence the noise generated. It is good piping practice to allow 10 pipe diameters of straight pipe upstream of the valve and 5 pipe diameters downstream. In designing the piping downstream of a pressure reducing valve or a valve that is handling a flashing liquid it is important to recognise the substantial increase in volume that is going to occur. If the piping is not adequately sized, then high velocities will occur, resulting in noise and vibration and erosion in the case of the flashed liquid. Good piping practice should be adopted in sizing pipe runs.
>
> **Pipe supports** - are a very important consideration particularly with control valves that can be susceptible to pipe loads and stresses. Pipe supports should be located in the vicinity of the control valve to support the valve and actuator weight and to absorb any pipe loads that might otherwise be transferred to the valve. The run of piping, the weight of the combined piping and fluid, and the possible expansion of the piping should be taken into consideration when selecting the location and type of supports used.
>
> **Installation** - One of the most common causes of problems with control valves is insufficient care taken at the installation stage. This can result in one or more of the following: -
> - The transference of pipe stresses to the valve, which can cause body distortion
> - The ingress of foreign matter into the valve trim resulting in the valve sticking or the trim becoming damaged
> - The valve being installed with flow in the incorrect direction.
>
> **Pre-commissioning** - checks should be carried out on all control valves following installation or major overall. The time spent on this will be more than adequately repaid at the commissioning stage when time is limited,

tempers are frayed, and the cost of delayed commissioning is considerable. Pre-commissioning should follow a set routine, which most manufacturers will set out in their instruction or installation manuals. Basically the object of a pre-commissioning check is to verify that the valve is capable of performing the function that is required from it but without the fluid being present.

Safety and Relief Valves - A safety/relief valve has been accurately set to operate at the desired set pressure and has passed the stringent factory leakage rate test. During installation it should be: -
- Stored under cover, in a vertical position, where valve cannot be knocked over until required
- Handled with care
- Not subjected to heavy shocks
- The internals kept clean
- The inlet and outlet flange covers removed at the last possible moment.

Proper installation requires careful consideration of the inlet piping and the outlet piping. If it is not possible to install the valve close to the pressure source, the intervening pipework and fittings should not create a pressure drop of more than 3% of the set pressure when the valve is discharging its full rated flow.

The outlet piping must be adequately supported. When a safety valve discharges, large reactive forces can be transmitted to the inlet piping (see Figure OP7). Loading applied to the outlet can cause misalignment of the valve and seat leakage. If the valve vents directly to atmosphere, protection must be provided to ensure rain, snow, or debris, do not accumulate in the outlet of the valve.

Always install the valve in a vertical position and where it will not be affected by vibration of the equipment it is protecting.

Figure OP7: Safety Valve with Braced Dual outlet supports (Anderson Greenwood Crosby)

Mechanical Key Interlocking

Valve operating procedures can be potentially dangerous if executed incorrectly, or in unsafe conditions. The scope for injury and/or damage is also significantly increased when high temperature, high pressure or a toxic/flammable product is present. While good practice begins with good design, both are inevitably hostage to the 'human factor'. 70% of reported incidents in the Oil and Gas industry worldwide are attributable to human error, accounting for in excess of 90% of the financial loss to the industry.

Modern processes are often highly automated, yet still require human intervention during essential maintenance procedures such as loading or unloading Pig Traps and the changeover of Pressure Relief Valves. Distributed control systems (DCS) cannot effectively regulate such procedures but mechanical key interlocking systems can.

The coded-card linear-key concept together with a range of modular key-operated interlocks will regulate operator execution of work procedures on any form of host process equipment. These can be used on every form of valve (including motorised and instrument valves) as well as access hatches/doors and electrical isolating switchgear.

Figure OP8: Mechanical key interlocks help industries take a disciplined approach to design and operating practice (Smith Flow Control)

Mechanical key interlocks work by controlling the sequence of events in which valve process activities are conducted.

Key interlocks are dual-keyed mechanical locking devices designed as integral-fit attachments to the host equipment and operate on a 'key transfer' principle. This limits actions to only those that produce a safe and desired outcome i.e. preventing a tanker from departing a loading/discharge station until the cargo hoses have been disconnected.

Typically key interlock systems are applied to valves, closures, switches or any form of equipment which is operated by human intervention. The 'OPEN' or 'CLOSED' status of an interlocked valve, or the 'ON' or 'OFF' status of an interlocked switch, can only be changed by inserting a unique coded key (see Figure OP8). Inserting the key unlocks the operating mechanism (e.g. hand wheel or push-button) enabling operation of the valve or switch. Operating the unlocked equipment immediately traps the initial (i.e. inserted) key.

When this operation is complete, a secondary (previously trapped) key may be released, thereby locking the equipment in the new position. This secondary key will be coded in common with the next lock (item of equipment) in the sequence. By this simple coded-key transfer principle, a 'mechanical logic' system is created which denies the scope for operator error.

In addition, keys may be customized to intelligent format by electronic tagging of individual keys and managed by system software that interfaces with the mainframe DCS system.

BVAA VALVE & ACTUATOR USERS' MANUAL - 6th Edition

Comid Valve Services

VALVE MAINTENANCE SERVICES

- ✓ Repair and Reconditioning of Industrial Valves
- ✓ Safety Relief Valve Repair, Service and Calibration
- ✓ Isolation, Control, Pressure Reducing Valves
- ✓ In-Situ Verification of Safety Valve Set Point (✓eritest)
- ✓ In-situ Valve Repairs
- ✓ On-Site Skills Team for Outage Work and Emergencies
- ✓ Established Supply Chain for Valve Procurement and Supply

BVAA

BSI / UKAS QUALITY MANAGEMENT 003

Contact Us
Comid Engineering Limited

Townfield Works
Greenacres Road
Oldham, OL4 2AB
www.comid.co.uk

UK Offices

Oldham
Lancashire
Tel: 0161 624 9592
Email: sales@comid.co.uk

Walsall
West Midlands
Tel: 01922 721649
Email: sales@comid.co.uk

Middle East Offices

Saudi Arabia
Jubail
Tel: +966 500120196
Email: jubail@valvserve.com

Bahrain
Manama
Tel: +973 3 9318725
Email: bahrain@valvserve.c

Maintenance and Repair of Valves and Actuators

Figure MR1: A good valve repair will focus primarily on procedures, quality control, fitness for purpose and documentation, rather than just aesthetics (Comid)

All the previous chapters essentially relate to new valves and actuators. This chapter considers the issues of maintenance and of repair. The nature of both processes is fundamentally different and it is important to understand these differences.

Maintenance is generally undertaken by the owner. The valve or actuator is almost always in situ and most likely the plant is operating. The objective of maintenance is to ensure that the valve and actuator continue to achieve an acceptable level of performance consistent with operating the plant safely and profitably. Maintenance is a scheduled activity with usually no critical time constraints.

Repair can be undertaken by the owner, by a third party specialist repair company or by the original valve or actuator manufacturer. The valve or actuator may be in situ, at a repair workshop or at the original manufacturing plant.

The objective of repair is to either correct a fault that has occurred or to return the valve and actuator to as near new conditions as is reasonably practical. Repair can be a scheduled or an unscheduled activity. It usually has to be performed under critical time constraints and in the event of unscheduled repairs costs of plant shut down can be huge.

Maintenance

The starting point for maintenance is to state the obvious. Each valve and actuator has been installed for a reason and considering what that reason is helps in assessing the requirements for maintenance. Valves and actuators will either have a safety function, a process function or a maintenance function in that they will be required to isolate so that activities can be carried out on other valves or equipment. Valves and actuators within these functions can probably be further categorised into critical and non-critical.

The valves in a plant can vary from tens to thousands and the task of simply assessing the requirements for maintenance can appear daunting particularly if the responsibility is given to a non-valve specialist. While valves and actuators are typically only a small percentage of the overall plant cost it should not be a task that is ignored or dealt with half heartedly. Their importance to safe and productive operation of the plant far outweighs their relative cost of purchase.

In addition to function valves and actuators should be broken down into type and into frequency of operation. Frequency of operation is important partly in that those that operate most frequently will be subject to the most wear of seating faces, packings and seals but also those that are not operated are very often not observed and are therefore allowed to deteriorate to the point when they will not function when called on to do so.

Figure MR2: The replacement of compression packings requires care, for example each ring should be fitted individually, with joints staggered by at least 90° (James Walker)

Manufacturers should provide appropriate maintenance instructions for their product. Those that do not should be considered to be of questionable competence. If the valve is covered by the Pressure Equipment Directive they are required to do so by law and it must be in the language of the user. Other directives also require manufacturers to provide relevant information to the user.

In general the requirement for maintenance activity on manual isolating valves is low. Mostly it relates to maintaining the functionality of the operating mechanism and the integrity of the stem or shaft seal. Some valve types such as lubricated plug valves do have the ability to adjust the seating faces, and from time to time may need to be re-stocked with lubricant (Figure MR3). Similarly most actuators are low maintenance although the filters on hydraulic systems need to be kept effective. Control valves will require more attention because of the number and sensitivity of items of equipment that make up the package. Isolating valves that are normally in the open position can usually be cycled to 90% open and back to 100% without affecting the process thus confirming the ability of the valve to move.

Figure MR3: A lubricated taper plug valve in the process of being loaded with grease, which is visible in the bore in the photograph below (Flowserve)

An essential feature of good maintenance is observation and listening. An awful lot of useful data about how the valve and actuator are performing can be gained this way. Feedback from the operators on valve operating torques in similar duties may highlight seat wear in one compared to the other. Stem threads that are obviously dry need lubrication. The start of initial gland leakage can often be seen or heard or smelt. The quicker the gland bolting is adjusted and the leak stopped the less damage is done to either the packing or the stem/shaft. Leaks from air supplies to actuators can either be heard or felt or found by bubble solution. The cumulative losses from the air supply system will certainly cost money and could have far more serious consequences if valves cannot be operated as intended. Noise in the form of check valve chatter or from vibration associated with fluid flow should not be allowed to continue indefinitely as they will cause damage to the valve.

External corrosion due to ambient conditions can seriously affect a valve or actuator's ability to function correctly. The most vulnerable items are bolting either body bonnet, gland or spring retaining bolting on safety valves or single acting actuators. Visual examination of these components should be made and appropriate repair to ineffective or damaged paint systems made.

Having analysed the population of valves into function, type and operations frequency and assessed the manufacturers' recommendations for maintenance, choices may have to be made based on what must be done as against the availability of resources to do this work. Operational history for a valve type in a particular plant is very helpful in determining the frequency that maintenance has to be carried out. Manufacturers specify the maintenance operations to be done but can only give guidance on frequency as each plant's condition is different.

An extremely useful function of modern actuators is their ability to provide feedback on the thrust or torque they are generating to operate the valve. This data can be collected over time giving an indication of changes to packing load on the stem/shaft or seating face sliding friction. These changes can be used to trigger appropriate maintenance or repair activities rather than at fixed intervals (see Figure MR4).

A key aim of maintenance should also be to identify valves and actuators that require either replacement or repair. Valve or actuator failures can be due to faulty design, materials or workmanship but equally they can simply have worn out due to adverse process conditions. It is the job of maintenance to make sure these unscheduled wear out failures are eliminated.

There are two issues that should be treated with great respect: -

- Stem leakage on a linear motion valve may no longer be prevented by tightening the gland bolting further. If the valve has a back seat this can be used to stop or reduce the leakage. Do not under any circumstances use the back seat to facilitate replacement of some of the packing rings while the valve is under pressure.

- If body bonnet joint leakage occurs do not attempt to stem the leak without contacting the manufacturer to obtain his advice and the appropriate bolt torques to be applied. If the leakage occurs during or within a few hours of commissioning it may be possible to stop it by simply re-tightening the bolting. If the leak has occurred after sometime it is most likely to have been caused by a failure of the gasket rather than a relaxation of the bolt stresses. Depending on the nature of the fluid the leak may also have damaged the bolting local to the leak site. For both reasons applying unspecified amounts of torque to the bolting hoping to seal the leak at best is likely to be ineffective and at worst could cause bolt failure and a far worse situation. Any valve with a continuing body bonnet joint leak should be scheduled for investigation of the reason for the leak and full repair at the earliest opportunity.

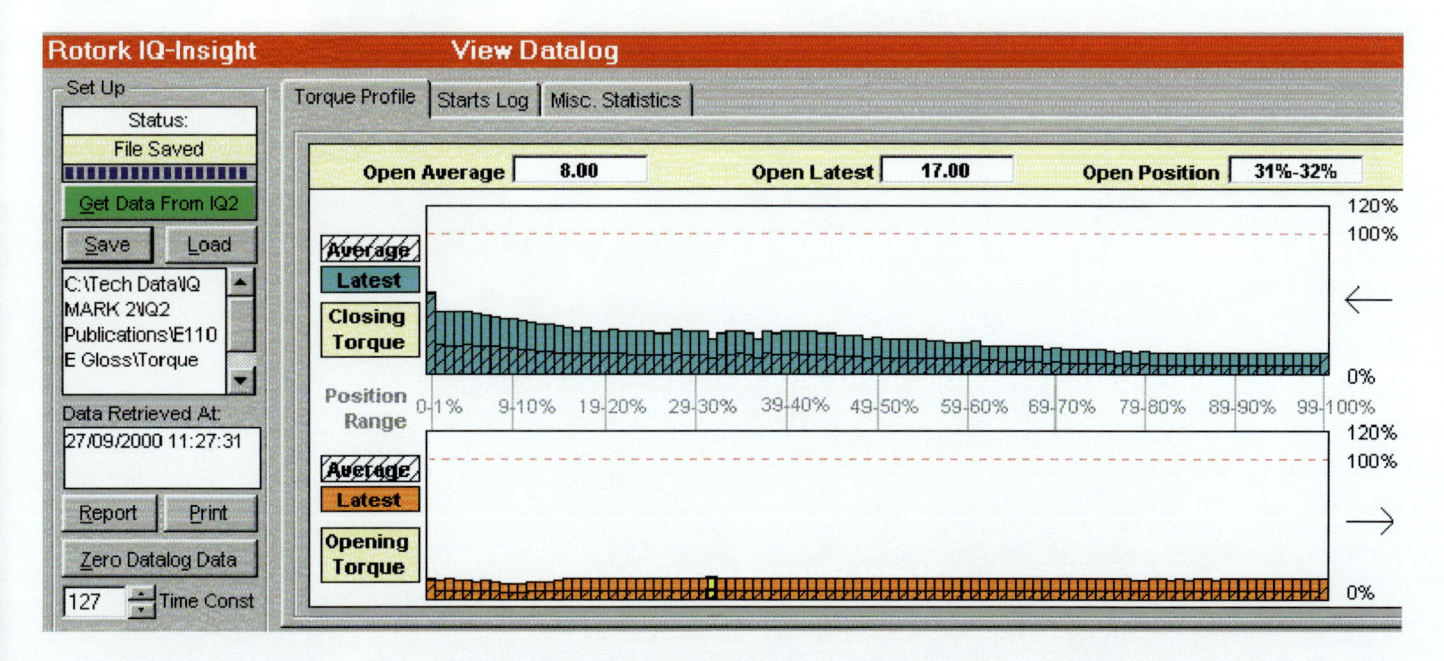

Figure MR4: Valve operating torque profile (Rotork)

SPE STEAM PLANT ENGINEERING LTD

Industrial Valve and Boiler Repair Specialists

- Industrial Valve Repair

- Supply of New & Reconditioned Valves from Stock

- Onsite Services

- Boiler Repairs & Coded Welding

- Boiler Annual & 5 year Insurance Preparation

- Industrial Pipework Design and Installation

- Feed Tank & Skid Unit, Design & Manufacture

- Full Range of Boiler Spares available from Stock

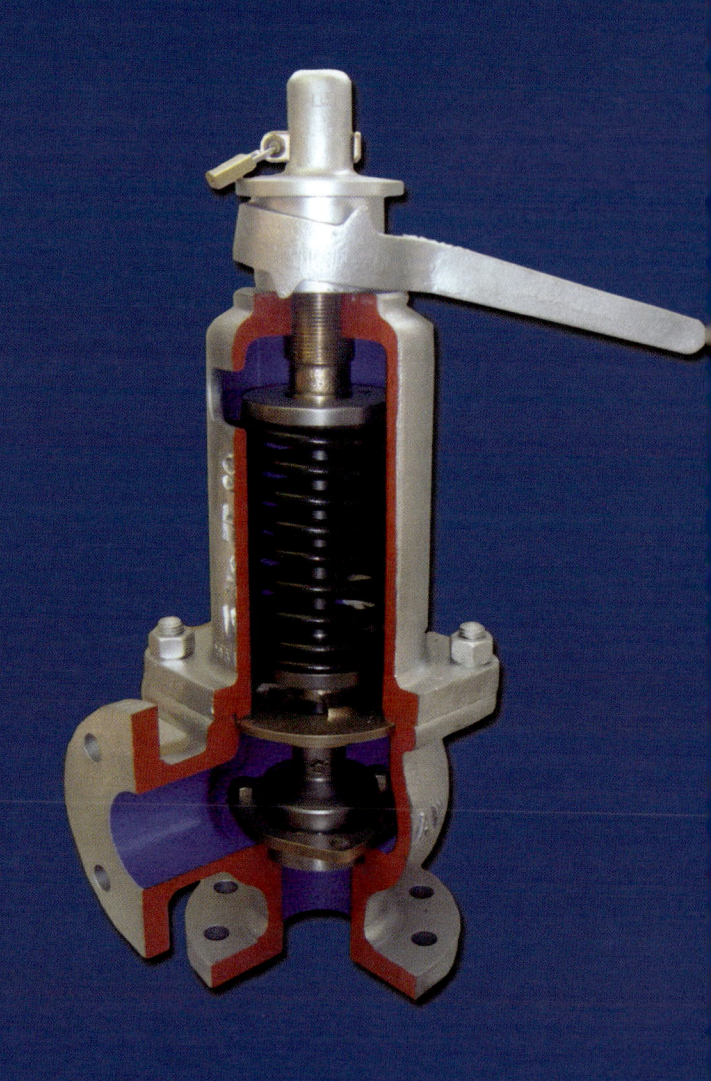

Steam Plant Engineering Ltd

Wartell Bank Industrial Estate   Kingswinford   West Midlands   DY6 7QJ

Tel: 01384 294936   Fax: 01384 295328   Email: sales@steamplant.net

www.steamplant.net

constructionline | CHAS | SAFEcontractor Approved | BVAA | Zurich Risk Services APPROVED SERVICE CENTRE Cert. No. CEN-019082-01 | ecITB | TOPOG-E DISTRIBUTOR

Repair

Repair or replacement becomes necessary once a valve or actuator is no longer capable of effectively fulfilling its function. Repair can be an unscheduled activity resulting from a failure or it can be a scheduled activity incorporated into a shut down of the plant.

The objective for a repair is to return the valve or actuator to as close to its as new condition as is reasonably practicable and that when it is returned to service it will perform at an acceptable level for at least the period up to the next planned shutdown.

working order particularly if the products are installed in a different country from that in which they were manufactured.

This issue is widely recognised by manufacturers who are generally keen to provide support for the products they sell. The larger companies establish their own repair organisations in the majority of countries and others use their authorised distributor to provide support to repair companies or alternatively they directly co-operate with an authorised repair company.

Another key issue is timescale and the availability of spare parts. Repairs usually have to be undertaken within a specified and critical time frame. Plant is shut down to facilitate the repair and delays in completion can be extremely costly. If the repair activity has been planned obvious spare parts can be ordered in advance such as body bonnet gaskets, gland packings, diaphragms, o-ring seals and other similar low cost items. The decision to order other more costly components in advance becomes a matter of judgement based on the importance of the valve, the likely immediate availability of the items and the potential to reuse the existing components. For the repair of large and complex items detailed planning well in advance is required between all the parties involved (see Figure MR6).

Figure MR5: The maintenance and repair of valves and actuators should be carried out under the right conditions by technically competent staff
(Above: Provalve, below: Rotork Site Services)

There are some important contractual issues that must be appreciated when considering the repair of a valve or actuator. Firstly unless the repair is necessitated by a failure under warranty of the valve or actuator the owner is responsible for specifying the repair activities to be undertaken. Very often the repair is carried out by a specialist repair company located close to the plant. Neither the owner nor the repairer is in full possession of the technical procedures, specifications and drawings the manufacturer used to create the original product. There is no requirement for the manufacturer to provide this information to either the owner or the repairer and it is likely that they will be reluctant to do so as they are divulging commercially sensitive information. This creates an obvious difficulty in trying to bring the valve and actuator back into

BVAA VALVE & ACTUATOR USERS' MANUAL - 6th Edition

is faced with the choice of either scrapping the valve and finding a replacement or accepting the reverse engineering of the component by the repair company. The re-conditioned valve shown in Figures MR7 and MR8 was turned around in approximately 1/10th of the lead time for a new valve.

Reverse engineering is the term applied when the original component is measured and the recorded dimensions are used by the repairer to produce a replacement component. The issue that is obvious is that the original component was damaged so not all dimensions can be measured. The not so obvious issue is that measuring the old does not give you the tolerance applicable to any given dimension or the surface finish or the radii of worn surfaces. As with most engineering the devil is in the detail and the detail is what made the valve work reliably in the first place.

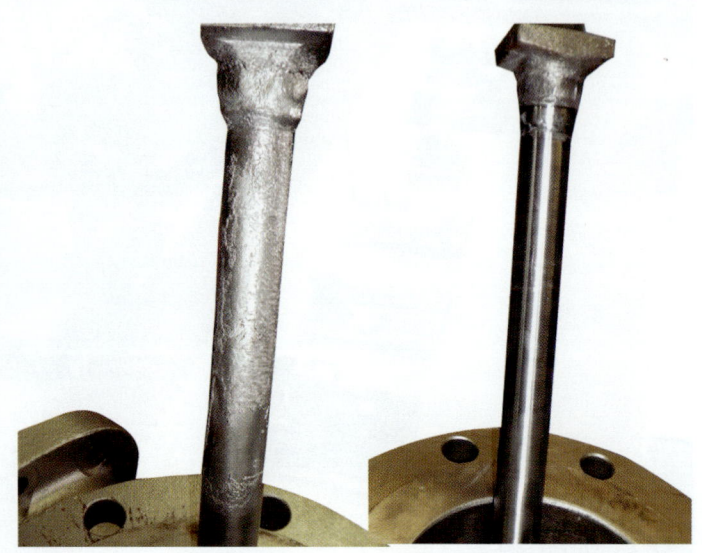

Figure MR6: For some valve types, complete overhaul is almost the only option and requires considerable logistics planning in order that end-users, owners, contractors and the valve supplier all complete their allotted task to schedule (Blackhall)

Figure MR7: Knowing when a part is beyond economical repair is also another vital skill. In this case the spindle was deemed irreparable and a replacement was manufactured to match the drive bush (Comid)

Many designs include allowances on seating faces so that a number of repair cycles can be carried out before parts must be replaced. The components can be re-machined to provide as new surfaces without loss of valve performance.

The repair becomes more complicated when components are damaged or worn to the extent that simply re-machining the component will not ensure a satisfactory repair. If the replacement new component is not available from the manufacturer within the critical time frame then the owner

It is therefore very important that if reverse engineering is part of the repair that the owner understands that this has been done and that the repairer is considered sufficiently component to do the work. It is strongly recommended that only original manufacturers' spare parts are fitted to safety critical items.

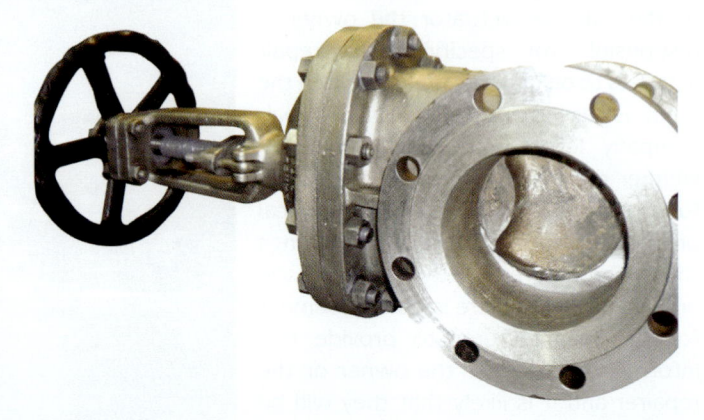

Figure MR8: A 6" Class 150 Wedge Gate Valve used on a sulphuric acid process. This overhaul and repair required the building up of seats with 'alloy 20' CN7M material and re-machining to correctly match seats and wedge (Comid).

EXCEL ENGINEERING SERVICES LTD

FISHER
Approved Repair Agent

Excel Engineering Services is a leading Industrial and Control Valve Supplier and Service Company in the UK in terms of quality, service and expertise. We offer a service to repair and refurbish all types of valves, fully reconditioned and tested to API 527 Standard.

We Can Offer the Following Services

- Full Valve Overhaul Service
- Control Valve Testing
- On-Site Valve Overhaul Service
- Valve Supply Service, Refurbished, Unused and New
- General Engineering Service
- Heat Exchanger Repair/Re-tube Service
- On-Site Heat Exchanger Repair/Re-tube Service
- Pump overhaul and flow test
- Safety Valve Re-certification
- Gauge Re-certification
- Testing Facility (Hydraulic, Nitro, Helium, O^2 Air)

Valve Refurbishment

We have a fully equipped workshop and machine shop that allows us to repair and overhaul all types of industrial and Control valves with a separate clean room for the testing and certification. We also provide an On Site Service with experienced Engineers all carrying a safety passport.

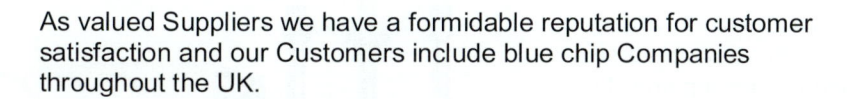

As valued Suppliers we have a formidable reputation for customer satisfaction and our Customers include blue chip Companies throughout the UK.

Heat Exchanger Refurbishment

We re-tube, overhaul or remanufacture any type of Heat Exchanger - utilising the latest tube removal and expansion technology.

Our service offers fast turnaround, competitive pricing and the ability to cater for all types of tubing material.

Facilities

Our Testing and Clean Room Facilities allow our multi-skilled, time served work force to test and certificate all Valve types. Fully automatic rigs allow us to test valves hydraulically to pressures up to 400 BAR.

ISO 9001, ISO 14001 and OHSAS 18001 accredited

Excel Engineering Services, Units 4-6, Burnhouse Industrial Estate, Whitburn. EH47 0LQ
Tel: +44 (0) 1501 744 588 Fax: +44 (0) 1501 744 589
Email: russellbrown@jbdtritec.co.uk and johnogilvie@jbdtritec.co.uk
www.excelengineeringservices.co.uk

JBDtritec

Code of Practice

These complex and important issues are being addressed within the BVAA Code of Practice for the Repair or Reconditioning of Industrial Valves and Actuators which defines the roles and responsibilities for the owner, repairer and manufacturer. The guidelines give recommendations on the important issues of: -

- **Decontamination** - The fluid that has been inside the valve is possibly hazardous. In that case the owner must decontaminate the valve and provide the repairer with a decontamination certificate and a Material Safety Data Sheet (MSDS) with specific COSHH data
- **Springs** - Valves and actuators containing springs should be disassembled strictly in accordance with the manufacturer's procedures as these components can release significant amounts of stored energy with dangerous consequences
- **Welding** - Welding processes shall be performed by qualified personnel only, using approved written procedures and controlled equipment. It should always be remembered that surfaces may have been contaminated by the fluid, and this could have an effect on the metallurgy of the welding process (See Figure MR8).
- **Documentation and Traceability** - Just as it is important to receive documentation about the materials of construction and testing of the valve or actuator when it was new it is equally important to know the extent of repair work that has been carried out and in particular details of any reverse engineered components supplied by the repairer. In the case of safety valves and control valves some repairers offer to manage the paperwork detailing the history of the valve from cradle to grave including scheduling future repairs and the availability of the necessary spare parts

Figure MR9: This valve has recently been cut out of a pipeline. Here it is mounted ready for re-machining of butt weld end profiles (Valvetek)

- **Flange faces** - The vast majority of work that is undertaken in repair workshops is for flanged valves. Screwed end or socket weld valves do not warrant the cost of repair and butt welding end valves are mostly repaired in situ. The refurbishment of the pipe flange gasket face to suit the gasket the owner is proposing is a crucial aspect of the repair. When butt welding end valves are repaired in the workshop the required weld profile must be provided by the owner together with any NDE requirements for the body ends (see Figure MR9).

It must always be remembered that a valve and a fluid powered actuator are items of pressure equipment. Actuators and instrumentation may have hazardous area classifications. Both owner and repairer have a responsibility to ensure that the product is put back into service in a safe condition (Figure MR1). The owner must supply the repairer with appropriate details regarding the product so that he can carry out the required work. In particular details of the pressure envelope materials, the pressure temperature rating of the product, hazardous area classification and the original factory testing must be supplied so that the repairer can conduct the appropriate tests at the end of the repair. The repairers' tests should be documented and ensure that the product is safe to be returned to service (Figure MR10).

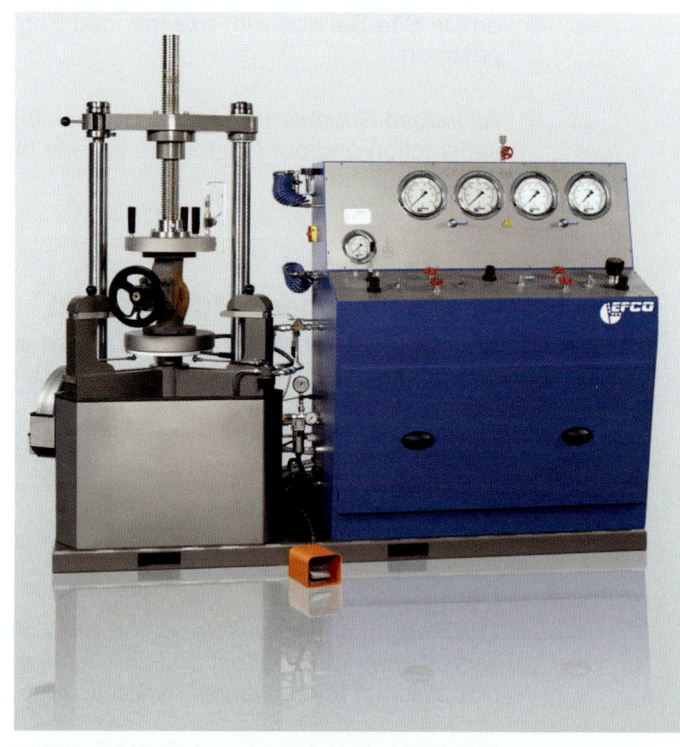

Figure MR10: Before returning to the owner, a valve must proven to be fit for purpose by testing (EFCO)

The exception to this is when the valve or actuator is repaired in situ and the ability to carry out the manufacturer's original test programme is restricted. In this case the owner and repairer must agree the test programme so that it can be demonstrated that the valve is safe to be returned to service. In situ repair uses specially designed equipment to enable the seating faces of top entry valves to be repaired (Figures MR11 and MR12).

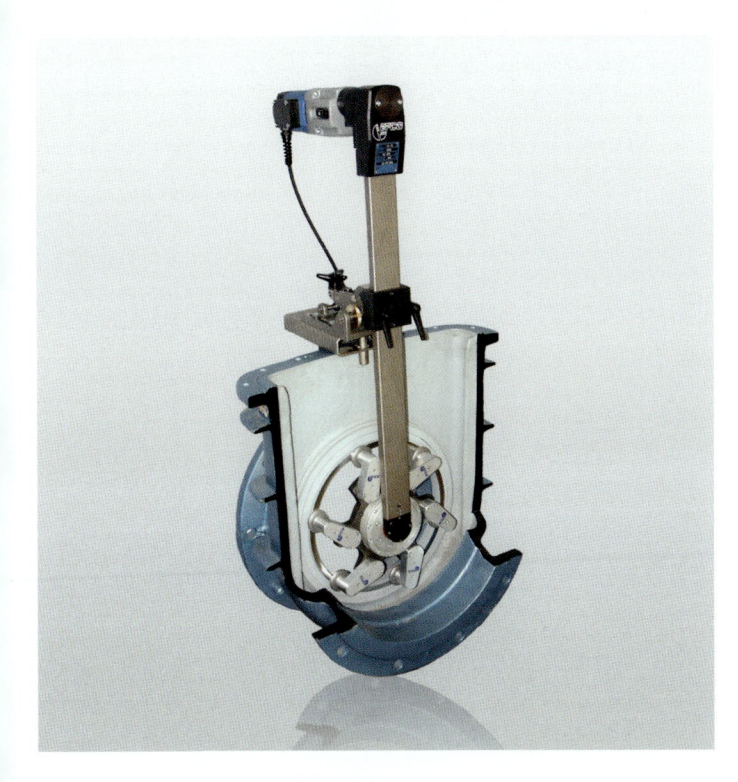

Figure MR11: Having the correct repair equipment is essential. This portable machine is for high-speed grinding of hard sealing faces in gate valves of various designs (EFCO)

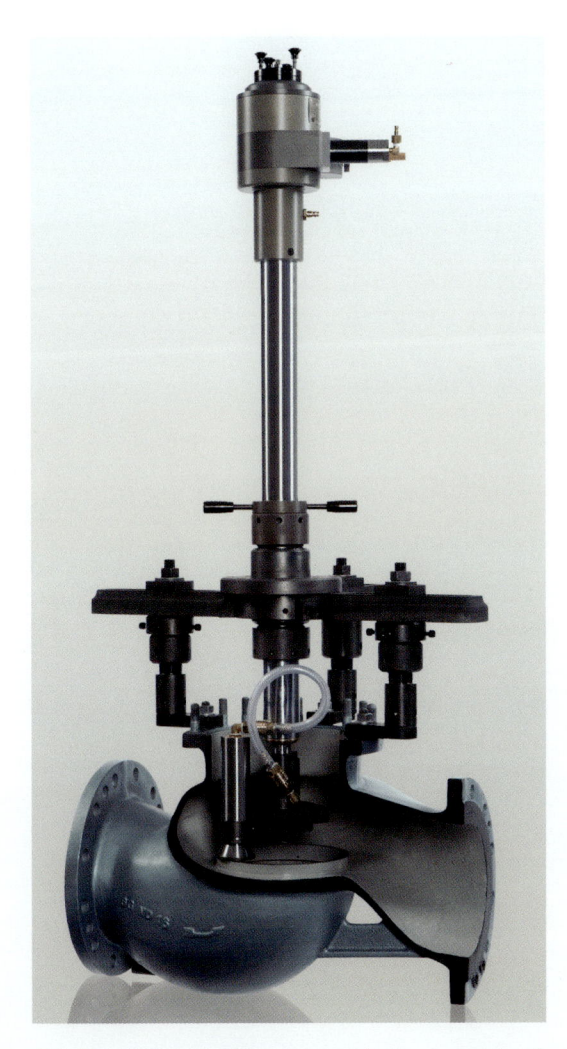

Figure MR12: High-speed precision grinding machine for the machining of build-up welding, sealing faces, conical faces and bores in valves (EFCO)

Unscheduled repairs resulting from valve failure create additional challenges. The owner may require that the cause of the failure be determined to assist either in his future selection of valves or in the way the plant is operated. In this case a competent person must be present during the disassembly of the valve so that all relevant evidence is collected regarding the cause of failure and a detailed report presented to the owner.

Failures typically fall into the following categories: -
- Through wall pressure envelope leakage or gross structural failure
- Body bonnet joint leakage
- Stem/shaft seal gross leakage
- Seat leakage
- Failure to operate in the case of isolation valves
- Failure to perform in the case of safety and control valves.

A careful analysis of the data collected from the failed valve will indicate whether process conditions have caused the failure or the manufacturer's design, faulty materials or faulty workmanship is responsible. The fact that the valve or actuator has failed does not prima facie imply the product manufacturer is at fault. It is in no-one's interest to ignore or hide the facts and responsible manufacturers will be keen to establish the cause of failure so that lessons can be learned and product improved.

Spare Parts

It is essential that the owner considers the availability of spares to facilitate a repair particularly an unscheduled one. The following terms are typical used: -
- **Commissioning spares** - These usually consist of one set of soft goods, comprising valve seats and actuator piston seals, for every ten valves and actuators per size supplied. These items could become damaged during the commissioning stage due to dirt in the lines
- **Two years spares** - These usually consist of a more comprehensive soft goods kit, including seats, gland seals, and body seals, one for every five valves per size supplied. A spare disc, plug, ball, or seal may also be recommended, depending on the quantity of valves of each size supplied
- **Long term spares** - In considering long term spares for valves and actuators, for which actual component recommendations will vary depending on the type of valve and actuator, the operational duty, and so on, serious consideration should be given to the purchase of complete valves, actuators, or even actuator/valve packages, so that in the event of major maintenance being required, the spare units can be used to replace the units requiring maintenance. The work can then be carried out in a suitable workshop under ideal conditions, with the minimum of disruption to plant operation.

Appendices

Appendix 1
British & International Standards for Valves & Actuators, Seals & Related Products

All British Standards are published by the British Standards Institution (BSI) and are available from BSI, 389 Chiswick High Road, London, W4 4AL (UK). Website: www.bsi-global.com. The members of the BVAA play a full and active part in the development of such standards via BSI standardization committees, currently designated 'PSE/18'.

These days many British Standards are implementations of standards published by either the International Organisation for Standardization 'ISO' or its European counterpart 'CEN' – the Committee for European Standardization. This means the content remains identical in each standardization body's version of the standard and in the UK these are prefixed 'BS EN' or 'BS ISO'. A similar arrangement applies in other countries. Occasionally standards are published as implementations of both EN and ISO and carry a 'BS EN ISO' prefix. Once again BVAA members participate in the meetings of CEN and ISO committees and working groups where these standards are developed.

The ISO 9000 series of standards have permeated most aspects of engineering and will be encountered frequently in the valve industry. ISO 9000 represents an international consensus on good quality management practices and consists of standards and guidelines relating to quality management systems and related supporting standards. ISO 9001 is the standard that provides a set of standardised requirements for a quality management system, regardless of the user organization's size or what it does. ISO 9000 will often appear as a normative reference in product and test standards, and is an important tool for helping to achieve compliance with European Directives.

European Directives

The UK is a member of the European Union (EU) and European legislation is now adopted throughout the EU in the form of Directives and Regulations. These EU directives are subject to the force of law through legislation in EU member states.

The basic purpose of EU Directives is to harmonise national legislation as the means of removing technical barriers to trade by the adoption of common regulations, thus promoting the creation of a single European market.

The European Community's Programme for the elimination of technical barriers to trade is formulated under the 'New Approach to Technical Harmonisation and Standards.' Its purpose is to harmonise national laws of EU Member States regarding design, manufacture, testing and conformity assessment of equipment to be placed on the market and put into service in EU member states. These New Approach directives provide a flexible regulatory environment that does not impose any detailed technical solution.

The UK has published laws, regulations and administrative procedures to comply with EU directives. Some of the EU directives that apply to valves and associated equipment include: -

Pressure Equipment Directive	97/23/EC
ATEX Directive	94/9/EC
Machinery Directive	2006/42/EC
Electromagnetic Compatibility Directive	2004/108/EC
Low Voltage Directive	2006/95/EC

The BVAA have published Guidelines to the interpretation of the requirement of the Directives and these are available by application to BVAA.

List of Standards

1 Valve Design

1.1 Materials

BS EN 1503-1:2000	Valves - Materials for bodies, bonnets and covers – Part 1: Steels specified in European Standards
BS EN 1503-2:2000	Valves - Materials for bodies, bonnets and covers – Part 2: Steels other than those specified in European Standards
BS EN 1503-3:2000/ AC:2001	Valves - Materials for bodies, bonnets and covers – Part 3: Cast irons specified in European Standards
BS EN 1503-4:2002	Valves - Materials for bodies, bonnets and covers – Part 4: Copper alloys specified in European Standards

1.2 Dimensions

BS EN 558:2008	Industrial valves - Face-to-face and centre-to-face dimensions of metal valves for use in flanged pipe systems - PN and Class designated valves
ISO 5752:1982	Metal valves for use in flanged pipe systems - Face-to-face and centre-to-face dimensions
BS EN 12982:2009	Industrial valves - End-to-end and centre-to-end dimensions for butt welding end valves

1.3 Terminology

BS EN 736-1:1995	Valves - Terminology - Part 1: Definition of types of valves
BS EN 736-2:1997	Valves - Terminology - Part 2: Definition of components of valves
BS EN 736-3:2008	Valves - Terminology - Part 3: Definition of terms

1.4 Installation and use

BS 6683:1985	Guide to installation and use of valves

1.5 Strength

BS EN 12516-1:2005/AC:2007	Industrial valves - Shell design strength - Part 1: Tabulation method for steel valve shells
BS EN 12516-2:2004	Industrial valves - Shell design strength - Part 2: Calculation method for steel valve shells
BS EN 12516-3:2002/AC:2003	Industrial valves - Shell design strength - Part 3: Experimental method
BS EN 12516-4:2008	Industrial valves - Shell design strength - Part 4: Calculation method for valve shells manufactured in metallic materials other than steel

1.6 General

BS EN 12351:2010	Industrial valves - Protective caps for valves with flanged connections

BS EN 12570:2000	Industrial valves - Method for sizing the operating element
BS EN 12627:1999	Industrial valves - Butt welding ends for steel valves
BS EN 12760:1999	Valves - Socket welding ends for steel valves
BS 5998:1983	Specification for quality levels for steel valve castings

2 Valve types

2.1 Butterfly

BS EN 593:2009	Industrial valves - Metallic butterfly valves
ISO 10631:1994	Metallic butterfly valves for general purposes
BS EN ISO 16136:2006	Industrial valves - Butterfly valves of thermoplastics materials (ISO 16136:2006)

2.2 Gate

BS EN 1171:2002	Industrial valves - Cast iron gate valves
BS EN 1984:2010	Industrial valves - Steel gate valves
ISO 5996:1984	Cast iron gate valves
ISO 6002:1992	Bolted bonnet steel gate valves
ISO 7259:1988	Predominantly key-operated cast iron gate valves for underground use
BS EN ISO 10434:2004	Bolted bonnet steel gate valves for the petroleum, petrochemical and allied industries (ISO 10434:2004)
BS EN 12288:2003	Industrial valves - Copper alloy gate valves
BS EN ISO 16139:2006	Industrial valves - Gate valves of thermoplastics materials (ISO 16139:2006)
BS EN ISO 15761:2002	Steel gate, globe and check valves for sizes DN 100 and smaller, for the petroleum and natural gas industries

2.3 Globe

BS 1873:1975	Specification for steel globe and globe stop and check valves (flanged and butt-welding ends) for the petroleum, petrochemical and allied industries
BS 5154 : 1991	Specification for copper alloy globe, globe stop and check, check and gate valves
BS 7350:1990	Specification for double regulating globe valves and flow measurement devices for heating and chilled water systems
ISO 12149:1998	Bolted bonnet steel globe valves for general-purpose applications
BS EN 13709:2010	Industrial valves - Steel globe and globe stop and check valves
BS EN 13789:2010	Industrial valves - Cast iron globe valves
BS EN ISO 21787:2006	Industrial valves - Globe valves of thermoplastics materials (ISO 21787:2006)

2.4 Check

BS 1868:1975/A1 : 1990	Specification for steel check valves (flanged and butt-welding ends) for the petroleum, petrochemical and allied industries
BS 7438:1991	Specification for steel and copper alloy wafer check valves, single-disk, spring-loaded type
BS EN 12334:2001/ A1:2004/AC:2002	Industrial valves - Cast iron check valves
BS EN 14341:2006	Industrial valves - Steel check valves
BS EN ISO 16137:2006	Industrial valves - Check valves of thermoplastics materials (ISO 16137:2006)

2.5 Ball

BS EN 1983:2006	Industrial valves - Steel ball valves
BS ISO 7121:2006	Steel ball valves for general-purpose industrial applications
DD CEN/TS 13547:2006	Industrial valves - Copper alloy ball valves
BS EN ISO 16135:2006	Industrial valves - Ball valves of thermoplastics materials (ISO 16135:2006)
BS EN ISO 17292:2004	Metal ball valves for petroleum, petrochemical and allied industries (ISO 117292:2004)

2.6 Plug

BS 5158:1989	Specification for cast iron plug valves
BS 5353:1989	Specification for steel plug valves

2.7 Water supply valves

BS EN 1074-1:2000	Valves for water supply - Fitness for purpose requirements and appropriate verification tests – Part 1: General requirements
BS EN 1074-2:2000/ A1:2004	Valves for water supply - Fitness for purpose requirements and appropriate verification tests – Part 2: Isolating valves
BS EN 1074-3:2000	Valves for water supply - Fitness for purpose requirements and appropriate verification tests – Part 3: Check valves
BS EN 1074-4:2000	Valves for water supply - Fitness for purpose requirements and appropriate verification tests – Part 4: Air valves
BS EN 1074-5:2001	Valves for water supply - Fitness for purpose requirements and appropriate verification tests – Part 5: Control valves
BS EN 1074-6:2008	Valves for water supply - Fitness for purpose requirements and appropriate verification tests – Part 6: Hydrants
BS 5163-1:2004	Valves for waterworks purposes - Part 1 : Predominantly key-operated cast iron gate valves - Code of practice
BS 5163-2:2004	Valves for waterworks purposes - Part 2 : Stem caps for use on isolating valves and associated water control apparatus

2.8 Industrial process valves

BS EN 1349:2009	Industrial process control valves

2.9 Thermoplastic valves

BS EN 28233:1992/ BS 2782-11 – Method 1131/ ISO 8233 : 1988	Thermoplastic valves - Torque - Test method (ISO 8233:1988)
ISO 8659 : 1989/ BS EN 28659 : 1992	Thermoplastic valves – Fatigue strength – Test method (ISO 8659 : 1989)

BS 6364:1984	Specification for valves for cryogenic service

2.10 Cryogenic valves

BS EN 13397:2002	Industrial valves - Diaphragm valves made of metallic materials
BS EN ISO 16138:2006	Industrial valves - Diaphragm valves of thermoplastics materials (ISO 16138:2006)

2.11 Diaphragm

BS EN 13942:2009	Petroleum and natural gas industries - Pipeline transportation systems - Pipeline valves (ISO 14313 : 2007 modified)

2.12 Pipeline

3 Safety

BS 759-1: 1984	Valves, gauges and other safety fittings for application to boilers and to piping installations for and in connection with boilers – Part 1 : Specification for valves, mountings and fittings
BS 1123: 2006	Fusible plugs for steam boilers and compressed air applications - Specification
BS 2915: 1990	Specification for bursting discs and bursting disc devices
BS 3463: 1975	Specification for observation and gauge glasses for pressure vessels
BS EN ISO 4126-1:2004/ AC:2006	Safety devices for protection against excessive pressure – Part 1: Safety valves (ISO 4126-1:2004)
BS EN ISO 4126-2:2003/ AC:2006	Safety devices for protection against excessive pressure – Part 2: Bursting disc safety devices (ISO 4126-2:2003)
BS EN ISO 4126-3:2006	Safety devices for protection against excessive pressure – Part 3: Safety valves and bursting disc safety devices in combination (ISO 4126-3:2006)
BS EN ISO 4126-4:2004	Safety devices for protection against excessive pressure – Part 4: Pilot operated safety valves (ISO 4126-4:2004)
BS EN ISO 4126-5:2004	Safety devices for protection against excessive pressure – Part 5: Controlled safety pressure relief systems (CSPRS) (ISO 4126-5:2004)
BS EN ISO 4126-6:2003/ AC:2006	Safety devices for protection against excessive pressure – Part 6: Application, selection and installation of bursting disc safety devices (ISO 4126-6:2003)
BS EN ISO 4126-7:2003	Safety devices for protection against excessive pressure – Part 7: Common data (ISO 4126-7:2004)
BS ISO 4126-9:2008	Safety devices for protection against excessive pressure – Part 9: Application and installation of safety devices excluding stand-alone bursting disc safety devices (ISO 4126-9:2008)

4 Marking

BS EN 19:2002	Industrial valves - Marking of metallic valves
ISO 5209:1997	General purpose industrial valves - Marking

5 Testing

BS EN 1267:1999	Valves - Test of flow resistance using water as test fluid
BS ISO 5208:2008	Industrial valves - Pressure testing of valves
BS EN ISO 10497:2010	Testing of valves - Fire type-testing requirements (ISO 10497:2004)
BS EN 12266-1:2003	Industrial valves - Testing of valves - Part 1: Pressure tests, test procedures and acceptance criteria - Mandatory requirements
BS EN 12266-2:2002	Industrial valves - Testing of valves - Part 2: Tests, test procedures and acceptance criteria - Supplementary requirements
BS EN 12569:1999/ AC:1999/AC:2000	Industrial valves - Valves for chemical and petrochemical process industry - Requirements and tests
BS EN 12567:2000	Industrial valves - Isolating valves for LNG - Specification for suitability and appropriate verification tests
BS EN ISO 15848-1:2006	Industrial valves - Measurement, test and qualification procedures for fugitive emissions - Part 1: Classification system and qualification procedures for type testing of valves (ISO 15848-1:2006)
BS EN ISO 15848-2:2006	Industrial valves - Measurement, test and qualification procedures for fugitive emissions - Part 2: Production acceptance test of valves (ISO 15848-2:2006)
BS EN 28233:1992/ BS 2782-11 – Method 1131/ ISO 8233 : 1988	Thermoplastic valves - Torque - Test method (ISO 8233:1988)
BS EN 28659:1992/ BS 2782-11:Method 1132/ ISO 8659 : 1989	Thermoplastics valves - Fatigue strength - Test method (ISO 8659:1989)

6　Performance

BS EN 13774:2003	Valves for gas distribution systems with maximum operating pressure less than or equal to 16 bar - Performance requirements
BS EN 14141:2003	Valves for natural gas transportation in pipelines - Performance requirements and tests
BS EN 15389:2008	Industrial valves - Performance characteristics of thermoplastic valves when used as construction products

7　Actuators

BS EN ISO 5210:1996	Industrial valves - Multi-turn valve actuator attachments (ISO 5210:1991)
BS EN ISO 5211:2001	Industrial valves - Part-turn actuator attachments (ISO 5211:2001)
BS EN 15081:2007	Industrial valves - Mounting kits for part-turn valve actuator attachment
BS EN 15714-1:2009	Industrial valves - Actuators - Part 1: Terminology and definitions
BS EN 15714-2:2009	Industrial valves - Actuators - Part 2: Electric actuators for industrial valves - Basic requirements
BS EN 15714-3:2009	Industrial valves - Actuators - Part 3: Pneumatic part-turn actuators for industrial valves - Basic requirements
BS EN 15714-4:2009	Industrial valves - Actuators - Part 4: Hydraulic part-turn actuators for industrial valves - Basic requirements

8　Steam traps

BS 6022:1983/ BS EN 26704:1991/	Automatic steam traps - Classification (ISO 6704:1982)
BS 6023:1981/ ISO 6552 :1980/	Glossary of technical terms for automatic steam traps
ISO 6552 : 1980/ BS 6023 : 1981	Glossary of technical terms for automatic steam traps
ISO 6553 : 1980/ BS EN 26553 : 1991	Automatic steam traps - Marking
ISO 6554 : 1980/ BS EN 26554:1991	Flanged automatic steam traps - Face-to-face dimensions
ISO 6704 : 1982/ BS EN 26704:1991	Automatic steam traps - Classification
ISO 6948 : 1981/ BS EN 26948:1991	Automatic steam traps - Production and performance characteristic tests
ISO 7841 : 1988/ BS EN 27841:1991	Automatic steam traps - Determination of steam loss - Test methods
ISO 7842 : 1988/ BS EN 27842:1991	Automatic steam traps - Determination of discharge capacity - Test methods
BS EN 26553:1991/ ISO 6553 : 1980	Automatic steam traps – Marking (ISO 6553:1980)
BS EN 26554:1991/ ISO 6554 : 1980	Flanged automatic steam traps - Face-to-face dimensions (ISO 6554:1980)
BS EN 26704:1991/ BS 6022:1983	Automatic steam traps - Classification (ISO 6704:1982)
BS EN 26948:1991/ ISO 6948 : 1981	Automatic steam traps - Production and performance characteristic tests (ISO 6948:1981)
BS EN 27841:1991/ ISO 7841 : 1988	Automatic steam traps - Determination of steam loss - Test methods (ISO 7841:1988)
BS EN 27842:1991/ ISO 7842 : 1988	Automatic steam traps - Determination of discharge capacity - Test methods (ISO 7842:1988)

9　Penstocks

BS 7775:2005	Penstocks for use in water and other liquid flow applications - Specification

10 Flanges

10.1 Design

BS 10:2009	Specification for flanges and bolting for pipes, valves, and fittings
BS EN 1092-1:2007	Flanges and their joints - Circular flanges for pipes, valves, fittings and accessories, PN designated - Part 1 : Steel flanges
BS EN 1092-2:1997	Flanges and their joints - Circular flanges for pipes, valves, fittings and accessories, PN designated - Part 2 : Cast iron flanges
BS EN 1092-3:2003/ AC : 2007	Flanges and their joints - Circular flanges for pipes, valves, fittings and accessories, PN designated - Part 3 : Copper alloy flanges
BS EN 1092-4:2002	Flanges and their joints - Circular flanges for pipes, valves, fittings and accessories, PN designated - Part 4 : Aluminium alloy flanges
BS 1560-3.2:1989	Circular flanges for pipes, valves and fittings (Class designated) - Part 3 : Steel, cast iron and copper alloy flanges – Section 2 : Specification for cast iron flanges
BS EN 1591-1:2001/ A1 : 2009	Flanges and their joints - Design rules for gasketed circular flange connections – Part 1 : Calculation method
BS EN 1591-2:2008	Flanges and their joints - Design rules for gasketed circular flange connections - Part 2 : Gasket parameters
DD CEN/TS 1591-3:2007	Flanges and their joints - Design rules for gasketed circular flange connections - Part 3 : Calculation method for metal to metal contact type flanged joint
DD CEN/TS 1591-4:2007	Flanges and their joints - Design rules for gasketed circular flange connections - Part 4 : Qualification of personnel competency in the assembly of bolted joints fitted to equipment subject to the Pressure Equipment Directive
BS EN 1759-1:2004	Flanges and their joints - Circular flanges for pipes, valves, fittings and accessories, class-designated – Part 1 : Steel flanges, NPS 1/2 to 24
BS EN 1759-3:2003/ AC : 2004	Flanges and their joints - Circular flanges for pipes, valves, fittings and accessories, class designated – Part3 : Copper alloy flanges
BS EN 1759-4:2003	Flanges and their joints - Circular flanges for pipes, valves, fittings and accessories, class designated - Part 4 : Aluminium alloy flanges
BS 3063:1965	Specification for dimensions of gaskets for pipe flanges
BS 3293:1960	Specification for carbon steel pipe flanges (over 24 inches nominal size) for the petroleum industry
ISO 7005-1 : 1992	Metallic flanges - Part 1 : Steel flanges
ISO 7005-2 : 1988	Metallic flanges - Part 2 : Cast iron flanges
ISO 7005-3 : 1988	Metallic flanges - Part 3 : Copper alloy and composite flanges

10.2 Bolting

BS EN 1515-1:2000	Flanges and their joints -Bolting - Part 1 : Selection of bolting
BS EN 1515-2:2001	Flanges and their joints -Bolting - Part 2 : Classification of bolt materials for steel flanges, PN designated
BS EN 1515-3:2005	Flanges and their joints - Bolting - Part 3 : Classification of bolt materials for steel flanges, class designated

11 Seals/Gaskets

11.1 Materials

BS EN 549:1995	Specification for rubber materials for seals and diaphragms for gas appliances and gas equipment
BS EN 751-1:1997	Sealing materials for metallic threaded joints in contact with 1st, 2nd and 3rd family gases and hot water – Part 1 : Anaerobic jointing compounds
BS EN 751-2:1997	Sealing materials for metallic threaded joints in contact with 1st, 2nd and 3rd family gases and hot water - Part 2 : Non-hardening jointing compounds
BS EN 751-3:1997	Sealing materials for metallic threaded joints in contact with 1st, 2nd and 3rd family gases and hot water - Part 3 : Unsintered PTFE tapes
BS 4249:1989	Specification for paper and cork/paper jointing
BS 4371:1991	Specification for fibrous gland packings
BS 5292:1980	Specification for jointing materials and compounds for installations using water, low-pressure steam or 1st, 2nd and 3rd family gases
BS 3381:1989	Specification for spiral wound gaskets for steel flanges to BS 1560
BS 6956-1:1988	Jointing materials and compounds - Part 1 : Specification for corrugated metal joint rings
BS 6956-5:1992	Jointing materials and compounds - Part 5 : Specification for jointing compounds for use with water, low pressure saturated steam, 1st family gases (excluding coal gas) and 2nd family gases
BS 7531:2006	Rubber bonded fibre jointing for industrial and aerospace purposes - Specification
BS EN 549:1995	Specification for rubber materials for seals and diaphragms for gas appliances and gas equipment
BS 7786:2006	Specification for unsintered PTFE tapes for general use
BS EN 12560-1:2001	Flanges and their joints - Gaskets for Class-designated flanges - Part 1 : Non-metallic flat gaskets with or without inserts
BS EN 12560-2:2001	Flanges and their joints - Gaskets for Class-designated flanges - Part 2 : Spiral wound gaskets for use with steel flanges
BS EN 12560-3:2001	Flanges and their joints - Gaskets for Class-designated flanges - Part 3 : Non-metallic PTFE envelope gaskets
BS EN 12560-4:2001	Flanges and their joints - Gaskets for Class-designated flanges - Part 4 : Corrugated, flat or grooved metallic and filled metallic gaskets for use with steel flanges
BS EN 12560-5:2001	Flanges and their joints - Gaskets for Class-designated flanges - Part 5 : Metallic ring joint gaskets for use with steel flanges
BS EN 12560-6:2003	Flanges and their joints - Gaskets for class-designated flanges - Part 6 : Covered serrated metal gaskets for use with steel flanges
BS EN 12560-7:2004	Flanges and their joints - Gaskets for class-designated flanges - Part 7 : Covered metal jacketed gaskets for use with steel flanges
BS EN 13090:2000	Means for resealing threaded joints of gas pipework in buildings

11.2 Dimensions

BS EN 1514-1:1997	Flanges and their joints - Dimensions of gaskets for PN-designated flanges - Part 1 : Non-metallic flat gaskets with or without inserts
BS EN 1514-2:2005	Flanges and their joints - Dimensions of gaskets for PN-designated flanges - Part 2 : Spiral wound gaskets for use with steel flanges
BS EN 1514-3:2005	Flanges and their joints - Dimensions of gaskets for PN-designated flanges - Part 3 : Non-metallic PTFE envelope gaskets
BS EN 1514-4:2009	Flanges and their joints - Dimensions of gaskets for PN-designated flanges - Part 4 : Corrugated, flat or grooved metallic and filled metallic gaskets for use with steel flanges
BS EN 1514-6 : 2003	Flanges and their joints - Dimensions of gaskets for PN-designated flanges - Part 6 : Covered serrated metal gaskets for use with steel flanges
BS EN 1514-7:2004	Flanges and their joints - Dimensions of gaskets for PN-designated flanges - Part 7: Covered metal jacketed gaskets for use with steel flanges
BS EN 1514-8:2004	Flanges and their joints - Dimensions of gaskets for PN-designated flanges - Part 8 : Polymeric O-Ring gaskets for grooved flanges
BS 7076-3:1989	Dimensions of gaskets for flanges to BS 1560 - Part 3 : Specification for non-metallic envelope gaskets
ISO7483:1991/Cor 1:1995	Dimensions of gaskets for use with flanges to ISO 7005
BS EN 13555:2004	Flanges and their joints - Gasket parameters and test procedures relevant to the design rules for gasketed circular flange connections
CR 13642 : 1999	Flanges and their joints - Design rules for gasketed circular flange connections - Background information

11.3 Quality Assurance

BS EN 14772:2005	Flanges and their joints - Quality assurance inspection and testing of gaskets in accordance with the series of standards EN 1514 and EN 12560

13 Pipework components

BS EN 1333:2006	Flanges and their joints - Pipework components - Definition and selection of PN
BS EN ISO 6708:1996	Pipework components - Definition and selection of DN (nominal size)
ISO7268:1983/Amd1:1984	Pipe components - Definition of nominal pressure

14 Leak detection

BS EN 14291:2004	Foam producing solutions for leak detection on gas installation

15 Lubricants

BS EN 377:1993/ A1 : 1996	Lubricants for applications in appliances and associated controls using combustible gases except those designed for use in industrial processes

16 Piping

16.1 Design

BS 806:1993	Specification for design and construction of ferrous piping installations for and in connection with land boilers
BS 1306:1975	Specification for copper and copper alloy pressure piping systems
BS EN 13480-1:2002/ A2 : 2008	Metallic industrial piping – Part 1 : General
BS EN 13480-2:2002	Metallic industrial piping – Part 2 : Materials
BS EN 13480-3:2002/ A3 : 2009	Metallic industrial piping. Part 3: Design and calculation
BS EN 13480-4:2002	Metallic industrial piping – Part 4 : Fabrication and installation
BS EN 13480-5:2002	Metallic Industrial Piping – Part 5 : Inspection and testing
BS EN 13480-6:2004/ A1 : 2005	Metallic industrial piping – Part 6 : Additional requirements for buried piping
PD TR 13480-7:2002	Metallic industrial piping – Part 7 : Guidance on the use of conformity assessment procedures
BS EN 13480-8:2007	Metallic industrial piping – Part 8 : Additional requirements for aluminium and aluminium alloy piping

16.2 Supports

BS 3974-1:1974	Specification for pipe supports – Part 1 : Pipe hangers, slider and roller type supports
BS 3974-2:1978	Specification for pipe supports – Part 2 : Pipe clamps, cages, cantilevers and attachments to beams
BS 3974-3:1980	Specification for pipe supports – Part 3 : Large bore, high temperature, marine and other applications

American Industry Standards

Many companies in the valve and actuator industry need to refer to, and comply with, American industry standards. These may include: -

API - American Petroleum Institute Valve Standards

API SPEC 6A	Specification for Wellhead and Christmas Tree Equipment
API SPEC 6D	Specification for pipeline valves
API STD 6C	Flanged steel gate and plug valves for drilling and production service
API STD 6D	Steel gate plug ball and check valves for pipeline service
API 526	Flanged steel safety relief valves
API 527	Commercial seat tightness of safety relief valves with metal - to metal seats
API 528	API standard for safety relief valve nameplate nomenclature
API 529	Cast - forged Steel plug valves flanged ends
API 594	Wafer - type check valves
API 595	Cast - iron gate valves flanged ends
API 597	Steel venturi gate valves flanged or butt welding ends
API 598	Valve inspection and test
API 599	Steel plug valves flanged or butt welding ends
API 600	Flanged and butt - welding - end steel gate and plug valves for refinery use
API 602	Compact design carbon steel gate valves for refinery use
API 603	150 - pound light - wall corrosion - resistant gate valves for refinery use
API 604	Ductile iron gate valves flanged ends
API 609	Butterfly valves to 150 psig and 150 F
API SPEC 6FA	Specification for fire test for valves
API SPEC 6FC	Specification for fire test for valves with automatic backseats
API BULL 6RS	Bulletin on referenced standards for committee 6, standardization of valves and wellhead equipment
API RP 11V6	Recommended practice for design of continuous flow gas lift installations using injection pressure operated valves first edition
API RP 11V7	Recommended practice for repair, testing, and setting gas lift valves
API RP 520 PT 1	Sizing, selection and installation of pressure relieving devices in refineries part 1- sizing and selection
API RP 574	spection of piping, tubing, valves and fittings first edition, replaces guide for inspection of refinery equipment Chapter XI
API RP 576	Inspection of pressure-relieving devices
API STD 608	Metal ball valves-flanged, threaded and welding end

The American National Standards Institute, (ANSI)

ANSI B1.1	Unified & American Screw Threads.
ANSI B2.1	Pipe Threads.
ANSI B16.5	Steel Pipe Flanges & Flanged Fittings.
ANSI B16.9	Steel Butt Welding Fittings.
ANSI B16.10	Face-to-Face Dimensions of Ferrous Flanged & Welding End Valves.
ANSI B16.20	Ring Joint Gaskets in Grooves for Steel Pipe Flanges.
ANSI B16.21	Non-Metallic Gaskets for Pipe Flanges.
ANSI B16.25	Butt Welding Ends.
ANSI B16.34	Valves - Flanged and Butt Welding Ends.

American Society of Mechanical Engineers (ASME)

ASME Boiler & Vessel Code:
Material Specifications (Section II)
Nuclear Power Plant Components (Section III)
Heating Boilers (Section IV)
Pressure Vessels (Section VIII)
Welding Qualifications (Section IX)

Manufacturers Standardization Society of the Valve & Fittings Industry (MSS SP)

MSS SP-6	Standard Finishes for Contact Faces of Pipe Flanges & Connecting End Flanges of Valves & Fittings.
MSS SP-25	Standard Marking System for Valves, Fittings, Flanges, & Unions.
MSS SP-54	Quality standard for steel castings – radiographic inspection method for valves
MSS SP-55	Quality standard for steel castings for valves, flanges and fittings
MSS SP-61	Hydrostatic Testing of Steel Valves.
MSS SP-84	Steel Valves – Socket Welding & Threaded Ends
MSS SP-86	MSS Guidelines for Metric Data in Standards for Valves, Flanges & Fittings.

Appendix 2
SI Units for Valves & Actuators

The data of greatest concern to the valve and actuator maker and users are those for:

  length;
  mass;
  temperature;
  pressure;
  volume or capacity;
  stress.

The SI units commonly used in catalogues, sales literature, etc., together with the respective SI coherent units, are set out in the accompanying table.

Quantity	SI unit commonly used	SI coherent or primary unit
Length	mm	m
Mass	kg	kg
Temperature	°C	K
Pressure	Bar (10^5 Pa)	Pa (1 Pa = 1 N/m^2)
Volume or capacity	1 (litre)	m^3
Stress (UTS, YP)	MPa or N/mm^2	Pa (1 Pa = 1 N/m^2)

Note that only one of the commonly used units, namely kilogram (kg), is in the coherent class of SI units.

Millimetre (mm) and litre (l) are decimal submultiples, and megapascal (MPa) is a decimal multiple of coherent SI units and are recognized as such within the SI system. The bar is also a decimal multiple of a coherent SI unit, but is not a selected multiple within SI. Nevertheless the bar is being retained because of its practical usefulness as a unit of pressure in some sectors of industry both in the UK and in Europe.

The SI unit of temperature is the Kelvin (K) and this unit is equally applicable to thermodynamic temperature and to temperature differences. It is however generally recognized that the term degree Celsius (°C) will continue in use as the everyday customary unit of temperature. The units of Celsius and Kelvin temperature interval are identical but the zero of the Celsius scale is the temperature of the ice point (273.15 K).

It is recognized that it will not always be practical to limit the usage of SI units to those mentioned above and other decimal multiples and sub-multiples will also be used. A list of SI prefixes used to form names and symbols of multiples of SI units is given in the table below.

Factor by which The unit is multiplied	Prefix Name	Prefix Symbol
10^{12}	tera	T
10^{9}	giga	G
10^{6}	mega	M
10^{3}	kilo	k
10^{2}	hecto	h
10	deca	da
10^{-1}	deci	d
10^{-2}	centi	c
10^{-3}	milli	m
10^{-6}	micro	μ
10^{-9}	nano	n
10^{-12}	pico	p
10^{-15}	femto	f
10^{-18}	atto	a

For full information on SI units and their application reference should be made to the relevant publications of the International Organisation for Standardization (ISO) or the British Standards Institution (BSI).

Appendix 3
Pressure Drop Through a valve and flow coefficient

Pressure drop

Liquid flow
Pressure loss through a valve can be expressed by the following equations

1. $\Delta h = \dfrac{Kv^2}{2g}$

2. $\Delta P = 0.000005\, Kv^2\, \rho$

3. $\Delta P = 1.389\, K \left(\dfrac{Q^2}{a^2}\right) G$

where
- Δh = head loss in metres of liquid;
- ΔP = pressure drop in bar
- K = resistance coefficient;
- v = mean flow velocity in m/s
- g = acceleration of gravity, 9.81 m/s².
- ρ = density of liquid in kg/m³.
- Q = flow rate in l/min;
- G = specific gravity of the liquid;
- a = nominal flow area of valve in mm².

Gaseous flow
In the case of compressible fluids some modifications may be required to the above formula and users are advised to consult the valve manufacturer.

Equivalent length
A convenient way of expressing valve resistances is in the form of an equivalent length of straight pipe which would offer the same amount of resistance as the valve. The value calculated from the following equation can then be added to that of the general piping circuit to obtain the total equivalent pipe length and the total head loss found by the use of established tables. The friction factor f used in the equation for equivalent length is the Darcy friction factor. Resistance coefficient K and friction factor f must be obtained from the valve manufacturer.

4. $L = \dfrac{dK}{1000\, f}$

Where
- L = equivalent length of pipe in m;
- K = resistance coefficient;
- d = diameter of pipe in mm;
- f = friction factor for pipe.

Flow Coefficient

The relationship between the pressure drop and flow rate through a valve is known as the flow coefficient. The internal geometry of the flow path through the valve determines the flow coefficient and in principle this is a unique number for each valve. However variations between different valves of identical design and size are so small that they can be ignored. Even variations between different manufacturers' designs result in only small variations in flow coefficient for the same basic valve type and DN. The formulae can be used to establish the pressure drop across a valve for a given flow rate or, alternatively the flow rate through a valve, which will generate a given pressure drop.

Flow coefficient values are usually based on testing with water. Typically a number of valves from a range are tested to determine the relationship between valve internal geometry and flow coefficient. For those valve sizes not tested the manufacture will interpolate between or extrapolate from the results obtained from the tests. Flow coefficients may not be valid for all conditions of flow. Hence, users requiring detailed information on the flow characteristics of a valve are advised always to consult with the valve manufacturer.

Liquid Flow
The basic formulae for liquid have already been given in the chapter on Valve Basics. In many parts of the world, including the UK and the USA, the flow coefficient in most general use is C_v (imperial units) while in Europe the coefficient K_v (metric units) is mostly used.

Compressible Flow
The flow coefficient for a compressible gas flow can also be calculated using a similar type of formula to that used for liquid flow. The fundamental difference between liquid and gas flows is the velocity at the controlling orifice. Gas flow velocities are generally much higher than for a liquid. The limiting factor in compressible flow is when the velocity at the controlling orifice reaches the speed of sound for the gas at the given inlet conditions. Under these conditions the flow is described as choked and the flow rate is called the critical flow rate. Further reductions in downstream pressure will not increase the flow rate. The ratio of $\Delta P/P_1$ which creates the choked conditions is known as the critical pressure drop. In many actual applications the critical pressure drop is exceeded but the flow rate will never exceed the critical flow rate.

The equations for flow coefficient take this into account using an expansion factor Y. This compares the critical pressure drop and the ratio of specific heats for air at 15.5°C with the actual pressure drop and ratio of specific heats of the flowing gas in the valve.

5. $Y = 1 - \dfrac{k_{air} x}{3 k_{gas} x_T}$ where $x = \dfrac{\Delta P}{P_1}$

and $x_T = \left(\dfrac{\Delta P}{P_1}\right)_{aircrit}$ and $x \leq \dfrac{k_{gas} x_T}{k_{air}}$

This gives the minimum value of Y = 0.6667 that is possible from Equation 5. Also included in the equations 6, 7, and 8 is the compressibility factor Z which compensates for the difference in properties of a real gas and those expected of an ideal gas. These equations make no compensation for piping factors associated with reducers or expanders upstream or downstream of the valve. For equations 6, 7 and 8

$x \leq \dfrac{k_{gas} x_T}{k_{air}}$ is also valid.

6. $C_v = \dfrac{Q}{417 P_1 Y} \sqrt{\dfrac{T_1 G Z}{x}}$ if the specific gravity G of the flowing gas is specified

7. $C_v = \dfrac{Q}{2240 P_1 Y} \sqrt{\dfrac{T_1 M Z}{x}}$ if the molecular weight M of the flowing gas is specified

8. $C_v = \dfrac{W}{94.8 P_1 Y} \sqrt{\dfrac{T_1 Z}{x M}}$ if the molecular weight M of the flowing gas is specified

where
W = gas flow rate in kg/hr.
Q = gas flow rate, m³/hr;
T_1 = temperature of gas in K (= °C + 273);
ΔP = pressure drop across valve in bar
P_1 = valve inlet pressure in bar absolute;
Z = compressibility factor
k = ratio of specific heats of the flowing gas
   (k_{air} = 1.4 at 15.5°C)
M = molecular weight of flowing gas;
G = specific gravity of flowing gas (air = 1);

To convert C_v to other flow coefficients, K_v and A_v, the following equations should be used:
K_v = 0.865 x C_v A_v = 2.4x 10^{-5} x C_v

Appendix 4
Steam Tables

Pressure	Saturation Temperature	Specific Volume of Saturated Water (vf)	Specific Enthalpy of Saturated Water (hf)	Specific Enthalpy of Evaporation (hfg)	Specific Enthalpy of Saturated Steam (hg)	Specific Volume of Saturated Steam (vg)
	°C	m³/kg	kJ/kg	kJ/kg	kJ/kg	m³/kg
bar absolute						
0.30	69.11	0.00102	289.3	2335.3	2624.6	5.230
0.50	81.34	0.00103	340.6	2304.8	2645.4	3.241
0.75	91.78	0.00104	384.5	2278.1	2662.6	2.218
0.95	98.20	0.00104	411.5	2261.4	2672.9	1.778
1.00	99.63	0.00104	417.5	2257.6	2675.2	1.694
1.01325	100.00	0.00104	419.1	2256.7	2675.8	1.674
bar gauge						
0.0	100.0	0.00104	419.1	2256.7	2675.8	1.674
0.1	102.7	0.00105	430.3	2249.6	2680.0	1.533
0.2	105.1	0.00105	440.8	2243.1	2683.8	1.414
0.3	107.4	0.00105	450.5	2236.9	2687.4	1.313
0.4	109.6	0.00105	459.7	2231.0	2690.7	1.226
0.5	111.6	0.00105	468.3	2225.5	2693.8	1.150
0.6	113.6	0.00105	476.5	2220.2	2696.7	1.083
0.7	115.4	0.00106	484.3	2215.1	2699.5	1.024
0.8	117.2	0.00106	491.8	2210.3	2702.1	0.971
0.9	118.8	0.00106	498.9	2205.6	2704.5	0.923
1.0	120.4	0.00106	505.7	2201.2	2706.9	0.880
1.1	122.0	0.00106	512.3	2196.8	2709.1	0.841
1.2	123.5	0.00106	518.6	2192.7	2711.3	0.806
1.3	124.9	0.00106	524.7	2188.6	2713.3	0.773
1.4	126.3	0.00107	530.6	2184.7	2715.3	0.743
1.5	127.6	0.00107	536.3	2180.9	2717.1	0.715
1.6	128.9	0.00107	541.8	2177.2	2718.9	0.690
1.7	130.2	0.00107	547.1	2173.5	2720.7	0.666
1.8	131.4	0.00107	552.3	2170.0	2722.3	0.644
1.9	132.6	0.00107	557.4	2166.6	2724.0	0.623
2.0	133.7	0.00107	562.3	2163.2	2725.5	0.603
2.2	135.9	0.00108	571.7	2156.7	2728.5	0.568
2.4	138.0	0.00108	580.7	2150.5	2731.3	0.537
2.6	140.0	0.00108	589.3	2144.6	2733.9	0.509
2.8	141.9	0.00108	597.6	2138.8	2736.4	0.484
3.0	143.8	0.00108	605.5	2133.2	2738.7	0.461
3.2	145.5	0.00109	613.0	2127.9	2740.9	0.440
3.4	147.2	0.00109	620.4	2122.7	2743.0	0.422
3.6	148.9	0.00109	627.4	2117.6	2745.0	0.404
3.8	150.4	0.00109	634.2	2112.7	2746.9	0.389
4.0	152.0	0.00109	640.8	2107.9	2748.8	0.374
4.5	155.6	0.00110	656.5	2096.5	2753.0	0.342
5.0	158.9	0.00110	671.1	2085.7	2756.8	0.315
5.5	162.1	0.00110	684.8	2075.5	2760.3	0.292
6.0	165.1	0.00111	697.7	2065.7	2763.4	0.272
6.5	167.9	0.00111	709.9	2056.4	2766.3	0.255
7.0	170.5	0.00111	721.6	2047.4	2769.0	0.240
7.5	173.0	0.00112	732.6	2038.8	2771.5	0.227
8.0	175.5	0.00112	743.2	2030.5	2773.7	0.215
8.5	177.8	0.00112	753.4	2022.4	2775.8	0.204
9.0	180.0	0.00113	763.2	2014.6	2777.8	0.194
9.5	182.1	0.00113	772.6	2007.1	2779.6	0.185
10	184.2	0.00113	781.7	1999.7	2781.3	0.177
11	188.0	0.00114	798.9	1985.5	2784.4	0.163
12	191.7	0.00114	815.2	1971.9	2787.1	0.151
13	195.1	0.00115	830.5	1959.0	2789.5	0.141
14	198.4	0.00115	845.1	1946.5	2791.6	0.132
15	201.5	0.00116	858.9	1934.5	2793.4	0.124
16	204.4	0.00116	872.2	1922.8	2795.0	0.117
17	207.2	0.00117	884.9	1911.5	2796.4	0.110
18	209.9	0.00117	897.1	1900.6	2797.7	0.105
19	212.5	0.00118	908.9	1889.9	2798.8	0.100
20	214.9	0.00118	920.2	1879.5	2799.7	0.095
21	217.3	0.00119	931.2	1869.3	2800.5	0.091
22	219.6	0.00119	941.8	1859.4	2801.2	0.087
23	221.9	0.00119	952.1	1849.7	2801.8	0.083
24	224.0	0.00120	962.1	1840.1	2802.2	0.080
25	226.1	0.00120	971.9	1830.7	2802.6	0.077
30	235.7	0.00122	1017.1	1786.2	2803.3	0.064
35	244.2	0.00124	1057.6	1744.8	2802.3	0.055
40	251.9	0.00126	1094.5	1705.6	2800.1	0.049
45	258.8	0.00127	1128.6	1668.3	2796.9	0.043
50	265.2	0.00129	1160.5	1632.4	2792.9	0.039
55	271.2	0.00131	1190.6	1597.6	2788.2	0.035
60	276.7	0.00132	1219.0	1563.8	2782.8	0.032
65	281.9	0.00134	1246.2	1530.7	2776.9	0.029
70	286.8	0.00135	1272.2	1498.3	2770.5	0.027
75	291.5	0.00137	1297.3	1466.4	2763.6	0.025
80	295.9	0.00139	1321.5	1434.8	2756.3	0.023
100	311.8	0.00146	1411.7	1311.0	2722.7	0.018
120	325.4	0.00153	1494.9	1187.5	2682.4	0.014

BVAA VALVE & ACTUATOR USERS' MANUAL - 6th Edition

Our art is releasing value through design

Our new range of PAM Valves designed specifically to take cost out of your business. For more information visit

www.pamvalves.co.uk

SAINT-GOBAIN
PAM UK

More than a work of art
Tough, long-lasting and affordable
A true work of genius

Eurostop Butterfly Valve   FH2-CE Hydrant   Eurocheck Non-Return Valve

Registered in England No. 56433. Registered Office: Stanton by Dale, Ilkeston, Derbyshire DE7 4QU England

Index

Actuators:

Actuating, linear motion valves	147
Actuating, rotary motion valves	148
Actuator control	166
Actuator, definition	21
Actuator duty types	166
Actuator power unit	179
Actuator sizing	158, 167, 177
Actuator torque characteristics	156, 158
Air fail open/ closed	157
Cam mechanism actuator	157, 176
Control accessories	159, 177
Control valve actuators	179
Control valve actuator terminology	179
Diaphragm actuators	155, 157
Drive coupling	185
Dual opposed	157
Electric	16, 151, 163, 181
Electric actuator advantages	168
Electric actuator disadvantages	169
Electro-hydraulic	16, 171, 172, 181
Fire-safe actuators	159, 177
Gas-over-oil (gas hydraulic)	16, 151
Gearbox, bevel	145
Gearbox, sizing	146
Gearbox, spur	144, 145
Gearbox, subsea	149
Gearbox, worm	143, 145
Gearboxes/ gearing	76, 87, 164
Helical spline, rotary	176
Hydraulic	16, 153, 171, 181
Hydraulic actuator advantages	177
Hydraulic actuator disadvantages	177
Hydraulic actuator types	175
Hydraulic motor	176
Piston actuator	155, 156, 173, 175, 180
Pneumatic	16, 125, 151, 152
Pneumatic actuator advantages	161
Pneumatic actuator disadvantages	161
Pneumatic actuator types	155
Pneumatic diaphragm actuator types	179
Pneumatic motor actuators	158
Pneumatic piston	155, 180
Rack and pinion actuators	156, 157, 175, 176
Rotary pneumatic	181
Scotch yoke	153, 156, 173, 175
Scotch yoke, canted	156
Self-contained	172, 177
Trunnion/ lever arm	157, 176
Vane actuators	157, 176

Valves:

5 basic types	19, 20
Ball	14, 15, 16, 19, 93
Ball, actuating	148
Ball, all welded	96, 97
Ball, anti blowout	97
Ball, anti static	97
Ball, body styles	95
Ball, cavity relief	98
Ball, ceramic	98
Ball, control valves	129
Ball, end entry	95
Ball, floating	94, 98, 129
Ball, full port	96, 97
Ball, lined	98
Ball, low temperature	99
Ball, notch/ v-notch	129
Ball, one-piece construction	94
Ball, reduced port	96, 97
Ball, rising stem	94, 95
Ball seats	29, 93
Ball, three-piece	95, 96
Ball, top entry	96, 183
Ball, trunnion mounted	94, 129
Ball, two-piece	95, 96
Ball, unidirectional	98
Butterfly	14, 16, 19, 101
Butterfly, actuating	148
Butterfly, concentric/zero offset	101, 129
Butterfly, conical geometry	106
Butterfly, control valves	129
Butterfly, double flange	101, 102
Butterfly, double offset	101, 105, 129
Butterfly, general purpose (concentric)	102
Butterfly, high performance	105
Butterfly, lined	103
Butterfly, lug type	101
Butterfly, triple offset	101, 106, 129
Butterfly, wafer type	101, 103, 105
Check	109, 186
Check, assisted opening/closing	113
Check, diaphragm	110
Check, dual wafer plate	109
Check, globe stop	113
Check, hinged flap	112
Check, lift	13, 110, 111, 112
Check, non-return	20
Check, nozzle	110
Check, selection considerations	109
Check, swing	13, 26, 109, 111, 113
Check, tilting rotating disc	110
Check, types	109
Choke	27
Control	19, 57, 123, 186
Control, ball	129
Control, globe	123, 124, 125, 127, 128
Control, linear rising stem	127
Control, plug, eccentric rotating	130
Control, rotary	127, 129
Control, rotating plug	129
Control, rotating plug in globe	133
Control, sizing and selection	130
Control, sliding gate	132
Control valve actuators	123
Control valve characteristics	130

Control valve types	127
DBB (see double block and bleed)	139
Diaphragm	14, 15, 19, 79, 148
Diaphragm seal	21
Diaphragm, straight through type	79, 80
Diaphragm, tee-style	81
Diaphragm, weir type	79, 80
Diverting	20
Double block and bleed (DBB)	71, 89, 107, 139
Double block and bleed, locking	140
Double isolation	139
Gate	19, 69
Gate, conduit	71
Gate, flexi wedge	70, 71
Gate, flexi wedge, actuating	147
Gate, knife	72, 73
Gate, parallel slide	72
Gate, parallel slide, actuating	147
Gate, reduced bore conduit (venturi)	73
Gate, sluice	71
Gate, solid wedge	70, 147
Gate, through conduit, actuating	147
Gate, through conduit, parallel expanding	71, 72
Gate, wedge	69, 70
Globe	14, 19, 75, 127, 148
Globe, angled	75, 128
Globe, oblique	75
Globe, straight	75
Globe, stop and check	77
Globe, three way	129
Globe, Y pattern	75
Iris	83
Isolating	19, 139, 190
Linear	67
Mixing	20
Mono flange	139
Multi end body	22
Needle	140
Non-return	109
Non-return, screw down (SDNR)	113
Penstock	73
Pinch	19, 83, 148
Plug	14, 15, 19, 87
Plug, actuating	148
Plug cock	13, 87
Plug, double isolation (DIPV)	90
Plug, eccentric	88, 89
Plug, expanding	89
Plug, full bore	88
Plug, lined design	88
Plug, lubricated	87, 88, 89, 190
Plug, multi-port	88
Plug, non-lubricated	87
Plug, parallel	87
Plug, sleeved design	88
Plug, spherical (ball)	93
Plug, Standard port	88
Plug, taper	15, 87
Plug, taper, pressure/dynamic balance	89
Plug, venturi style	88
Pressure regulators/ regulating	123
Pressure sustaining	124
Pressure relief	115
Reduced bore valve	23
Regulating	19, 179
Relief	115, 116, 187
Relief, thermal	116
Safety	19, 115, 116, 187
Safety, balanced or unbalanced	120
Safety, direct spring	117, 119
Safety, full or semi Nozzle	120
Safety, open or closed bonnet	120
Safety, pilot operated	119
Safety relief valve	115, 116
Safety, steam	13, 14
Safety, torsion bar	119
Safety, types of	117
Safety, weight-loaded	119
Solenoid/ solenoid pilot	159
Solid wedge, actuating	147

General:

Acceptance test	23
Accumulation	116
Accumulators	160, 174, 175
Adjustable stops	160
Airset	123
Allowable differential pressure	22
Angle pattern body	22
ANSI	23
ANSI/FCI 70-2	130
Anti-blow out design	23
Anti-cavitation/ low noise trim	128
Anti-static design	23
Autonomous shut down	64
API	23, 115
API 607 fire test	47
API 622 packing test	56
Aseptic service	81
ASME	23, 115
ASTM	23
ATEX (Potentially Explosive Atmospheres)	49
Austenitic	26, 57
Backlash	185
Back pressure	116
Back seat	21
Back-up ring, sealing	58
Bellows, balanced	120, 121
Bellows, seal	21, 34, 35, 76, 127
Bio-Pharmaceutical	80, 81
BISVMA	17
Blowdown/ring	116, 117, 119
Body	20, 25
Body bonnet /cover gasket	20, 55
Body end	20, 21
Body styles	21
Bolting	43
Bonnet	20, 25, 33, 127, 191
Bonnet, bushing	20
Bonnet, extended	99
Bonnet, pressure seal	33, 56, 69
Boundary layer lubrication	57, 58
Bredtschneider joint	56

Bridgeman joint	56
BS 6755 Part 2 fire test	47
BSI	23
BSI Kitemark	46
BSI PSE/18 Committee	46
BVAA (British Valve & Actuator Association)	17, 23
Butt welding end	22, 30, 31, 196
BX type ring joint	56
Bypass	21, 179
Cage, types	127
Capillary end	22, 30
Centre-to-end dimension, CTE	22
Centre-to-face dimension, CTF	22
Coating	23
Compression end	22, 30
Carbon steel	26
Carbon and low alloy steels (corrosion)	42
Cast iron	26
Cavitation	131
CEN	23
CE Marking	45, 48, 51
Ceramics	29
Chatter	110, 113, 121, 190
Chemical Resistance of rubber/plastic linings	41
Chemical Vapour Deposition (CVD)	29
Chromium	26
Class	22
Cobalt alloys	28
Cold differential pressure	117
Commissioning	186, 197
Conformity assessment (PED)	49
Controlled Safety Pressure Relief System (CSPRS)	120
Copper alloys	28
Corrosive attack	39
Corrosion resistance tables	41
Corrosion resistant alloys	42
COSHH	196
Creep	26, 31
Cv values	109, 132
Cyrogenic service	28, 99, 107, 111
Declaration of Conformity	48, 50
Declaration of Incorporation	50
Decontamination	196
Dewrance & Co.	13
DIN	23
Direct acting	179, 180
Directives, European	48
Discharge characteristic	186
Distributed Control Systems (DCS)	187
DN (nominal size)	22
Documentation	196
Double acting	155, 158, 172, 174, 175, 180
Duplex	26
Duty cycle	166
Elastomeric seats	94
'Electrical Equipment' (Low Voltage Directive)	52
Electrical Equipment Safety Regulations	52
EMC (Electromagnetic Compatibility/ Directive)	51
EMC, 'Apparatus'	51
Emergency shutdown (ESD)	62
EN	23
EN 12570	144
EN 15714 Actuator standard	166
EN ISO 5210/5211	152, 185
End connections	29
End-to-end dimension, ETE	22, 64
Environmental considerations	63
Environmental protection	163
Equilibrium pressure/ operation	115
Essential Safety Requirements (ESRs)	48
ETFE fluoropolymer	15, 29
Ethylene-propylene, seal	57
EU	23, 48
European Norms (ENs)	48
Exercising	62
Ex marking (ATEX)	50
Expanded graphite	55, 57
Face-to-face dimension, FTF	22, 64
Ferritic	26, 57
Field reversible	180
Filter regulator	123
Finite Element Analysis (FEA)	152
Fire and explosion hazards	42, 105
Fire type test, EN ISO 10497	46, 65, 89, 93, 98, 106
First valves	13
Flange covers	187
Flanged end	22, 30
Flap	13
Flashing	130, 186
Flow characteristics	34
Fluid sealing materials	55
Flurocarbons, seals	57
Flutter chatter	121
FMEDA (Failure Modes Effect & Diagnostic Analysis)	137
Fugitive emissions, EN ISO 15848	47, 56
Full bore valve	23
Galling	29, 72
Gaskets	30, 32, 55
Gland sealing	34, 56
Gland flange	21
Graduated travel indication plate	179
Greenhalgh, J	17
Greenhalgh, M	9
Grey Iron	26
Gunmetal (red brass)	28
Hackworth, Timothy	13
Handwheel	179
Hard facing	20
Harmonised standards	48, 51
Hastelloy®	27
Hazardous areas/ratings/ classification	63, 164
Hazardous chemicals	77
Helium (test)	36
Hindle, Joshua	17
HIPPs (High Integrity Pipeline Protection)	135
HTHP (High Temperature High Pressure)	135
Hydraulic power supplies	172
Hydrogen embrittlement	40
Hydrogen sulphide	42
IEC	23
IEC 60534	125
IEC 61508	135, 136
IEC 61511	136

Inconel®	27
Industrial revolution	14, 16
Inside screw	143
Installing, actuators	184
Installing, flanged-end valves	184
Installing, thread-end valves	184
Installing, welded-end valves	184
Installation & Operation	183, 186
Intergranular corrosion	39
Intrinsically safe	164
IP rating	163
ISA	23
ISO	23
ISO 9001	45, 65
JIS	23
Joukowsky	113
Kammprofiles	55
Key Interlocks	187
Lantern ring	21, 34, 57
Lens rings	55
Limit setting	186
Limit switches	159
Lining	23
Lip seal	57
Live loading	34
Low pressure recovery trim	131
Low Voltage Directive (LVD)	52
Lug type body	22
Machinery Directive	50
Maintenance & Repair	65, 189
Maintenance instructions	190
Manganese	26
Manual operation	143, 144, 165, 172
Manual override	143, 160
Manufacturers' instructions	185
Market Growth & Consolidation	17
Marking	36
Martensitic	26, 57
Materials	25
Metal ring joints	56
Metal seats	148
Molybdenum	26
Mounting brackets	185
MSS	23
Maximum allowable pressure, PS	22, 48
Maximum allowable temperature, TS	22
Maximum travel	23
NACE	23, 27, 39, 42
NDE (Non-Destructive Examination)	196
Newcombe, Thomas	13
Nickel	26
Nickel alloys	27, 28
Niobium	39
Nitrile, seal	57
Nitrogen	26
Non-rotating	77
Nordstrom, Sven	14
Normalising	26
Notified Body (PED)	49
NPS	22
Nuclear service	70
Oblique pattern body	22
Obturator	19, 20
Operating mechanism	21
Operating time	69, 75
Operating torque	23, 69, 144, 152, 190
O-rings	57
Outside screw	143
Over pressure	116
Packing	21, 33
Packing chamber	21
Packing gland	21
Packing set	57
Partial stroke testing	173
Pearson, George	17
PEEK	29, 93, 148
Peglers	14
Pig traps	187
Pigging	136
Piping	186
Pitting corrosion	41
Plunkett, Dr.	15
PN	22
Position indication	160
Position limit switch	144
Positioner	125, 179
Positive material inspection (PMI)	26
Pre-commissioning	186
Pressure accessories (PED)	49
Pressure equipment (PED)	49
Pressure Equipment Directive 97/23/EC (PED	26, 36, 48, 65, 115, 190
Pressure seal ring	20
Pressure testing	35
Probability of Failure on Demand (PFD)	137
Process considerations	62
Production test	23
PS, Maximum allowable pressure	22, 48
PTFE fluoropolymer	15, 29, 55, 56, 57, 70, 88, 93, 103, 148, 175
p/t ratings	31, 32
pup weld ends	184
Quenching	26
Rapid gas decompression (RGD), sealing	58
Rate torque	167
Recent developments	135
Reliability	64
Repair	193
Repair, Code of Practice	196
Reverse acting	179, 180
Reverse engineering	194
Ring joints	55
Risk Reduction Factor	136
RTJ flanges	56
Running torque	147
Safety Accessories (PED)	49
Safety Component (Machinery Directive)	51
Safety Instrumented Function (SIF)	136
Safety Instrumented System (SIS)	136
Safety Integrity Level (SIL)	135, 136, 173
Safety Related System (SRS)	136
Saunders, P.K.	14, 15, 17
Savery, Thomas	13
Screwed end	29

Seal / joint integrity	32	Torque output	185
Seal ring bushing	21	Torque rating	145
Seal weld	20	Torque sensing	165
Seating surface	20	Toxicity	42
Seat leakage/ test	36, 111, 121	Traceability	196
Seat leakage classifications	130	Transgranular cracking	40
Seats, metal	16	Travel	22, 165
Shaft	21	Trim	20, 28
Shell	20	TS, Maximum allowable temperature	22
Shell leakage test	36	Type test	23
Socket welding end	22, 30, 31	Tungsten Carbide	27, 29, 93
Soft seal	21	Turbulence	186
Seating torque	144	UKAS	45
Sequence tightening	184	Union nut	20
Set pressure	115, 116, 117	Unseating torque	147
Shock waves	186	Valve and actuator testing and adjustment	185
SIL rating	136	Valve application matrix	61
Silicon, seals	57	Valve layout and position	186
Single acting	155, 172, 174, 175, 180	Valve operating torques	143
Sizing selection chart, gas/ vapour	131	Valve selection techniques	61
Sizing selection chart, liquid	131	ValveUser magazine	17
Smart Valve Monitoring	137	Velocity control trim designs	127
'Sniffing' (fugitive emissions)	47	Vena contracta	79, 131
Sound Engineering Practice (SEP, PED)	49	Vent	139
Sound pressure level	132	Venturi effect	110
Sour environment	42	VMA	23
Spares	193, 197	Wafer type body	22
Special class ratings	32	Water hammer	113
Specification sheet	65	Welding	196
Spigot end	22, 31	Welding end	22
Spiral-wound gaskets	55, 56	Wireless technology	17, 140, 141
Spheroidal Graphite (SG) Iron	26	Worm	(see gearboxes)
Spring opposed	180	Yoke	21
Spring return	158, 174	Yoke bush	21, 143
Stainless steel	26, 57	Yoke sleeve	21, 77, 143
Stall torque	167	Zinc	28
Standardisation	17, 45, 65	Zones 0, 1, & 2 (ATEX)	50
Standards list	199		
Standards, performance	46		
Stellite®	29		
Stem	21		
Stem leakage	191		
Stem nut	21, 143		
Stem shaft seal	33		
Straight pattern body	21		
Strength test	35		
Strength torque	23		
Stress corrosion cracking (SCC)	40		
Sulphide stress cracking (SSC)	40		
Supplementary Loaded	120		
Switchboxes	159		
Technical Construction File (EMC)	51		
Tempering	26		
Thermal cycling	34, 57		
Thrust rating	145		
Thermoplastic polymers	29		
Threaded end	22		
Titanium	25, 29		
Top entry	77, 80		
Top works	81		
Torque characteristic	158, 167		
Torque limitation	165, 186		

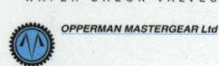

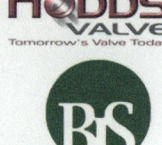

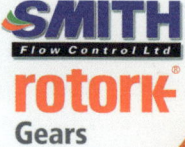

What do over 140 world class companies...
...have in common?

..Superb Backing!

The British Valve & Actuator Association
Professional Support For The Process Flow Control Industry
9 Manor Park, Banbury, Oxon. OX16 3TB (UK)   Tel: (0) 1295 221270   Fax: (0) 1295 268965

www.bvaa.org.uk